THE MAN WHO
SAVED GEOMETRY

T0292842

THE MAN WHO SAVED GEOMETRY

The Multidimensional Mind of Donald Coxeter

SIOBHAN ROBERTS

Princeton University Press
Princeton and Oxford

Published by Princeton University Press
41 William Street, Princeton, New Jersey 08540
99 Banbury Road, Oxford OX2 6JX

press.princeton.edu

All Rights Reserved
ISBN (pbk.) 9780691264745
ISBN (e-book) 9780691264752
Library of Congress Control Number: 2024938462

British Library Cataloging-in-Publication Data is available

Editorial: Diana Gillooly and Whitney Rauenhorst
Production Editorial: Terri O'Prey
Cover Design: Benjamin Higgins
Production: Erin Suydam

Cover illustration by Christian Gralingen

This book has been composed in Sabon

1 3 5 7 9 10 8 6 4 2

FOR
Doris Schattschneider

Marjorie Senechal

Glenn Smith

I could be bounded in a nutshell and count
myself a king of infinite space.

—WILLIAM SHAKESPEARE, *HAMLET*, ACT II, SCENE 2
(AS CITED BY COXETER REGARDING
"THE FINITENESS OF TRIANGLES,"
INTRODUCTION TO GEOMETRY)

CONTENTS

FOREWORD

SOME PERSONAL REMINISCENCES OF DONALD COXETER

DOUGLAS R. HOFSTADTER
CENTER FOR RESEARCH ON CONCEPTS AND COGNITION
INDIANA UNIVERSITY

It is a great honor to have my name linked with that of Donald Coxeter.

As a mathematics and physics student in the 1960s and 1970s, I often ran across the intriguing name H. S. M. Coxeter. I knew that this man's books were world famous, had heard that they were elegant and concise, and, on flipping through them once or twice, had even seen that they were filled with beautiful, enticing diagrams. But somehow, I had other things on my mind and I paid them little heed. When, decades later, I finally came under the spell of Coxeter's words, images, and ideas, I fell in love with geometry.

What eventually launched me on a collision course with geometry was a spectacular course on complex analysis that I took at Stanford University way back in 1962. This course was given by a young professor named Gordon Latta, who hailed from Toronto, the city in which English-born Donald Coxeter eventually settled. Latta, without doubt the best mathematics teacher I ever had, was extremely visual in his teaching, and he conveyed the depth and power of calculus in the two-dimensional arena of complex numbers in an inimitable fashion. One image from that course stuck with me for three decades—that of a circle turning the complex plane inside out, flipping the finite disk inside the perimeter into the infinite region outside the perimeter, and vice versa.

One fateful morning in 1992—thirty years after Latta's course—I woke up with that image of circular inversion in my head, for God knows what reason, and in particular with the vague memory that any circle outside the disk was carried, by this strange but lovely operation, into a circle inside the disk (and vice versa). This weird geometric fact, which I knew Latta must have proven, struck me as so marvelous that I immediately decided to try to prove it myself. Actually, I wasn't entirely sure that I was remembering the statement correctly, and this made my idea of proving it a little dicier. Indeed, my first attempt, rather ironically, showed that a random circle did *not* become another circle! However, my sense of mathematical aesthetics insisted that this statement had the ring of truth, and compelled me to try again. The second time around, I caught my dumb mistake (the center doesn't go to the center!) and proved that circles indeed remain circles when flipped inside out by circles.

This small but joyful excursion into inversion was the tiny spark that ignited a forest fire in my brain, and over the next few months, as geometric imagery started cramming my head fuller and fuller, I knew I needed an external guide. Where else to turn but to the person whose name for me was synonymous with the word "geometry"—H. S. M. Coxeter? I bought a copy of the thin volume he had written with Samuel Greitzer, called *Geometry Revisited*, and went through it from beginning to end, absorbing the ideas with passion. Some of them, as it happened, I had already invented on my own, but by far the majority were brand new to me and served as springboards for countless geometrical forays that I made over the next several years. Thanks to Coxeter and Greitzer, I was flawlessly launched on one of the richest and happiest explorations in my life.

Somewhere around six months into my geometrical odyssey, I used a chain of analogies to make a discovery that excited me greatly, and I wrote up the story of this discovery in a short essay. I wanted to find out if my discovery was new or old, so I decided to seek the reaction of a number of geometers whose books I admired. First and foremost was Donald Coxeter, and so I took the plunge and sent him my essay along with a cover letter. Not wishing to impose, I tried to be very brief (a mere ten pages!), but felt I at least had to tell him how much his book had meant to me. In a most cordial and prompt reply, he suggested I take a look at a couple of books he had written on projective geometry, and so, without hesitation, I purchased them both.

The older of the two was a concise opus entitled *The Real Projective Plane*, and I have to say that reading this was another dazzling revelation to me. As Coxeter points out in his preface, the restriction to the real plane in two dimensions makes it possible for every theorem to be illustrated by a diagram. And not only is this *possible*, but in the book it is *done*. By itself, this simple fact makes the book a gem. Moreover, Coxeter strictly adheres to the philosophy of proving geometric theorems using geometric methods, not using algebra. This means that a reader of *The Real Projective Plane* comes to understand projective geometry through the ideas that are natural to it, building up an intuition totally unlike the intuition that comes through formulas. I am not impugning what is called the *analytic* style of doing geometry; I am just saying that coming to understand projective geometry using the *synthetic* style was among the most gratifying mathematical experiences I have ever had. I will never forget the many nights I spent in bed reading Coxeter's monograph with only a tiny reading light perched on it (in fact, inside it), in order not to wake up my wife, who had nothing against my infatuation with geometry but who seemingly couldn't sleep a wink if even a single photon impinged on her eyelids.

I cannot resist quoting a sentence in the preface to *The Real Projective Plane*. It says this: "Chapter 10 introduces a revised axiom of continuity for the projective line, so simple that only eight words are needed for its enunciation." I think Donald Coxeter must have felt not only pleased but also proud as he wrote this down, because he was so in love with simplicity, elegance, and economy of means. Here is the eight-word definition to which he was referring: "Every monotonic sequence of points has a limit." What a delight! As you probably can tell, my copy of *The Real Projective Plane* is one of my most lovingly read and most prized possessions.

Speaking of doing geometry with a minimum of photons, I have to relate one of the most absurd and yet enriching geometrical experiences I have ever had. Somewhere in my many readings on geometry, I came across a vignette about a famous nineteenth-century German geometer—probably Steiner, Plücker, von Staudt, or Feuerbach—who was so suspicious of the insidious dangers supposedly lurking in diagrams that he insisted on teaching his students geometry in a pitch-dark room, using words and words alone to convey all the ideas. When I first read about this, I was nonplussed, thinking it to be among the silliest notions I had ever heard of. But perhaps precisely because it was so silly, this scene kept bouncing around in my head for a long time, and eventually, years later, when I myself was teaching a course on triangle geometry that often met at my house at night, I couldn't resist pulling down all the shades, turning off all the lights, and trying out this technique myself. The room became absolutely pitch dark, so dark that the students couldn't even see my arms move when I traced geometric shapes in the air. All they ever knew about were my spoken words, not my physical gestures. And what theorem did I prove to them in that darkest darkness of night? None other than the gleaming jewel known as Morley's theorem, which states that the "taboo" trisectors of the three angles of a random triangle join each other at the corners of an equilateral triangle floating somewhere inside the random triangle. Did they see it in their mind's eyes? I am sure they did! And what proof did I relate to my assembled students? Well, naturally, it was the one I had found in the pages of Coxeter and Greitzer's little volume and had made my own, although of course I had to adapt it to fit my brave new light-free, diagram-free circumstances.

This whole episode may seem like an exercise in utter craziness, but in retrospect, I don't think so. Quite the contrary, it was an unforgettable exercise in visualization without vision. One has to remember that some of the greatest of all mathematicians have been blind, and yet that didn't stop them from making astounding discoveries. I was reminded of this as I perused Coxeter's famous book *Introduction to Geometry*, chock-full of literary quotes (the index includes Aeschylus, Aristophanes, Plato, Shakespeare, Goethe, Lewis Carroll, H. G. Wells, Dorothy L. Sayers, and even Tom Sawyer), and found the following sentence, which he took from E. T. Bell's book *The Development of Mathematics*: "Euler overlooked nothing in the mathematics of his age, totally blind though he was for the last seventeen years of his life."

There is a vast difference, I feel, between having no diagrams before one's eyes and having no diagrams inside one's head. They are not the same thing at all; indeed, internal imagery is indispensable. For that reason, one of the most regrettable and baffling tendencies in the mathematics of the twentieth century was a mad stampede toward obliteration of the visual and even the visualizable. Donald Coxeter, however, as everything he wrote vividly demonstrates, was among the people who most systematically opposed this madness.

I will never forget how, at age fifteen or so, I came across the book *General Topology* by John L. Kelley. This austere volume, the first treatise I had ever seen on "rubber-sheet geometry," that mysteriously alluring branch of mathematics I thought was populated by Möbius strips and distorted doughnuts, did not, in its hundreds of pages, contain a single diagram; instead, it was filled with incredibly dense and prickly notation using all sorts of arcane symbols (many of which, I realized years later, stood for rather simple, bland words, but were used in their place for the dubious sake of maximal symbolic compression). Being young and naïve and in love with mathematics, and not yet having had the experience of struggling with it, I merely thought to myself, "Oh, so this is the kind of thing I will have mastered in just a few years! Won't that be wonderful!" I wasn't dismayed in the least by the prospect of reading long and picture-free works of mathematics, and writing such things myself; it struck me as a natural part of the process of reaching the mythical status known as "mathematical maturity."

Within a few years, however, I discovered that I personally could not survive in such an arid atmosphere. Diagrams (or at least mental imagery that could be thought of as personal, inner diagrams) were the oxygen of mathematics to me, and without them I would simply die. And thus, when the air of abstraction for abstraction's sake became too thin for me to breathe, I wound up with no choice but to bail out of graduate school in mathematics. It was a terrible trauma. If, at that crucial moment in my life, someone had suggested that before abandoning mathematics, I take a look at geometry, I might have discovered the works of Donald Coxeter and followed a very different pathway in life.

In 2000, several years after my correspondence with Donald Coxeter, I went to the University of Toronto to give two colloquia in the Physics Department. After the first (a talk describing the key role played by analogies in physics), a very thin and well-dressed elderly gentleman walked up and softly said to me that he was Donald Coxeter. You could have knocked me over with a feather. At the time, he was ninety-three years old! We walked out to an informal reception together and ate cookies and chatted for a little while. Mentally speaking, he was completely at the top of his game, and we talked in a lively fashion about the importance of analogies in both math and physics. I was deeply touched by his presence at my lecture.

But the capper came at my second physics colloquium. Just as I started speaking, I spied Donald Coxeter once again in the audience. And after I had finished, we once again met and chatted for a little while. This time, after we had touched on the family of geometries about which I had written to him some eight years earlier, the conversation somehow veered to the topic of Coxeter's vegetarianism and his incredible daily exercise program, which at that time he was still religiously following.

How honored I felt that this great man, this icon of twentieth-century mathematics, had come to hear me not just once but twice, and had presented himself to me as if he were an admirer of mine rather than the reverse. The logic was simply upside down. Moreover, here was someone who for almost his entire life had stuck to a moral principle that I, too, had found central: the sacredness of life, whether that of humans or that of "lower" creatures. Altogether, the message that came straight to me was that this was a human being entirely without pretension, the kind of person that I had grown up hearing described as a "mensch"—the best kind of person that exists. I had the privilege of meeting this marvelous mensch face-to-face on only those two occasions, but they remain indelibly imprinted on my mind.

This concludes my personal reminiscences of Donald Coxeter, but I would like to add a few words about Siobhan Roberts's book. I have never met Siobhan, but we have corresponded a little bit. What I know of her comes almost entirely from reading her words about Donald Coxeter, and what emerges loud and clear is that she understands the man's spirit very deeply. She understands what drove him, and she knows just how to put into words the fire that always inhabits a great mathematician's soul. I hope that Siobhan's book will bring to many people not only a sense for the beauty of mathematics itself, but also a sense for how the very human love of hidden patterns and symmetries can result in a hundred years of exultant exploration.

PART I

PURE COXETER

CHAPTER 0

INTRODUCING DONALD COXETER

Tell me something is impossible
and I will set about it immediately.

—H. S. M. COXETER

On a cold and crystalline night in January 2002, the geometer Donald Coxeter sat waiting for the formalities to begin at a reception put on by the Royal Society of Canada, a club of distinguished scientists modeled after the Royal Society in Britain. Coxeter, age ninety-four, sat near the fireplace in the library of the University of Toronto president's mansion, holding in one hand a glass of red wine tilted dangerously toward a spill, and in the other an exploding pastry. "This cream puff is not very sensible," he said, fastidiously dressed—as he always was, even for breakfast—in a suit and tie. He waited contentedly, the elder Genius among geniuses.[1]

Donald Coxeter was a man whom most admirers only ever knew as old. Encountering Coxeter in his tenth decade, fledgling mathematicians were often taken aback by his preternaturally ancient appearance, the patina of time at once smoothing and wrinkling his face with a certain cosmic glow. The standard joke among his longtime colleagues was that Coxeter had looked equally ancient a quarter century before.[2] In the memory of his children, he was always balding, and what hair he had was gray. His great-grandchildren found him a frightening presence and avoided his company.[3] Michel Broué, director of the Institut Henri Poincaré in Paris, became acquainted with Coxeter while a student in the 1960s, but only by the coattails of Coxeter's reputation. "I was amazed to hear he was still alive. I thought he had lived in the nineteenth century," Broué recalled. "His name was everywhere. He was such a legend."[4]

At the Royal Society gathering, between the advances of fans and well-wishers, Coxeter—never one to waste an idle moment when he could instead pounce on a geometry problem—gestured toward the middle distance and asked, "What shape is that table?" It seemed like a trick question. Anyone could see the table was round, it was a circle. But Coxeter begged to differ: "If I were suspended from the ceiling looking down upon the table," he said, "then it would be a circle." From Coxeter's coordinates across the room, however, his perspective was slanted and transformed. "I see it as an ellipse," he declared, adding as a footnote that he had written a paper on this exact subject, titling it poetically "Whence Does a Circle Look like an Ellipse?"[5]

This was quintessential Coxeter, ruminating about the romance of shapes—ellipses and circles, hexagons and icosahedrons. Coxeter's definition of his discipline, often recited, was this: "Geometry is the study of figures and figures. Figures as in shapes"—triangles, cubes, dodecahedrons—"and figures as in numbers."[6] He delighted in the geometry of frothy bubbles, porous sponges, the cells of honeycombs, the buds on pineapples, and sunflowers. During his professorial days, Coxeter picked towering sunflowers from his garden, taller than the diminutive man himself, and toted their yellow-rayed faces along on the city bus to the University of Toronto, where he employed them as teaching devices. He dabbed a dot of glossy red nail polish on each of the sunflower's seeds, highlighting the geometrically perfect golden ratio of their graceful whorl—a phenomenon known as phyllotaxis.[7] (For further discussion of phyllotaxis, see appendix 1.)

Coxeter was also known to be both instructive and entertaining in revealing the hidden symmetry of an apple. Around the dinner table with colleagues gathered for the American Math Society conference in 1981, he asked: "Did you know that apples do not have cores?" They thought he was pulling their leg, until the hostess, Marjorie Senechal, a mathematics professor at Smith College, procured an apple and placed it before him with a knife, as requested. He filleted the fruit into thin horizontal sections, demonstrating that there was no stem-to-stern core, but rather elongated pods of seeds suspended within. The pièce de résistance occurred when he reached the center of the apple and sliced through its equator. There lay its secret symmetry—not nature's sloppy attempt at spherical symmetry, as suggested by an apple's exterior, but rather perfect fivefold symmetry, hidden at the apple's heart: the apple seeds were arranged in a five-pointed star. Everyone around the table gasped when they saw it. "It just shows," said Senechal, "that he was looking everywhere, and looking deeply. Coxeter delighted in the geometry of everyday objects, and, because he was so curious and astute, he found symmetries and regularities in these objects that the rest of us never suspected."[8]

Everyday patterns grabbed Coxeter's attention, played in his mind, and provoked his geometer's passion for over eighty years (he made his first discoveries at age thirteen, and was still practicing, still pulling books from his library for yet another paper, at age ninety-six).[9] The renowned futurist and innovator Buckminster Fuller captured Coxeter's century-spanning stewardship of classical geometry with this dedication in his book on the geometry of thought:

> By virtue of his extraordinary life's work in mathematics,
> Dr. Coxeter is *the* geometer of our bestirring
> twentieth century, the spontaneously acclaimed
> terrestrial curator of the historical
> inventory of the science of
> pattern analysis.
> I dedicate this work with particular esteem for him
> and in thanks to all the geometers of all time
> whose importance to humanity
> he epitomizes.[10]

For a figure of such majestic status—perfectly pedigreed at Cambridge and Princeton, muse to such titans as Fuller and M. C. Escher, and masterminds the likes of Douglas Hofstadter and John Horton Conway—Coxeter was first and foremost a humble, hands-on geometer who appreciated the feel of his shapes and models, turning them in his fingers, peering through

Coxeter, King of Geometry.

One in a series of "cartoon Coxeters" drawn by the geometer David Logothetti.

their corners with x-ray vision to get a reading on their intrinsic symmetrical properties. Above all, he valued visual input to feed his vivid geometric intuition. As a geometer, as a number of mathematicians have commented, "Coxeter could really see things."[11]

The honorifics only continue. He was reverentially called the "King of Geometry."[12] However, while his contributions to geometry were formidable, in person he was never one to wield his status with even a trace of blustery bravado—Coxeter was modest, self-effacing, and soft-spoken.[13] Others likened him to a modern-day Euclid, the greatest classical geometer of the twentieth century. And, he was considered the man who saved geometry from near extinction in a mathematical era characterized by its penchant for all things algebraic and austere.[14] In the twentieth century, jungles of symbols and equations, a tangle of subscripts and superscripts, overtook mathematics, leaving a dearth of diagrams and shapes.

Coxeter's obsession with geometry was motivated exclusively, almost with an elitist bent, by beauty. And yet, classical geometry is not merely a paean to the beauty of patterns and shapes. It is also intensely practical. While we no more notice geometry and its crucial impact on our lives than we notice the curve of the Earth when walking upon it, geometry is everywhere and its reach is infinite. Geometric algorithms produce the computer-designed curves of a Mercedes-Benz, animated films such as Pixar's *The Incredibles,* and the fluid contours of detergent bottles.[15] László Lovász, mathematician-in-residence at Microsoft, discerned an important application in Coxeter's last paper (delivered at a conference in Budapest when he was ninety-five), addressing the properties of four mutually touching circles. In the field of computer algorithms, the elementary classical interest in four mutually touching circles is "a hot topic," said Lovász. "It's a central topic in the geometric representation of graphs. These geometric representations are related to issues in data-mining programming." Data mining is the technology of finding patterns in massive amounts of raw information. It powers e-commerce engines such as eBay, and the American government's surveillance software MATRIX (Multistate Anti-TeRrorism Information eXchange). Amazon.com exploits this technology when you buy or search for a book and the site prompts you with recommendations—when you click into your shopping cart Coxeter's book *The Beauty of Geometry,* you learn that customers who bought this book also bought Coxeter's bestseller *Introduction to Geometry,* his *Regular Polytopes,* and *Famous Problems of Geometry and How to Solve Them* by Benjamin Bold. "Each customer is a data point," Lovász explained, "spending this much money here and that much money there, and so you get a set of points associated with each particular visit to the site. You get a huge number of points, because there are a huge number of customers. This generates points in some space that is higher-dimensional than three." Patterns amass on graphs in these multiple dimensions and become computerized geometric representations of who buys what.[16]

The inadvertent applications of Coxeter's pure geometry go on and on and on, appearing in linear programming, modem technology, and immunology, to name but a few.[17] Most often the applications involve mathematical tools that Coxeter invented, which in time have revolutionized the way mathematicians and scientists create and investigate. Coxeter pioneered tools that are now called "Coxeter groups" and "Coxeter diagrams"—tools that shed new light on symmetry, and deepen its study. Symmetry underpins all mathematics—an equation being an expression of perfect balance. And symmetry describes the forces of nature—everything from the smallest spec of a subatomic particle, to a sunflower, to the shape of the universe and the hypothetical parallel universes that mirror our own.[18]

Mathematicians today can't say enough good things about these Coxeterian innovations. They are "one of the pillars of mathematics,"[19] "part of the substrate . . . the air we breathe"[20] and almost as essential as numbers themselves.[21] Papers have been written on why Coxeter groups pop up so much, why they are such a versatile and omnipresent tool that can be deployed in such a diversity of domains in both mathematics and science. They crop up even in our existential search for the shape of the universe. The physics of superstring theory, the much-lauded "theory of everything," rests on the concept of supersymmetry. Some physicists conjecture that infinite-dimensional symmetries will be important in unraveling the puzzle of string theory.[22] "[A]nd if so," said Ed Witten, the "pope of strings," at Princeton's Institute for Advanced Study, "maybe it will be helpful to understand the Coxeter groups."[23]

△ ⬠ ◆ ◉ ⬡

Symmetry was the center of gravity for Coxeter's geometry—he was incessantly searching for the symmetries of shapes. Coxeter was a classical geometer, the classical goal of geometry being not so much to prove theorems but to discover gemlike geometric objects. He explored and enumerated diverse species of geometric configurations, and uncovered how they relate to one another through their symmetrical properties. Prefacing *Introduction to Geometry*, Coxeter stated: "The unifying thread that runs through the whole work"—and indeed that ran through his whole life and career—"is . . . in a single word, *symmetry*."[24]

Louis Kahn's geometrically inspired National Assembly Building in Dhaka, Bangladesh.

Etymologically the word breaks down to *sym*, meaning "together," and *metry*, meaning "measure," and implies that different parts "measure together."[25] Symmetry is ubiquitous, with faces, feet, and much of the human body being approximately symmetrical. The music of Bach has many symmetrical qualities, as does the art of Leonardo da Vinci, the metrical rhythm and rhymes of poetry, and the designs by architects such as Louis Kahn.[26]

Symmetry, generally speaking, occurs when two halves of a whole are each other's reflection in a mirror (bilateral symmetry). Examples of symmetry abound in the chemistry of life, though often the symmetry of life is "chiral" or "handed," meaning the halves or mirror images are different. The spearmint molecule and caraway molecule are chiral twins—one molecule is the mirror reflection of the other, and with that minor difference the molecules have considerably different effects on our taste buds.[27]

The chirality of pharmaceuticals (drugs often being compounds of left-handed and right-handed molecules) was demonstrated with tragic consequences in the use of Thalidomide[28]: one of its molecular geometries was therapeutic as a sedative quelling morning sickness, but the mirror opposite caused unexpected birth defects.[29] Mirror symmetry is not the only type of symmetry. There is also rotational symmetry (a pinwheel), or translational symmetry (repeating rail ties spaced equally apart), and then many combinations thereof (a trail of footprints is symmetrical in the sense that it is produced by "glide reflection," a composition of translation and reflection).[30]

In geometry, a finite object is symmetrical if it looks the same after being subjected to a geometric change (a rotation, or reflection), called a symmetry operation, or a transformation. The sphere can be rotated and reflected in an infinite number of ways and always remain exactly the same; a sphere is invariant under an infinite number of symmetry operations.[31] However, these infinite symmetries are predictable and thus hold less allure than the shapes with discrete symmetries that Coxeter preferred to investigate.* A square, for example, has only eight symmetries, eight precise ways in which its position can be moved or changed, all the while leaving the square looking the same (see chapter 6). This mathematical study of symmetry is systematized in "group theory." The meaning of a group in mathematical terms is distinctly different from the everyday meaning of the word. While a group in ordinary language can be defined as a number of people or objects located, gathered, or classed together based on some distinguishing characteristic, a group in mathematical terms is the set of eight actions—the symmetry operations—that preserve the square's appearance.[32]

A Coxeter group is a tool for exploring the world of group theory. But Coxeter investigated shapes more complex than the square—he liked shapes with a complexity analogous to that of an exquisite crystal. He studied how the facets of a crystal, the angles between its corners and edges align "just so" and make it a highly symmetric object. A Coxeter group pertains to these finite symmetries, the finite number of rotations that preserve a crystal's appearance. "It is this finiteness," said Ravi Vakil, a geometer at Stanford, "within the infinite group that makes some sort of magic happen mathematically."[33]

Coxeter followed in a tradition of classical geometers who extended the investigation of symmetries into multiple dimensions,[34] where shapes rotate and reflect upon themselves, replicating their properties in the hall of mirrors that is hyperspace. These multidimensional shapes are called polytopes. Coxeter's preoccupation with polytopes was so conspicuous that during his stint at Princeton in the 1930s he earned the nickname "Mr. Polytope."[35]

Polytopes, meaning "many shapes," are a broad class of geometric figures whose subsets of families, related like cousins by their symmetries, live in

---

*Astronomer Fritz Zwicky (1898–1974) was notorious for calling people "spherical bastards" if he found them uninteresting and dislikable—no matter which way he considered these people, they were equally offensive.

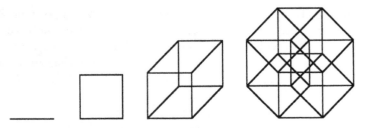

The dimensional progression from a point to a hypercube.

various dimensions.[36] The two-dimensional polytopes are called polygons (meaning "many angles"—*gon* derives from the Greek word *gonu* meaning "knee," a knee often being bent at an angle).[37] Everyone is acquainted with some of the regular polygons, having equal sides: we test our geometrical skills in grade school with an equilateral triangle; the square, as Coxeter said, "confronts us all over the civilized world"; then there's the Pentagon Building; the hexagonal snowflake; the eight-sided octagon of a stop sign; and the twelve-sided dodecagon of the old Canadian nickel or British three-penny bit.[38]

The three-dimensional polytopes are called polyhedra (meaning "many surfaces"—*hedron* is an Indo-European word meaning "seat," so a polyhedron has many seats, or surfaces on which one could sit).[39] The most famous polyhedra are the Platonic solids: the tetrahedron, octahedron, icosahedron, cube, and dodecahedron. Analogous figures exist in higher dimensions—the fourth dimension, for example, contains the simplex (the 4-D analog to the tetrahedron), and the hypercube (the 4-D analog to the cube). And in higher dimensions still, polytopes morph into more and more complex cousins of the originals, some continuing to infinity.[40]

Coxeter's house was a veritable zoo of polytopes, overtaking every available surface. He hung posters of higher-dimensional polytopes as art on his wall; he had polyhedral lamps and polyhedral bookends. Polyhedra—made of cardboard, wood, marble, plastic straws, string and sticks, plaster, soldered wires, and stained glass[41]—filled the china cabinet, lurked among plants, encroached on the window seat, on the fireplace mantel, on side tables, and sometimes the dining room table.* Coxeter's book *Regular Polytopes* became a best-seller and a mathematical classic, the geometrical analog

---

*Many of Coxeter's models were sent to him as gifts from strangers, fans from afar, such as George Odom, a resident of the Hudson River Psychiatric Center, in Poughkeepsie, New York. Odom sent so many models over the years that Coxeter, though grateful, eventually ordered him to stop. As Odom recalled, Coxeter begged him: " 'Please! NO MORE MODELS!' "

of Darwin's *Origin of Species*. With his Coxeter groups, Coxeter did for polytopes what Darwin did for organic beings[42]—he classified and quantified their very existence.

Around the time Coxeter chose classical geometry for his career, however, circa 1930, the classical tradition—of hands-on visual reasoning, using antiquated treasures such as triangles, circles, and polyhedra as specimens of study—was exiting its golden age. Historian E. T. Bell pronounced in 1940: "The geometers of the 20th century have long since piously removed all these treasures to the museum of geometry where the dust of history quickly dimmed their luster."[43] Geometry was being recast, like a remake of a cinematic classic, in an abstract and dry format. Geometry was being subsumed by algebra and analysis—it was all equations and no shapes, like prose without poetry.[44]

The eminent German mathematician Hans Freudenthal (1905–90) lamented classical geometry's dethronement in a 1971 essay, "Geometry Between the Devil and the Deep Sea." For a long time, he said, mathematics

George Odom with one of his models, which he sent to Coxeter.

was synonymous with geometry, but today it is rejected as not firmly enough rooted in reality. Freudenthal countered with a "haphazard" list of questions that evoke the singular mind-set of a geometer, some of them very much related (on varying levels) to the space in which we live.

Why does a rolled piece of paper become rigid?

How do shadows originate?

What kind of curve is the terminator on the moon?

What is the intersection of a plane and a sphere, or two spheres?

Why can the radius of a circle be transferred six times around the periphery?

How come a beautiful star arises by this construction?

Why is the straight line the shortest?

Why do congruent triangles fit to cover the plane and why do congruent pentagons in general fail to do so?

How can people measure big distances on the earth, the diameter of the earth, and distances of celestial bodies?

What is the shortest path for a light ray to travel from one point to another while touching a mirror?

How does a kaleidoscope work?

If a cube is split into six square pyramids with their vertices in the center and these pyramids are turned outside upon the corresponding faces, why does a rhombic dodecahedron arise?

Why can a table with four legs wobble, and what is the difference with a table with three legs?

Why does a door need two hinges, and how can we add a third?

And finally the old question: why does a mirror interchange right and left though not above and below?[45]

Another of Freudenthal's questions matched a geometric trick Coxeter demonstrated whenever he found the chance: "Why does a tied paper ribbon show a regular pentagon?" Coxeter's instructions for folding a five-sided polygon are simple. "The figure of a pentagon with diagonals can be neatly displayed," he said in his best-selling book *Introduction to Geometry*, "by tying a simple knot in a long strip of paper and carefully pressing it flat."[46] It is easy enough to do. Tear a 2 × 15 inch strip of paper with a ruler to keep the edge straight. Loop the ends as if beginning to tie two shoelaces. Slide the edges together until they jimmy into place, meeting flush with the fold, and then press the woven strips flat. There you have a very practical pentagonal bookmark.

Geometer Walter Whiteley, director of Applied Mathematics at York

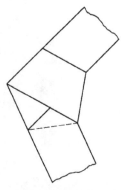

A tied paper pentagon, from Coxeter's *Introduction to Geometry*.

University, in Toronto, asked similar questions in his course "Introduction to Geometries." Do the tracks of a bicycle indicate it was traveling forward or backward? Why does a piece of paper fold along a straight line? Whiteley called all this "Learning to See Like a Mathematician." And in a paper titled "The Decline and Rise of Geometry in 20th Century North America," Whiteley warned that if this visual perspective had met its demise during the dark ages of classical geometry, the consequences would be profound and far-reaching. Should classical geometry ever become extinct, he reckons a "geometry gap" would haunt Western civilization for generations to come.[47] Without classical geometry, as without Mozart's symphonies or Shakespeare's plays, our culture, our understanding of the universe, would be impoverished and incomplete. Donald Coxeter did much to save us from such a loss.

According to mathematical folklore, the shift away from classical geometry manifested itself most dramatically with a statement by a bombastic French mathematician, who declared:

"*À bas Euclide! Mort aux triangles!*"—"Down with Euclid! Death to triangles!"[48]

Legend has it that this war cry came from one Nicolas Bourbaki. He believed mathematical education in France was falling behind the international standard. He wanted to overhaul the entire structure of mathematics. In so doing, he sought to stamp out the use of diagrams. Bourbaki endeavored to write an algebraic encyclopedia of mathematics without a single picture. This aversion to shapes was defended as serving the interest of purity: all mathematical results were to be reached by reason alone—by rationality—rather than by the corruptible visual sense. According to Bourbaki, our

"General Bourbaki," as depicted in a *Scientific American* article, 1957.

visual perception of the world was unreliable, our eyes leaving us victim to subjectivity and error.[49]

More than forty years later, when reminded of Bourbaki and the "Death to Triangles!" rant, Coxeter was cool and calm, with the retrospection of old age. "Everyone is entitled to their opinion," he said. "But Bourbaki was sadly mistaken."[50] Coxeter had by then become geometry's apostle. He ignored the fads and fashions, and through steadfast rear-guard action, simply persevering with the shapes he loved, he preserved the classical tradition of geometry and sustained it through its lean years.[51] For this he has become a hero for many mathematicians the world over.[52]

Even if you have never before heard of Coxeter, let alone his Coxeter groups or Coxeter diagrams, you will nonetheless find it hard to resist a tumble down the Coxeterian rabbit-hole into his geometrical wonderland. There is something marvelous in witnessing Coxeter's seemingly esoteric and arcane obsession, his intense focus and all-consuming passion for classical geometry that lifts him above the humdrum of the everyday and makes his life take flight.[53] It's easy to get hooked by his devotion and stoicism in fighting for his beloved geometry. And, by following Coxeter as an ambassador and interpreter in these foreign parts, you may find yourself viewing the world from a new and illuminating lens—Coxeter's perspective—seeing a hypertext reality where everything takes on shades and shapes of geometry. For example, not long ago a billboard at a small-town Di$count car rental lot beckoned with a truly Coxeterian double entendre. Rising out of the cultural wasteland of fast-food joints, car dealerships, and gas stations lining the main road into town, the billboard read:

Front Street North, Belleville, Ontario, January 2002.

CHAPTER 1

MR. POLYTOPE GOES TO BUDAPEST

Geometry will draw the soul towards truth.

—PLATO, *THE REPUBLIC*

Bursts of white light lit up the splendidly restored auditorium of Hungary's Academy of Sciences as Donald Coxeter inched toward the lectern, leaning only slightly on his cane. Photographers had descended upon the academy, located on the east bank of the Danube in Budapest, to capture a few shots of the president of the Republic of Hungary, Ferenc Madl, who was there making a rare public appearance at the opening ceremony of the János Bolyai Conference on hyperbolic geometry, in July 2002. But afterward, the photographers stayed to snap a few shots of Coxeter as well.[1]

Flash from the cameras reflected off his pale pate and the bejeweled turtle brooch pinned to his lapel. Well into his nineties, Coxeter still traveled the international conference circuit. He had been invited to give the opening lecture at this event, commemorating the two hundredth birthday of Hungary's sainted Bolyai, who, with his discovery of non-Euclidean geometry in 1823, changed forever our perception of space.[2]

When a long-retired mathematician is asked to give an address at a conference, his audience would be forgiven in assuming that he'll provide an autobiographical synopsis of his career. Coxeter, however, wrote a scholarly paper, months and months in the preparation. Titled "An Absolute Property of Four Mutually Tangent Circles,"[3] it addressed a topic tangentially related to Descartes' Circle theorem, one of Coxeter's favorites.[4] As he was announced to the audience, Coxeter shuffled the pages of his talk, and readied his visual aids—numerous transparencies and a geometrical model, a cubic nexus of multicolored straws. Three hundred or so mathematicians awaited his presentation, a discrete group of individuals more than willing to forfeit July's summer sun for the somnolent glow cast by the lecture hall's overhead projector. Most were a fraction of Coxeter's age. Many, the organizers included, had been skeptical that he would be able to make the journey. A similar number no doubt wondered whether he could possibly have anything left to profess.[5]

Coxeter began slowly, enunciating meticulously with his lingering British accent: "The absolute property of four mutually tangent circles that I am describing seems to have been discovered by Mr. Philip Beecroft, of Hyde Academy, Cheshire, England, and published in *The Lady and Gentleman's Diary* . . . In Beecroft's own words, [the theorem states,] 'If any four circles be described to touch each other mutually, another set of four circles of mutual contact may be described whose points of contact shall coincide with those of the first four.' "[6] He waded into an examination of what he believed to be his new proof—a simple, elegant proof—of Beecroft's theorem, delineating the four mutually tangent circles, a_1, a_2, a_3, a_4, and another set, b_1, b_2, b_3, b_4. "This figure makes the theorem almost obvious," he said, fixing his transparency into position, "but for the sake of completeness it seems desirable to

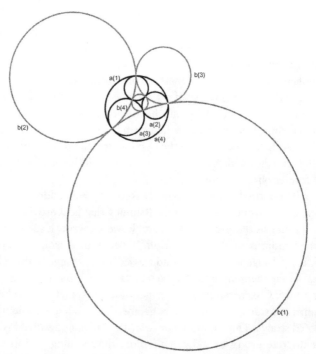

The diagram of four mutually tangent circles used by Coxeter in his Budapest talk.

consider further details." He proceeded, pausing here and there, whistling lightly under his breath, as he often did to focus his concentration.[7]

△ ▢ ◈ ☺ ⬡

When Donald Coxeter barely squeaked by his ninety-fifth birthday, his doctor diagnosed that he was in that final and waning stage of life warranting palliative care—there was no cure for what ailed him. He was creaky and tired, acute pointy parentheses wrinkling around his subtle smile. He had weathered cancer of the prostate and the right eye, and a heart attack, and now suffered chronic digestive troubles. Nonetheless, against doctor's orders, he was determined—ever the obstinate optimist—to make the trip to the Bolyai conference.[8]

In an attempt to be as prudent as possible in planning his journey, Coxeter tended with grandfatherly gumption to life-and-death details such as buying travel health insurance (he was refused), and determining what to do with his brain should he die while away. Coxeter's brain functioned so impressively over the years that he had received a request for its mass of synapses to be seized swiftly, no more than twelve hours' postmortem, in order to undergo scientific research at McMaster University, in Hamilton, Ontario,[9] where a specimen of Einstein's well-traveled and well-dissected brain resides.[10] As Coxeter recalled, McMaster's Dr. Sandra Witelson rang and asked with pardonable clinical insensitivity: "Dr. Coxeter, when you die can we have your brain?" He took it as a compliment and agreed.[11]

On the day of departure, Coxeter sat at his cluttered kitchen table and set about the task of testing his hearing aids. He snapped his fingers from one ear to the other, then tried a ticking pocket watch. "Dead as a doornail! I get the best results when I don't have any hearing aid in at all!" he concluded, at once confused and bemused. He glanced out the window to find the airport limousine waiting at the sidewalk. "Oh, bother!" he cursed (that was the extent of his cursing, for a minor mishap or a major flood in the basement). "How very, very awkward. I'm not ready!" Coxeter checked and rechecked that his passport and airline ticket and envelope stuffed with Hungarian money were safely stowed in his briefcase. He packed his hearing aids and the Tupperware container full of dead and fresh batteries ("A terrible nuisance that they were somehow mixed together"). He gathered all the parts of his electric shaver and stowed his high-altitude antiembolismic socks in the waistband of his daughter's skirt, like Kleenex at the ready under the cuff of a sleeve.[12]

Coxeter's daughter, Susan Thomas, a retired nurse, was his escort. With the chauffeur patiently standing by outside, Coxeter checked for his passport yet again, then snapped his briefcase shut. He inched down the stairs, taking each step two feet at a time, and finally passed the cuckoo clock in the foyer embossed with the motto "Delay Not the Hour Flies." He shuffled along the front walk, and slid his stiff, angular body, not an ounce of body fat for cushioning, onto the limo's leather seat. He was off, venturing forth on one more journey into what he termed "the wild wicked world"—a world, according to the classical definition of "geo-metry," which he had spent more than three-quarters of a century measuring.[13]

One day after Coxeter arrived in Budapest, he attended a welcoming luncheon at his posh Hyatt hotel. There to greet him was the conference organizer, András Prékopa. A member of the Hungarian Academy of Sciences, and a professor of mathematics and operations research at Rutgers University, Prékopa had never before met Coxeter. When he did he shook his hand and announced with a beaming smile: "Dr. Coxeter is *the* world's greatest living classical geometer. No question!"[14]

Later that evening, relaxing in the hotel lobby, Coxeter met with another fan, Texan Glenn Smith, a self-described "geometry groupie," who makes a successful living in the sesame business. Smith brought to Budapest the geometric model for Coxeter's presentation—constructed by special order—as well as an antique set of wooden geometric solids, circa 1850, which he had purchased during a stopover in London. Smith always travels with models in his suitcase that can be assembled and disassembled like LEGO; it's how he kills time in airports and keeps himself company in hotel rooms.[15]

Even with his hobbyist's perspective, Smith had a cogent argument for Coxeter's designation as the savior of classical geometry. "Coxeter so understood the importance of geometry that he stuck with it. He went out on a hilltop—when all the rest of us were down in the valley—and he saw what was out in front of us and how important geometry was going to become, and he led us out of the darkness. We've been in a dark age," said Smith. "And I think we're still trying to come out of that age. The more we investigate geometry, the better off we all will be." Smith also provided an interesting way of explaining the importance of geometry in the world. "Geometry is at the root of everything, whether we recognize it or not. If you take everything and strip it down—start out with the universe and galaxies and stars and planets and solar system and the Earth, then the Earth is organized into countries and countries become communities and communities are made of families, families are made up of people, people have organs, organs have cells, molecules, atoms, subatomic—strip all that away, and at every stage there are certain geometries or configurations of patterns. If you study those patterns, you will see them almost wherever you go, they will always exist. That's the nice thing about geometry, about polytopes or polyhedra—we could be anywhere in the universe and have the same thoughts. In other words, geometry is not particular to this planet we live on."[16]

"What I told my children when they were young," he continued, "is that you need to learn geometry because if you are ever picked up by a flying saucer, you'll need to show the aliens that you know geometry. They will know geometry for sure. You'll need to be able to make a tetrahedron like this"—he placed his right hand on his forehead and his left hand on his right elbow, forming the frame of a tetrahedron. "If you see somebody from another planet, do that and they'll know you have some intelligence, and they won't treat you like an insect and pull off your arms and legs."[17]

Coxeter, not long before, had articulated much the same sentiment when speaking of the Platonic solids: "I don't think they were invented. I think they were discovered. Somebody on a different planet, with the right kind of mind, would find the same thing."[18] That evening in Budapest, Coxeter added as a footnote: "It was Plato's idea that everything that is true has always been true and people simply reconstructed true things by thinking about them."[19]

△ ▢ ◈ ⊕ ⊗

Researching the family tree of geometry, tracing the ancestry from Thales who begat Pythagoras who begat Plato, is comparable to retelling tales from the Bible, since most of what is known about these single-name ancients comes from unattributed or biased sources, anecdotes passed down and spun together to form a grand mythology.[20]

The five regular polyhedra, for example, the mainstays of geometry, are also called Platonic solids even though they were known before Plato (427–347 BC). But Plato took a special interest in these solids and left us the earliest surviving description in his book *Timaeus* (the sequel to his *Republic*).[21] In Scotland, a complete set of five carved out of stone have been attributed to Neolithic people dating back some 4,000–6,000 years. In fact, according to George Hart's Web-based *Encyclopedia of Polyhedra*,[22] hundreds of stone spheres have been found with carved edges roughly corresponding to the regular polyhedra, and ranging in material from sandstone to granite and quartzite. Ornate bronze dodecahedra by the dozens, dating to Roman times from the second to fourth century, have been unearthed across Europe, in the United Kingdom, Belgium, Germany, France, Luxembourg, the Netherlands, Austria, Switzerland, and Hungary. Their function has not been confirmed—perhaps candle stands, flower stands, staff or scepter decorations, surveying instruments, leveling instruments, finger ring–size gauges, or just plain geometric sculpture.[23]

Coxeter liked to note that a pair of icosahedral dice of the Ptolemaic dynasty reside in one of the Egyptian rooms of the British Museum in London, and that excavations on the Monte Loffa, near Verona, extracted an Etruscan dodecahedron, revealing that this figure was enjoyed as a toy at least 2,500 years ago.[24] Known for meticulously sourcing his ideas, Coxeter provided perhaps the best summation of origins: "The early history of these polyhedra is lost in the shadows of antiquity. To ask who first constructed them is almost as futile as to ask who first used fire."[25]

In the ancient Greek tradition, geometry was elevated beyond its practical Egyptian and Babylonian usage (5000–500 BC) to the rank of science.[26] The Greek word *mathemata* translated to "science of learning"—and mathematics in those days essentially comprised geometry.[27] Geometry was the purest measure of truth and the highest form of knowledge, with schools dedicated to its study. The Pythagorean School, which became part of the zeitgeist,[28] was attended by citizens of all social strata, especially the upper class. Women disregarded a law forbidding their presence at public meetings and flocked to hear Pythagoras speak.[29] The ingenuity of the Pythagorean theorem—stating that the square of the hypotenuse of a right-angled triangle is equal to the sum of the squares of the other two sides—provided early affirmation of the direct relationship between number and space. However, when Pythagoras squared the hypotenuse, he did not do "modern" mathematics, multiplying the hypotenuse by itself. Rather, he literally constructed a geometrical square on top of the hypotenuse. Likewise, the sum of two squares being equal to a third meant that the two squares could be physically cut up and reassembled to form the third square.[30]

Pythagoras (569–475 BC) believed that mathematics was religion, capable of purifying the spirit and uniting the soul with the Divine. He made the study of geometry part of a liberal education, probing theorems in an intellectual manner.[31] His heir was Plato, who proclaimed, "God ever geometrizes."[32] And when Plato started his own school, the Academy, the sign hanging over the entrance indicated he did not suffer geometrical fools gladly: "Let none ignorant of geometry enter my door."[33]

Plato, too, held that mathematics was the finest training for the mind, the secrets of the universe being embedded in number and form. He believed the ideal geometrical shapes—circles, spheres, squares, cubes—did not exist in reality but only in a higher realm of their own, independent from the physical world; a sphere in the physical world was only an approximation of the perfect form of a sphere.

"The ideal notion is the mathematical concept," said mathematical physicist Sir Roger Penrose. "A mathematical concept or mathematical structure, in a certain sense, conjures itself into existence. Mathematicians tend to think of mathematics as having its own existence . . . of mathematical notions and mathematical truths as having a timeless existence. And mathematicians are somehow explorers in that world." The notion that mathematical structures contain an inviolable reality of their own is somehow reassuring. The human mind operates with significant margin of error, so often imprecise, inconsistent, and selective in its judgments. In mathematics, there exists logical rigor, an absolute purity. Plato's world of mathematical forms provided a methodology that modern science has followed ever since—scientists propose models of the world, and the models are tested against observations from previous or new experiments.[34]

Plato himself had a model of the world, based on his namesake solids. In his book *Timaeus*, four interlocutors gather to discuss cosmology and natural science. The main character, Timaeus, constructs a story for the creation and composition of the universe. As one Plato biographer, A. E. Taylor, recounted, "What Timaeus is really trying to formulate is no fairy tale, but, as we shall see, a geometrical science of nature." In devising his theory of everything, Plato paired the classical elements with the five regular solids.[35] These shapes, Plato said, were "forms of bodies which excel in beauty,"[36] their beauty residing in the criteria they meet for being "regular," or uniform. First, each solid's surfaces are all the same regular polygon—a shape with all sides and all angles equal (the equilateral triangle, the square . . .). The classification of the Platonic solids as "regular" also depends on a second criterion: the same number of regular polygon faces must meet in the same way at each corner, or vertex.[37]

There are three Platonic solids constructed solely with the equilateral triangle. The simplest is the tetrahedron, composed of four equilateral triangles, three at each of its four vertices. In his scheme of the elements, Plato chose the tetrahedron, due to its simplicity and sharp corners, to represent fire, the fiercest and most basic of the elements—with its "penetrating acuteness . . . the pyramid is the solid which is the original element and seed of fire."[38] The octahedron is built from eight equilateral triangles, four at each vertex, and Plato considered it symbolic of air, because this solid spins nicely in the wind (or by blowing on it) when you hold it between finger and thumb.[39] The icosahedron has twenty equilateral triangles, five at each vertex, which combine to make it the roundest of the regular polyhedra. As a result, Plato associated the icosahedron with a drop of water, "the densest and least penetrating of the three fluid elements."[40]

The cube, Plato assigned to earth: "for earth is the most immovable of the four and the most plastic of all bodies, and that which has the most stable bases must of necessity be of such a nature."[41] Thus four of the five convex regular polyhedra symbolized the four elements: fire, air, water, and earth. "The discrepancy between four elements and five solids did not upset Plato's

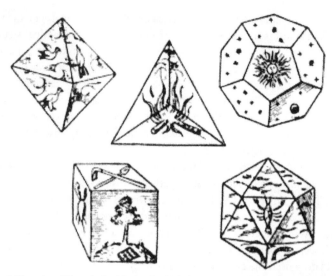

Kepler's Platonic solids etched with the classical elements, from *Harmonice Mundi*, 1619.

scheme," Coxeter noted. "He described the fifth as a shape that envelops the whole universe."[42] The dodecahedron, with twelve pentagonal or five-sided faces, was the model of the universe as a whole. "There remained a fifth construction," said Plato, "which God used for embroidering the constellations on the whole heaven."[43] Plato's scheme demonstrated considerable prescience, because the Platonic solids, even though they did not turn out to be the exact elements of all existence, are in many ways elemental, or fundamental, components of the universe, emerging on both microscopic and macroscopic dimensions in the most unexpected places—a recent cosmological hypothesis revisited Plato's notion that the universe might be dodecahedral; and in astrochemistry, the shape of the Nobel-winning C_{60} molecule is a truncated icosahedron. (See chapter 10 for C_{60} and chapter 12 for the dodecahedral universe.)[44]

The Fabergé-egg feature of Platonic solids—what makes them such exquisite treasures—is the fact that only five regular solids can physically exist.* This cunning act of geometric sorcery is explained by the solids' regularity

---

*Another awe-inspiring feature of the Platonic solids is their interconnectedness. The dodecahedron, with twelve faces and twenty vertices, is the mate, or dual, of the icosahedron, which has twenty faces and twelve vertices. Similarly, the cube, with six faces and eight vertices, is the dual of the octahedron, which has eight faces and six vertices. The fact that these solids are dual to one another has the result that they also share their symmetries.

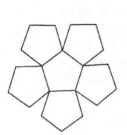
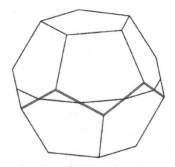

The elasticized popping dodecahedron, from *Introduction to Geometry*.

(faces all the same regular polygon, with the same grouping of polygons around each vertex). It is best appreciated by constructing the Platonic solids for oneself, piece by piece—simply taping the component polygons together. "Any intelligent child who plays with regular polygons (cut out of paper or thick cardboard, with adhesive flaps to stick them together) can hardly fail to rediscover the Platonic solids," said Coxeter. "They were built up that 'childish' way by Plato himself."[45]

Models of these highly symmetric solids can also be constructed from "nets," made by tracing a flat pattern of adjoined component polygons. Coxeter provided instructions in his book *Introduction to Geometry*[46] for a springy dodecahedron model made by fitting together two nets, folded into "bowls" of pentagons that are then strung together by an elastic band. When assembled, the dodecahedron model becomes alive—animated by its crude spring-release system, it can be pushed flat and stored in a book, but when not compressed by sufficient weight it spontaneously pounces back into shape. During class, Coxeter made a stunt of pretending to have lost his dodecahedron model. "Oh, bother!" he'd mutter mid-lecture. "Now, where is my dodecahedron?" He'd look around, opening a book or lifting a stack of papers and then—POP!—there it was, springing into being.[47] (Endnote 47 contains illustrated instructions for constructing a popping dodecahedron.)

△ ▱ ◈ ⊕ ⊛

Euclid (365–25 BC) proved there are only five Platonic solids.[48] And given the above-mentioned restrictions, only three regular polygons (the equilateral triangle, square, and pentagon) can be used in the construction of the Platonic solids. This is because the sum of polygon angles that meet at a vertex must be less than 360° in order to form a convex solid. This can be proved algebraically, or by physically putting the component polygons to-gether and discovering what works. For example, if you try to fit three, four, or five triangles around a vertex, there is still a gap, and the triangles then can be folded down to meet one another, forming a corner of the respective solid (the tetrahedron, octahedron, or icosahedron). All other options with the equilateral triangle would not work: two triangles around a vertex can-not possibly meet at all edges to form a solid, while six triangles add up to 360° exactly, thus leaving no gaps and forming a flat tiling, and seven, eight, or more triangles overlap or meet in accordion-like folds.[49]

Euclid's seminal contribution to geometry was his book *The Elements*. But Euclid was not the author of *The Elements* so much as its editor. He compiled and organized the fundamentals of geometry, work done by Thales, Pythago-ras, and other predecessors. Euclidean geometry in general, to loosely define it, encompassed the study of familiar shapes, their areas and angles, and filled thirteen books. The first book covered triangles; the next, rectangles; fol-lowed by circles, polygons, proportion, similarity; with four books on number theory, and one each on solid geometry and pyramids, culminating with the properties of the majestic five regular polyhedra—here Euclid placed the Platonic solids on a pedestal and gave his proof that there are only five.[50]

By the middle of the nineteenth century, Euclid's *Elements* had been the bible of mathematics for two millennia. Arabian mathematicians and authors, providing one of few sources of information on Euclid's life, translated his name as "Uclides," *ucli* meaning "key" and *des* meaning "measurement"—Euclid was the "key of geometry."[51] And the Euclidean framework was as-sumed to be the geometry of the real world. Immanuel Kant's philosophy still dominated metaphysical beliefs, and in his *Critique of Pure Reason* he asserted that the Euclidean system was "a priori"—meaning "prior to experience," based on synthetic, theoretical deduction rather than empirical observation, or, as Kant translated it, "an inevitable necessity of thought."[52]

In 1847, Oliver Byrne, a mathematics schoolteacher and Queen Victoria's surveyor of the Falkland Islands, published a beautiful new edition of Eu-clid's *Elements*, with color diagrams replacing equations (this in addition to the simple line drawings of previous editions).[53] Byrne's book, *The First Six Books of the Elements of Euclid*, stated on its title page, "Colored Diagrams and Symbols Are Used Instead of Letters for the Greater Ease of Learners." In the preface, Byrne elaborated: "The arts and sciences have become so ex-tensive, that to facilitate their acquirement is of as much importance as to ex-tend their boundaries. Illustration, if it does not shorten the time of study, will at least make it more agreeable. This work has a greater aim than mere illustration; we do not introduce colors for the purpose of entertainment . . .

but to assist the mind in its researches after truth, [and] to increase the facilities of instruction."[54]

Euclid, then, was enjoying continued popularity, but there were undercurrents of dissent. In *The Elements*, Euclid had outlined his exalted five postulates, and the first four were simple enough:

1. A straight line may be drawn between any two points.
2. A piece of straight line may be extended indefinitely.
3. A circle may be drawn with any given radius and an arbitrary center.
4. All right angles are equal.[55]

But the fifth postulate—the parallel postulate—was unlike the others, and allegedly Euclid himself had been hesitant to include it in his *Elements*. "His reluctance to introduce it," Coxeter observed, "provides a case for calling [Euclid] the first non-Euclidean geometer!"[56] It stated:

5. If a straight line crossing two straight lines makes the interior angles on the same side less than two right angles, the two straight lines, if extended indefinitely, meet on that side on which are the angles less than the two right angles.[57]

Coxeter deemed it "unnecessarily complicated."[58] Indeed, since Euclid's time, the parallel postulate had dogged mathematicians, and annoyed them. It was not intuitively obvious and required mathematicians to suspend disbelief; it stumped them because it could in no way be verified by experience.

Another way of expressing the parallel postulate is to say that, given a line and a point not on the line, every line through the point will meet the line, except in one "freaky case": when the two lines are parallel to each other. But, as Jeremy Gray, historian of mathematics at the Open University, pointed out, who is to say what happens to two parallel lines when extended to infinity, or off 10^{10} light-years away, where strange things might alter the laws of space? Maybe parallel lines could meet somewhere in the "vaguer cluster," said Gray. Regardless, it is impossible to check. "So it's a very strange statement," he said. "It's a blot. Because it's a leap of faith unlike all the other postulates."[59]

Over the years, most mathematicians ignored this blot, for if they didn't the reign of Euclidean geometry threatened to collapse like scaffolding with one faulty strut. Some mathematicians, the more daring, courageous, and foolhardy—Greek, Arab, Islamic, and eventually Western mathematicians— tried and failed to prove the parallel postulate using the other four postulates. As the failures accumulated, these attempts of geometrical derring-do only continued, forming a procession of doomed parallel postulators throughout history.[60] The predicament was decried in the mid–eighteenth century as "the scandal of elementary geometry."[61]

Hungary's János Bolyai (1802–60) was one of the adventurers who went in search of geometry's Holy Grail. He first tried to prove the fifth postulate, with no success. He then wondered whether the postulate was perhaps false. Bolyai became infatuated, convinced he was closing in on the chase for geometry's mercurial axiom. His efforts dismayed his father, Farkas Bolyai, who himself had exercised self-destroying due diligence with the parallel postulate.[62] "I have traveled past all reefs of this infernal Dead Sea," he told his son, "and have always come back with broken mast and torn sail." He tried desperately to disabuse János of his interest.[63]

> You must not attempt this approach to parallels. I know this way to the very end. I have traversed this bottomless night, which extinguished all light and joy of my life. I entreat you, leave the science of parallels alone . . . I thought I would sacrifice myself for the sake of the truth. I was ready to become a martyr who would remove the flaw from geometry and return it purified to mankind. I accomplished monstrous, enormous labours . . . I turned back when I saw that no man can reach the bottom of this night. I turned back unconsoled, pitying myself and all mankind. Learn from my example: I wanted to know about parallels, I remain ignorant, this has taken all the flowers of my life and all my time from me.[64]

His son, however, ignored the warnings:

> I am determined to publish a work on parallels as soon as I can put it in order, complete it, and the opportunity arises. I have not yet made the discovery but the path that I am following is almost certain to lead to my goal, provided this goal is possible. I do not yet have it but I have found things so magnificent that I was astounded . . .[65]

Eventually Farkas relented and encouraged his son to publish whatever he had as soon as possible, lest the ideas pass to someone else. "There is some truth in this," János agreed, "that certain things ripen at the same time and then appear in different places in the manner of violets coming to light in early spring."[66] They published János's findings in 1832, as an appendix to a book on geometry his father had long been preparing.[67]

János's findings proved that the fifth postulate was not a theorem—not a consequence of Euclid's first four postulates—by showing that there are geometries in which Euclid's first four postulates hold true but the fifth does not. He had discovered a consistent and self-contained system of geometry that differed from Euclid's in its properties of parallelism; in Bolyai's non-Euclidean geometry, there are infinitely many lines through a given point that do not meet a given line. With this, Bolyai had performed a seemingly impossible feat.[68] He had discovered a new geometry—"one of the most momentous discoveries ever made," said Gray—but the world simply ignored it. By the time János Bolyai died, in 1860, he had received no recognition for his discovery of non-Euclidean geometry.[69]

With Bolyai's discovery, there were then two types of geometry, Euclidean and non-Euclidean, each rooted in the classical tradition. As a classical geometer, Coxeter carved a unique, surprisingly productive, and far-reaching career from Euclidean geometry, elevating it to complex and hyper-dimensional levels, and he made forays in the non-Euclidean realm as well.[70] Thus, Coxeter was what you might call a modern classical geometer, according to Sir Michael Atiyah, one of the finest mathematicians of our day: "Coxeter's geometry was classical flat geometry, geometry of ordinary space. Then he moved into variations on that, with group theory. And this brings geometry into touch with modern algebra in lots of interesting ways. He was the master of that bridge," said Sir Michael. "But Coxeter stayed in the old world. He didn't become a modern geometer. He didn't embrace modern geometry as a whole. He stayed very close to the spirit of classical geometry . . . He was a virtuoso in that area. Quite unique. He's almost the last classical geometer more than the first modern geometer."[71]

Since Bolyai's time, many more types of non-Euclidean geometry have been discovered. Geometry, broadly speaking, is anything that shares the general ideas of Euclidean geometry. If a few rules are changed, however, then a slightly different "non-Euclidean" geometry results. There is a seemingly infinite diversity of geometries, either classical or contemporary in origin—each logical systems unto themselves and devised for a specific purpose.[72] Some of them, Coxeter waded into headlong (especially projective geometry); some, he approached in spurts (such as topology, also known as "rubber-sheet" geometry, with the four-color problem, regarding the theory of maps); and other areas he touched on scarcely at all (modern curved complex geometry, fractal geometry, and taxicab geometry*).[73] The different geometries evolved slowly, like a genealogy, responding to ideas of the times, and sometimes pushing the envelope. For example, the study of knots

---

*Taxicab geometry measures distances by vertical and horizontal steps—east–west and north–south increments—the way taxis traverse city blocks, rather than by the shortest distance between two points, as the crow flies. Fittingly, the distance units in taxicab geometry are known as the "Manhattan metric."

required the development of topology, which in turn required the development of metric spaces. Whereas differential geometry, the study of curved surfaces via calculus, originated in the mid-1800s and was found to be relevant (along with non-Euclidean geometry) at the turn of the next century in the space-time geometry of Einstein's relativity theory.[74] Many of these branches were "beyond my powers," Coxeter once admitted,

> There are so many branches of the subject in which I am almost as ignorant as the proverbial man in the street. I must ask you to forgive me if I concentrate on my own favorite branches, and I must take the risk of offending various geometers who will ask why I have not dealt with algebraic geometry, differential geometry, symplectic geometry, continuous geometry, metric spaces, Banach spaces, linear programming, and so on . . . Thus there are many geometries, each describing another world: wonderlands and Utopias, refreshingly different from the world we live in.[75]

A different non-Euclidean geometry from Bolyai's, for instance, occurs when you assume there are no parallel lines at all—every pair of lines intersects. One way to illustrate this mind-bending geometry is with a query that Coxeter entertained (one long a part of geometry folklore): If you had your pilot's license and flew ten hours due south, then ten hours directly west, and then ten hours due north, how could it transpire that you would find yourself right back at your starting place?[76]

Flummoxed disbelief is the usual reaction to this question, because the directions are envisioned in the flat Euclidean plane. Coxeter demonstrated this warped perspective in 1957 with a grainy black-and-white television appearance on a Canadian news magazine. In comparing the "nature of space" and alternative geometries to Euclidean, he made use of two blackboards—a standard flat blackboard on the wall, and a swiveling globe of the world painted black. First, Coxeter said, consider an ordinary triangle in the plane. He gestured to his triangle drawn on the regular chalkboard—a traditional Euclidean triangle with angles summing to 180°. Another kind of geometry, he continued, moving toward his globe, is geometry on the surface of a sphere. And then, beginning at the North Pole, he chalked lines on the globe running due south, then traveling due west, and finally due north, leading directly back to his starting point and forming a triangle with his path—a triangle constructed from three 90° angles.[77] So if we choose, Coxeter concluded, "we can find a triangle having right angles at each vertex, and the sum of the three is 270°." This is spherical geometry, one example of a non-Euclidean geometry. Non-Euclidean geometry exists in worlds where, tinkering with

Coxeter demonstrating non-Euclidean spherical geometry with a 270° triangle on a globe.

qualitative and quantitative factors, the angles of a triangle sum to more or less than the traditional Euclidean 180°. It is simply a matter of experiment; mathematicians invent new geometries and then it is left to the physicists to figure out which of these geometries, if any, apply in the real world.[78]

When Bolyai's non-Euclidean geometry eventually gained attention, people began asking, "Which geometry is valid in physical space—Euclidean or non-Euclidean?"[79] Bolyai's new geometry had exposed a firmly entrenched misunderstanding about the nature of space. For ages mathematicians had believed that Euclidean geometry was the one and only logical account of the way the world could be. But, as Bolyai announced: "All I can say now is that I have created a new and different world out of nothing."[80]

△ ▢ ◆ ❂ ❂

Wading into his talk at the commemorative Bolyai conference, Coxeter twirled in his fingers Smith's model, a nexus of multicolored straws—a skeleton of a cube surrounding the skeletons of two interlocked tetrahedra. As he proceeded with his lecture, a rumble of unease stirred in the audience, the skeptics straining to hear. "Louder! Louder, please! We cannot hear!" cried Coxeter's daughter. His microphone wasn't working. Neither was his hearing aid.

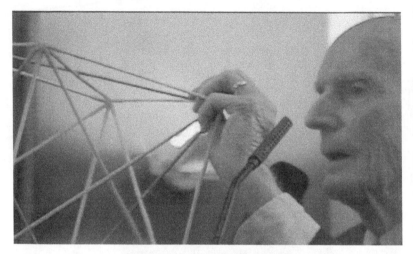

Coxeter at the microphone in Budapest.

Oblivious to the predicament as it was being resolved, Coxeter carried on, the audience scribbling bouquets of tangent circles into notepads on their laps.[81]

Coxeter was hardly a showman. He was a gentleman geometer who held his audience's attention with the beauty and elegance of his work. He was known for being birdlike, both in appearance and style—very delicate, very precise, very spare. "You have to know what's important," said Gray, who likened Coxeter to a violin rather than an entire orchestra. "He's not going to rhapsodize, he's not going to tell you that this is a huge big deal . . . he's not going to write you any advertising copy."[82] Coxeter meets the measure of an elegant and beautiful practitioner also because his mathematics flourishes in the minds of other mathematicians. When a piece of mathematics is called beautiful or elegant, it is presented in a way that conveys understanding, and one litmus test for understanding is whether other mathematicians can do something with it, fitting it nicely into the bigger picture. "It becomes elegant because it opens something up," said Gray. "The elegance is in the power it conveys to do something that couldn't, or hadn't, been done before."[83]

In the end, Coxeter's talk went over well, and it proved relevant—it was related to the hot applied topic of data mining. "His proof [of an absolute property of four tangent circles] is not an earth-shaking discovery," said Karoly Bezdek, the secretary of the Budapest conference committee. "But his proof is the simplest one, the ideal proof for Beecroft's theorem. Nowadays many mathematicians publish with very complicated proofs. It is important to have simple proofs that we can digest and really learn from. It's an art to discover the right proof."[84] The conference organizer, Prékopa, was very pleased as well: "It is amazing that somebody who is 95 years old can invent new scientific results of such depth and present them at a meeting. I wish I could be such a fresh-minded person, and interested and active. Coxeter gets distracted and falls asleep during some of the other talks," he noted (many an audience member was caught nodding off), "but he always wakes up when he's interested."[85]

One widely accepted mathematical truth is that mathematics is a "young man's" game. "Young men should prove theorems, old men should write books," said the legendary G. H. Hardy, a professor of Coxeter's at Cambridge who penned *A Mathematician's Apology*, a lament for his waning mathematical prowess.[86] Hungarian mathematician Paul Erdös (1913–96) is the usual counterexample. Erdös was a prolific problem solver to the end of his life, publishing more than one thousand papers, more than any mathematician in history. He was "the man who loved only numbers," as the title of Paul Hoffman's biography proclaimed.[87] A close friend and collaborator, Ron Graham (introduced to Erdös by Coxeter in 1958), recalled that Erdös "was completely dedicated to, as he would say, 'taking a peek into *The Book*'—'*The Book*'* was this hypothetical book of the Almighty that contains all the best possible proofs, all the gems of mathematics that you can present in a page or two. Erdös really lived mathematics."[88]

Donald Coxeter is an equally good counterexample disproving the stereotype of a mathematician's "best-before date."[89] Coxeter's only professional regret, articulated at the end of his days, was that he had not collaborated with Erdös[90]—his Erdös Number was 2 (as was Einstein's).[91] A person who coauthored a paper with Erdös gained an Erdös Number 1; a person who coauthored a paper with such a person has an Erdös Number 2, and so on, forming an international nexus of Erdös's 485 coauthors. Erdös had no real home base and traveled the world with his battered Mexican leather briefcase of worldly possessions, landing on the doorstep of welcoming or unexpecting mathematicians. Upon arriving at his destination Erdös would announce: "My brain is open!" (the title of another Erdös biography, by Bruce Schechter).[92] His visits were so intensive that Graham often joked, "We had Erdös over for a month last weekend."[93] After Erdös squeezed all the mathematical juice from his host he moved on to his next stop.[94]

---

*Galileo Galilei (1564–1642) also referred to a "grand book" of the universe, and to the importance of geometry in gleaning knowledge of its contents. In *The Assayer* he wrote: "Philosophy is written in this grand book, the universe, which stands continually open to our gaze. But the book cannot be understood unless one first learns to comprehend the language and read the characters in which it is written. It is written in the language of mathematics, and its characters are triangles, circles, and other geometric figures without which it is humanly impossible to understand a single word of it; without these one is wandering in a dark labyrinth."

Left to right: Unidentified woman, Coxeter, Branko Grünbaum, Paul Erdös, circa 1965.

Coxeter had plenty of opportunity to become an Erdös 1. He and Erdös often crossed paths. One day in 1935, when Erdös was at Cambridge, he rang Coxeter and asked him a question regarding a problem he was working on about parallelotopes.[95] Coxeter worked on Erdös's problem for a few days, but it didn't lead to collaboration. They met numerous times thereafter, in London, Toronto, and elsewhere. In 1965, Coxeter noted in his diary: "while shaving I solved Erdös's problem (of the dancing girls and boys)." But still no collaboration.[96]

For Erdös's sixty-eighth birthday, Coxeter dedicated a talk in his honor on "a symmetrical arrangement of eleven hemi-icosahedra,"[97] and the two bounced ideas back and forth in correspondence—a letter from Erdös, always written with a fountain pen, typically began with a brief pleasantry, promptly launching into pages of mathematical proposition, suppositions, equations, conclusions, and a diagram. Coxeter sent Erdös problems he might appreciate, and Erdös contributed ideas to a few of the problems Coxeter was working on,[98] but upon Erdös's death in 1996, Coxeter settled for an Erdös Number 2.[99]

◁ ◻ ◇ ◉ ✦

Coxeter's daughter Susan, not being the least bit mathematically inclined (or even empathetic), wasn't so impressed with her father's intellectual longevity, and in general she ran hot and cold on his status as a mathematical legend. Having solved the problem of the microphone malfunction during his lecture, Susan settled in and read her novel. And at the end of his talk, when Coxeter hightailed it to the loo, Susan gave her evaluation. "To think," she said. "We've come all this way to talk about circles touching circles when there are so many more important things going on in the world. Dad would hate to be equated with Elvis Presley, but Elvis gave people some moments of joy, happiness, inspiration. And if that's what Dad's work does for these people, that's wonderful. Personally, I get more from Elvis Presley."[100]

The day wound down with a reception in the Academy of Sciences ballroom. Conference-goers stood nibbling on a dinner buffet, and scrounged for a miscalculated supply of desserts. Coxeter found one of few seats in the house, a majestic dais elevated above the crowd. A steady stream of admirers stopped by, bowed at his side, and gave him praise. Ernest Vinberg, from Moscow State University, introduced himself and thanked Coxeter for long ago writing a letter to his Soviet-era PhD committee, reassuring them that Vinberg's field of study—Coxeter groups—was not politically suspect (after perestroika Vinberg's PhD was finally conferred, and he proceeded to do a second PhD, also on Coxeter groups).[101] Daina Taimina, a senior research associate at Cornell, approached Coxeter to show him her crocheted model of the hyperbolic plane, and to tell him that his *Introduction to Geometry* was a blessing—"it saved me," she said—when she started teaching high school geometry in Latvia in 1975.[102]

At the close of the festivities Coxeter plodded back to his hotel, the scorching July sun retreating over the Danube. Just then John Ratcliffe, from Vanderbilt University, in Nashville, Tennessee, caught up with him on the sidewalk. Ratcliffe told Coxeter he had two copies of his *Regular Polytopes*—one at work and another in his study at home for late-night consultations. "This is the modern-day Euclid's *Elements*," said Ratcliffe, pulling a copy of the book from his attaché case. "It's like the Bible for me. I refer to it all the time."[103]

All in all, it was a jubilant day for Coxeter. He had managed the trip, delivered an apropos presentation, and been showered with adulation. "It was very satisfactory!" he said, never one for hyperbole in language. Susan deposited her father in his hotel room and withdrew for some time on her own. Coxeter climbed out of his suit jacket, undid his shirt and tie, sat on the edge of the bed, and sipped on some champagne from the minibar.* After the high of the day he was stung by melancholy (as he was a few times during the

---

*His usual bedtime elixir, to fortify his constitution, was a stomach-curdling mixture of Kahlúa coffee liqueur, peach schnapps, sometimes a splash of vodka, and soy milk.

Budapest conference, prompted by a documentary camera following him the entire trip). He thought of how he could have been a better husband, father, and grandfather, spending less time on his work. He thought of the recent invitation he passed up to return to his alma mater, Trinity College, Cambridge, as a newly minted emeritus fellow—a mark of honor that allowed him, if he chose, to live out the end of his days in a room in Great Court, kept company by all his old haunts. He thought of his childhood governess May Henderson, whom he had been known to confess he loved more than his mother. When he was in his late sixties, Coxeter planned to pay May a surprise visit on a trip home to England. He was devastated to find she had died of cancer only two weeks before he arrived. May had taught Coxeter French and Latin, multiplication, division, and quadratic equations. Little did she know, way back then, what a fine mathematical mind she was molding.[104]

CHAPTER 2

YOUNG DONALD IN WONDERLAND

Beside the actual universe I can set in imagination
other universes in which the laws are different.

—J. L. SYNGE, *KANDELMAN'S KRIM*

Coxeter's mother, Lucy Gee, a portrait and landscape painter, had grudgingly relinquished her freedom in favor of her husband's wish for progeny, and their only child, Donald, was born in London, England, on February 9, 1907.[1]

Lucy was a stern and scrawny woman, unphotogenic, her sallow complexion relieved by demure and engaging brown eyes. She preferred sports jackets, knee breeches, and harlequin-diamond stockings to constraining Victorian dresses. She was a decent painter (attending the Royal Academy of the Arts) who jealously guarded her creative realm. Coxeter's father, Harold Samuel Coxeter, was a hobby sculptor and baritone singer. Fittingly, their family home at 34 Holland Park Road, in the Royal Borough of Kensington and Chelsea,[2] sat in a high-society artists' colony whose denizens had included Henry James. "The weather is hideous, the heaven being perpetually instained with a sort of dirty fog-paste, like Thames-mud in solution," ranted the American writer. "At 11 a.m. I have to light my candle to read!"[3]

Harold, a robust, white-haired, flushed-faced man, was an autodidact and read voraciously. If there were ever questions of general knowledge discussed at a party, Coxeter's father always trumped the other guests with trivia.[4] Harold earned his living as a manufacturer of surgical instruments. As a boy he had wanted to become a doctor, but he dutifully joined Coxeter & Son Limited, a family business started in 1836 by his grandfather. The company became well known for inventing a mechanism that anaesthetized surgical and dental patients with a continuous flow of oxygen and laughing gas—a promotional flier boasted that Coxeter & Son won many a "prize medal" at international exhibitions.[5]

Initially, Lucy and Harold named their child simply Donald. Donald himself came to wish it had been left at that, though the dithering that followed formed a tale he often enjoyed telling. The birth certificate recorded his first name officially as MacDonald, after his father's father. His mother added "Scott" in homage to a renowned relative—the British architect Sir Giles Gilbert Scott, designer of the iconic windowpaned red phone booth, as well as the Bankside power station (now the Tate Modern), and the University Library at Cambridge. Then a meddling godparent suggested the boy ought to have his father's name as well. His first name thus became Harold, making him H. M. S. Coxeter, or, as one quick-witted observer pointed out, a ship in Her Majesty's fleet. A simple rotation of the names produced Harold Scott MacDonald Coxeter.[6]

One of the earliest surviving portraits of Donald, painted by his mother, shows him at about three years old, dressed in a frilly collared shirt and knickerbockers, a swag of blond curls hanging at his shoulders. With his feet dangling from a velvet-upholstered bench, he is seated, smacking at the keys of a grand piano, a gift to Lucy from Harold upon Donald's birth. The house was also furnished with a billiards table for Harold in the living room. A cavernous study to the rear was Lucy's, though it provided the setting for Donald's first exposure to geometry (even if subconscious) as he crawled across the oak floor, a symmetrical herringbone pattern of rectangles—in geometric terms, a tiling or tessellation of the Euclidean plane.[7]

At about the same age that Donald posed for his mother at the piano, he demonstrated the first signs of his interest in numbers, staring intently at the financial pages of his father's *Times*, columns upon columns of numerals. Within a few years, his precocious intellect unequivocally showed itself— first with music, not mathematics. Before he was ten, Donald was an accomplished pianist. He learned from his father's friend, musician and composer Ernest Galloway, who played live music in the silent movie theaters to make ends meet. He often dropped by the Coxeters' for impromptu musical ensembles, after which he taught Donald how to play and compose. Donald penned stacks of arrangements. One piece, written over five months when Donald was sixteen, titled *Magic*, contained several movements and was composed as incidental music for G. K. Chesterton's play of the same name. He even went to the trouble of indicating when the curtain should rise and the music fade in and out. Most arrangements he wrote as Christmas or birthday gifts, with dedications to his mother or father.[8]

△ ▢ ◈ ◉ ⊗

Some eighty years later, Coxeter discussed the link between his early mathematical and musical inclinations. "I was interested in the structure of the

The first sheet of Donald's score for *Magic*.

notes of music," he said. "That was somewhat mathematical. I think it's clear that one has to regard [music] as being mathematical: the 12 semitones in the octave and the 8 diatonic notes and how they are different."[9] As for the aesthetic analogy between mathematics and music, Coxeter admitted he could hardly do better than quote his Cambridge mentor G. H. Hardy in describing the intersection of the two arts: " 'There is a very high degree of *unexpectedness* combined with *inevitability* and *economy* . . . A mathematical proof should resemble a simple and clear-cut constellation, not a scattered cluster in the Milky Way.' Similar words might well be used as advice to composers, with 'mathematical proof' replaced by 'piece of music.' "[10]

Coxeter itemized several parallels between these two "precise arts"— replacing a D♯ with a D♭ in a piece of music and switching a plus and minus sign in mathematics would be equally disastrous (whereas, he noted in contrast, "A painting or a piece of sculpture would not be essentially changed if a few of its daubs of paint or lumps of clay had been differently placed, and Keats's "Ode to a Grecian Urn," with its awkward line, 'O Attic shape! Fair attitude! . . .' would be positively improved if the word 'attitude' could be replaced by one that did not clash with 'Attic.' ").[11] He also remarked on historical similarities. The development of music was hampered by lack of notation until the Middle Ages; so was much of mathematics until the invention of Arabic numerals. "In this respect Geometry was exceptional," he said, "because its essential ideas are so simple that they can be adequately expressed in words, especially when accompanied by sketched diagrams. It is pleasant to see how closely the illustrations in the oldest manuscripts of Euclid's *Elements* resemble those that we draw today."[12]

In illuminating how mathematical ideas are "inherent in music itself," Coxeter went on to discuss rhythm, and the time signature, "a fraction whose denominator is a power of 2 while its numerator indicates the number of beats in a bar—usually 2, 3, 4, or 6 (like the period of the rotational symmetry of a crystal)."[13] Harmony provided another example, the pitch of a note being determined by its frequency, or the number of vibrations per second. Here, Coxeter co-opted the words of English physicist Sir James Jeans: " 'It is found to be a quite general law that two tones sound well together when the ratio of their frequencies can be expressed by the use of small numbers, and the smaller the numbers the better is the consonance . . . This was known to Pythagoras 2,500 years ago; he was the first, so far as we know, to ask the question, 'Why is consonance associated with the ratios of small numbers?' "[14] Coxeter continued, commenting that, "One is tempted to see some significance in the fact that the agreeable harmonics, 3, 4, 5, 6, 8, 10, 12 correspond to the numbers of sides of regular polygons that Euclid was able to construct with his chosen instruments, the straight edge and the compass . . . whereas the dissonant harmonics—7, 9, 11, 13—correspond to polygons that cannot be so drawn."[15]

And of course, Coxeter's analysis truly sings when he addressed the point of comparison closest to his domain:

> Most mathematics depends for its appeal on some aspect of symmetry. Symmetry is likewise a guiding principle in musical composition. In a fugue, for instance, the second occurrence of the main theme is usually in a different but related key. Such a transposition is analogous to the geometric operation of translation or parallel displacement. Again, Bach's trick of inverting a theme is analogous to reflection in a mirror. One of the most interesting transformations in elementary geometry is the dilative reflection, which combines a reflection with a dilation or steady increase of size. This has its musical counterpart in Bach's *Wedge Fugue*.[16]

The pleasure Coxeter experienced from writing music transferred naturally into mathematics. "I got the same kind of euphoria from a successful piece of mathematical rediscovery," he said, "that I formerly did in writing a piece of music."[17] Lucy sought an evaluation of her son's musical talent, taking him to see British composer Gustav Holst. "I don't know how she got to him," Coxeter recalled, "but she took me along and I showed him some of the music I had written, and I played a little bit on the piano. On the whole he thought it was rather poor." They received much the same response from a visit to Irish composer C. V. Stanford, who said: "Educate him first."[18]

△ ▢ ◈ ❂ ❀

Donald's joy in music and math was his salvation from an anxious environment at home. His parents' common passion for the arts was not enough to make them a happy couple. For starters, his father wanted more children, but Lucy did not. Harold attended Royal Psychological Society meetings to cope with the marital difficulties. There he befriended a German divorcée, Rosalie Gabler, who along with her daughter, Katie, six years older than Donald, became family friends of the Coxeters'. Relations did not improve, however, and Donald's parents took steps toward ending their marriage. The indignity of divorce was great in that day—"It was taboo," Coxeter recalled. Divorce was rare and required proof of adultery. The shock was enough to send Donald headlong into alternative worlds of his own creation, taking refuge in music, mathematics, and make-believe.[19]

Coxeter's first teacher was his dearly loved nanny, May Henderson. In confessing he loved May more than his mother, Coxeter explained it was perhaps due to the long periods of time he spent separated from his mum

Young Donald at work.

during the First World War. He and May lived outside London, at the Coxeters' weekend cottage in the south of Kent, near the border of Surrey.[20] There they were at a safe remove from Germany's zeppelins, the passenger airships doing double duty as bombers.[21]

May's curriculum distracted Donald, and inspired him. Her introductory lessons in French and Latin moved him to do something many children fantasize about but few actually execute: He created his own language—Amellaibian. He filled a 126-page notebook detailing the construction of this language and the imaginary world where it was spoken. The inhabitants were called " 'bainia,' a spherical kind of fairy," who drew life energy from batteries made of cork, wood shavings, paraffin, lanoline, white paint, Vaseline, and cloth, all sedimented together, layer upon layer, within glass casings of various shapes. He wrote this novella in impeccable uppercase letters. It contained vocabulary lists ("The Terminations of Amellaibian Words," neatly divided into verbs, nouns and pronouns, adjectives and adverbs), maps, histories, genealogies, short stories, and a section called "Fairies' Birthdays and Other Events." Much of the narrative chronicled the fairies' romantic adventures and happy unions. Gradually, the treatise turned mathematical, with

Amellaibian Numbers.

In Amellaibian, if two numbers are hyphened together, they are multitlied (as CLARRIF-GLANTÓ = 4 × 10 = 40); and if they are put as two seperate words they are added as (GLANTÓ CLARRIF = 10 + 4 = 14). In the first case, the second word of the number, the multiplier, is always GLANTÓ, GLANTIN, GLANTINI, GLANTINÓ, GLANTININ, etc.; and in the second case, the first word of the number is.

1 = CLARRIN.	11 = 10 + 1 = GLANTÓ CLARRIN.
2 = CLARRINT.	12 = 10 + 2 = GLANTÓ CLARRINT.
3 = TENTIÓ.	13 = 10 + 3 = GLANTÓ TENTIÓ.
4 = CLARRIF.	14 = 10 + 4 = GLANTÓ CLARRIF.
5 = TENTIF.	15 = 10 + 5 = GLANTÓ TENTIF.
6 = CLARRIS.	16 = 10 + 6 = GLANTÓ CLARRIS.
7 = CLARRISAI.	17 = 10 + 7 = GLANTÓ CLARRISAI.
8 = CLIARIAI.	18 = 10 + 8 = GLANTÓ CLIARIAI.
9 = TENTIN.	19 = 10 + 9 = GLANTÓ TENTIN.
10 = GLANTÓ.	20 = 2 × 10 = CLARRINT-GLANTÓ.

21 = 2 × 10 + 1 = CLARRINT-GLANTÓ CLARRIN.
22 = 2 × 10 + 2 = CLARRINT-GLANTÓ CLARRINT.

38.

A page from Donald's primer on his invented language, "Amellaibian."

pages and pages dedicated to weights and measures, formulas, equations, and Amellaibian magic numbers—any number that factored into Donald's favorite number at the time, 250.[22]

While Donald was enrapt with his Tolkien-esque fairy tale (predating Tolkien), Harold and Lucy were in the midst of their divorce. After May Henderson left to get married, Donald's parents sent him to the coeducational St. George's boarding school, twenty kilometers north of London, to shield him from any nasty indiscretions during the divorce proceedings.[23] His father dropped by the house to see him off: "Donald was such a dear and looked so nice in his school things and was so very glad to see me. He's so sensible, and brave about school. But he will be dreadfully homesick."[24]

"I was incarcerated at boarding school," Coxeter later said. "The headmaster was something of a freak. He took a sadistic interest in caning the bottoms of boys who behaved badly. My father explained to me later that he got sexual pleasure from beating boys on their bums."[25] Coxeter and his one friend at boarding school, John Flinders Petrie, son of Egyptologist and adventurer Sir William Matthew Flinders Petrie, suffered under the headmaster's cane quite a lot. Donald also found his parents' visits disconcerting, since they came separately, alternating weekends. Epidemics of measles or chicken pox often quarantined the schools, preventing his parents from taking him on outings. He looked forward to walking as far as he could down the school's driveway, to the edge of the quarantine, to meet his mother or father when they arrived. They spent most of their visit in a room at the back of the school, usually reserved for practicing the piano. Once during a visit with his father, Donald suddenly burst into tears, uncontrollable weeping. His father tried to comfort him, but he was unconsolable, overcome with the trauma of his parents' separation. "It was too much at that adolescent age," Coxeter recalled. "I was weeping at the collapse of my family."[26]

Harold's announcement of his plans to remarry only made matters worse. Donald and his mother, and everyone among family and friends, assumed his new wife would be Rosalie Gabler. In hindsight, family members speculated as to whether Harold had occupied her bed, or insinuated the pretense of a romantic liaison, in moving along his divorce.[27] But Rosalie was too old to bear Harold the children he wanted, so he proposed to her daughter, Katie, instead. Rosalie sent Katie off to Munich for a year in an attempt to cool the affair, or at least test its mettle. The May-September couple married in 1922, when Katie was twenty-one and Harold forty-three. For Donald it was a double blow. Katie had been the object of his first crush.[28]

Donald idolized his father, the pivot point of his life and the dominating influence. But as Harold began his new family, becoming a father to three daughters in rapid succession (Joan, Nesta, and Eve), Donald was no longer the sole focus of his attentions, even during their father-son visits. "I took Donald to town yesterday—morning at the office and afternoon at a cinema," wrote Harold to Rosalie in 1924. "He's a queer kid and not very easy. He practically ignores Joan [his eldest half sister]—not that he is the least unfriendly or jealous, I think, but simply that she doesn't interest him, not being mathematical or—so far—particularly musical."[29]

At the uneasy and vulnerable time of adolescence, Donald's broken family hurt him to the heart—into his nineties Coxeter's memories of his parents' broken marriage stayed with him as one the greatest tragedies of his life, bringing him to tears even then. As a boy he was deeply sensitive and idealistic, clever and solitary, and his parents' failings were horribly disappointing. They had read him the Bible every night before bed (they explored Quakerism for a time), and were his closest playmates. Photographs show Donald and his mother and father on picnics with only adults for company, and sometimes a chicken or a dog. Donald hardly knew how to relate to other children and didn't make many friends at St. George's. He was ridiculed and bullied for his brainy peccadilloes, girls tittering as they kicked at his shins under the desks. "One boy had a grudge against me because I was a weakling," Coxeter recalled. "That's why I dreaded the break between early lessons and late lessons. All the children went out to play games and I hid under the teacher's desk, to avoid this teasing boy."[30]

These desperate circumstances worked as a catalyst. During his incarceration at St. George's, Donald experienced his formative encounter with geometry.[31] The only problem being that just as Donald fell in love with the jewels of classical geometry, the tradition was falling decidedly out of fashion.[32]

The golden age of classical geometry had been the middle of the nineteenth century. The science of triangles and circles, to be sure, was founded in the era of the great Euclid, Pythagoras, and Archimedes, but theorems on triangles and the like accumulated until the central cache of knowledge amassed in the 1800s, and to a dwindling extent the beginning of the twentieth century.[33]

"Geometry sometimes has had to fight for its existence," observed Jeremy Gray. "You might think, if you were not a mathematician, that of course geometry is important. Ask people what mathematics is and they would probably say geometry in the first minute of their answer. But in some math departments there is a feeling that, 'Oh we've left that behind . . . Who wants to prove things about triangles?' "[34] The classical visual and intuitive approach with circles and triangles and polyhedrons had come to be thought of as "playing," "tinkering with toys,"[35] and "second-rate math"—amusements for idling away daydreams and Sunday afternoons.[36]

The luster of Euclidean geometry first began to fade after the dethronement of the parallel postulate in the mid 1800s. After that, "logical worries" crept into Euclidean geometry as a whole. It had been the paradigm of truth, the bedrock of empirical pursuit; three-dimensional Euclidean geometry had been a concrete perspective that jibed well with the scientific rationalism of the Victorian era, and it neatly explained the real space in which we lived. Or so mathematicians thought. Then it came time to face reality: geometers had been constrained to one space, instead of roaming two, or more, for millennia.[37]

Mathematicians began questioning the veracity and reliability of Euclid's *Elements* as a whole. His dependence on diagrams drew the most criticism. And Euclid's original work included not only an abundance of diagrams, but also a collection of intentional geometrical fallacies made believable by convincing but flawed figures—Euclid thought it was a good exercise for the students to find the errors in reasoning.[38] The lesson now appeared not so constructive: diagrams are deceptive. And as a result, the pendulum swung to the opposite extreme; the trend became mathematics without the disinformation of pictures, without appealing to the corruptible visual sense.

To restore the faith, mathematicians took to formalism, a term spoken pejoratively among disheartened classical geometers. Formalism embraced a systematic method, enforcing the logic of geometry axiom by axiom—pictorial geometry was rendered abstract and algebraic through the power of sheer deduction; it was all numbers and equations.[39] This rigorous rote method was no way to teach geometry to children, however, and the consequences at the grade school level were grim—this mobilized the Association for the Improvement of Geometrical Teaching, founded in England in 1871.[40] But there was no stopping the march of modern mathematics. The eminent David Hilbert, at Göttingen—known for his Hilbert space—was the beacon of formalism, the prime mover pushing modern geometry toward a more formalist and abstract style.[41] He published his book *The Foundations of Geometry* in 1899, the first thoroughly systematic study of Euclidean space, supplanting Euclid's axiomatization—with this work Hilbert is said to have made the greatest impact on geometry since Euclid. And, Hilbert famously underscored the arbitrary nature of visual space, and the importance of keeping geometrical terms abstract, with this catchy remark: "One must be able to say at all times—instead of points, straight lines, and planes—tables, chairs, and beer mugs."[42]

△ ▢ ◈ ⊛ ⬡

While the non-Euclidean revolution changed the foundation of geometry and "kicked up a fuss," as Gray described, "It was never the fuss that the fourth dimension caused—Woooo woooo! The fourth dimension!!"[43] Flouting tradition by stretching the boundaries of three-dimensional Euclidean space into higher realms—into hyperspace, or what's called *n*-dimensional geometry—was also a trend very much in the air at the latter part of the nineteenth century, as the restrictions on rigid three-dimensional Euclidean geometry loosened.[44]*

Edwin A. Abbott concocted a fictional world with inhabitants who contemplated higher dimensions in his book *Flatland, A Romance of Many Dimensions*, published in 1884 under the pen name A. Square. The main character, Square himself, is visited in his two-dimensional land by an alien creature named Sphere, from three-dimensional Spaceland. Square gets acquainted with this odd creature, and in so doing becomes convinced Sphere is not a "burglar or cut-throat, some monstrous Irregular Isosceles," who by disguising himself as spherical had gained entry to his house with plans of stabbing him with his acute angle. Square comes to believe that there is in fact life in Spaceland after he learns to visualize three-dimensional entities.[45]

In relaying his experiences, Square aimed to be instructive and encouraging, yet when he voiced his views about higher dimensions in *Flatland* he was imprisoned for heresy. And indeed, an unorthodox mystical bent, popular in spiritualistic and theosophical circles, manifested itself in the proliferation of writings on the fourth dimension.[46] English mathematician Charles Howard Hinton was notorious for his mystical leanings, but he also devised a very down-to-earth bridge, a mental aid, to expedite the crossing into hyperspace—a system of multicolored cubes, which he explained in an essay called "Casting Out the Self."[47] Coxeter encountered Hinton's work at about the age of thirteen. He absorbed Hinton's book, *The Fourth Dimension*, during the ensuing years, and it opened wide this gosling geometer's insatiable appetite for polytopes.[48]

Another influence on the young Donald was H. G. Wells's science fiction

---

*As early as 1827, German mathematician August Möbius (1790–1868) hypothesized that a trip through a fourth spatial dimension could transform an object into its mirror image. In 1909, *Scientific American* issued a call for explanations of the fourth dimension through an essay contest, and many essays explored similar mirror reversals.

novel *The Time Machine*: "Space . . . is spoken of as having three dimensions, which one may call Length, Breadth, and Thickness," wrote Wells. "But some philosophical people have been asking why three dimensions particularly—why not another direction at right angles to the other three? . . . Well, I do not mind telling you I have been at work upon this geometry of Four Dimensions for some time."[49] Published in 1895, *The Time Machine* was still a best-seller in 1920, just when Donald was off to boarding school.

Also in the air at the time was the fallout from the total eclipse of the Sun on May 29, 1919. During the eclipse, British astronomer Sir Arthur Eddington had measured the bending of starlight by the Sun, confirming Albert Einstein's theory of general relativity, a geometrical theory postulating that the presence of mass and energy generates gravity, and that gravity has the effect of "curving" space and time—gravity = space-time geometry, a continuum with three dimensions of space and a fourth dimension of time. The media event made Einstein immediately famous.[50] The London *Times* headline proclaimed, REVOLUTION IN SCIENCE, NEW THEORY OF THE UNIVERSE. Two days hence the *New York Times* answered with, LIGHTS ALL ASKEW IN THE HEAVENS/MEN OF SCIENCE MORE OR LESS AGOG OVER RESULTS OF ECLIPSE OBSERVATIONS/EINSTEIN THEORY TRIUMPHS.[51]

With the fabric of space changing before his eyes, Coxeter experienced a personal geometric epiphany. His formative encounter with the study of shapes and space occurred when he wound up in his school's sickbeds, lying next to his friend John Petrie. Woozy with the flu and surrounded by the sick-bay smells of antiseptic, freshly laundered sheets, and the coal fireplace burning at the end of the room, he and John lay there musing about the mysteries of the world.

"How do you imagine time travel works?" John asked.

"You mean as in *The Time Machine*?" replied Donald. After thinking for a moment, he answered John's question. "I suppose one might find it necessary to pass into the fourth dimension."

The fourth dimension that most intrigued Donald, however, was spatial,

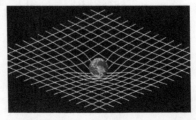

Einstein's general relativity, espousing a "spacetime geometry," is often summarized as follows: Matter tells spacetime how to curve, spacetime tells matter how to move.

not temporal. Having recently studied Euclid in math class, the two boys pondered for a while why there were only five Platonic solids, and then whiled away the time imagining how to stretch the Platonic solids into the fourth dimension.[52] Even more intriguing than the reason why the teenaged Donald and John reveled in such a heady exercise, is the question of how, exactly, they set out with their mental machinery and traveled into hyperspace.

Put the question of how to think in four dimensions to John Horton Conway and he jokingly snaps back: "None of your business! That's personal!"[53] Conway occupies the John von Neumann Chair of mathematics at Princeton University. Coxeter called Conway "a great friend!" whenever he mentioned him, even though they engaged only on a professional level.[54] Conway has a boyish dishevelment about him, but all the same he is an imposing presence, statuesque even, a grinning Archimedes of the twenty-first century. He is best known for inventing the Game of Life, surreal numbers, and his "Conway group" or "Conway's constellation"—a beastly group of sporadic symmetries (sporadic, because they do not fit into any classification scheme). Conway calls these "white hot" discoveries that had him walking around in a world all his own for weeks. And he characterizes such a piece of mathematics by his "Hotspur property,"[55] in reference to a character in Shakespeare's *King Henry IV*. In act 3, Glendower says, "I can call spirits from the vasty deep . . ." To which Hotspur replies: "Why, so can I, or so can any man; But will they come when you do call for them?"[56]

Conway calls himself an honorary student of Coxeter's. He never studied with the great man, but much of Conway's work is Coxeterian in nature. And Conway is considered by some to be Coxeter's successor—they held in common a wide-ranging mathematical curiosity and a profound geometric spirit.[57] In 1957, when Conway was a teenager in his first year at Caius College, Cambridge, he sent Coxeter a fan letter.

Dear Professor Coxeter,

Over the past year or so my copy of your edition of Ball's "Mathematical Recreations" has accumulated an astonishing number of notes and some corrections. Most of these can hardly be said to be suitable for publication in later reprints, but one or two may seem important.[58]

The letter went on for five pages; tiny scrawls interrupted by a very surehanded rendering of a four-dimensional cube, or a hypercube. Conway had discovered that by labeling the vertices of a hypercube in a certain way he could derive a number of magic squares.* Eventually Conway signed off, with a query about a four-dimensional polytope:

> My absolutely last remark is a question. Where can I find the requisite information required to draw {5,3,3}, or do I have to work out the details for myself? I should be very thankful if you could supply me with some accessible information.
>
> Yours hopefully,
> J. H. Conway[59]

A while later at Cambridge, Conway made an earnest effort to train himself to think in four dimensions. He did not expect to see *the* fourth dimension, as if it were a physical reality. Time is most often thought of as the fourth dimension, but higher dimensions can measure any value or feature of existence.[60] The fourth dimension could be temperature or wind direction, the fifth dimension could be the rate of interest on your credit card, and the sixth dimension could be your age, and on and on and on as you please. Each characteristic measured adds another "dimension"—the dimensions become coordinates, a navigational tool that quantifies our existence, our position in the world. Being geometers, Conway and Coxeter naturally preferred contemplating a fourth dimension in terms of space.[61]

In attempting to visualize a fourth coordinate or dimension in space, Conway built a device that allowed him to see with "double parallax"—in addition to the displacement that occurs horizontally when you look at an object by closing one eye and then the other, he tried to train himself to see vertical parallax. If he could experience both horizontal and vertical parallax, he would have four coordinates for every point in space, and thus would be seeing four dimensions. In his attempt to do so, Conway donned a recycled motorcycle helmet, adapted with a flat visor and cheap, old war-surplus periscopes. The periscopes were bolted to the visor (not very well; they rattled when he walked) and extended from his right eye up to his forehead and his left eye down toward his chin. The only name Conway had for the helmet was "that damned contraption" because it was rather uncomfortable—his nose was pressed up against the visor, as a child's to a toy shop window at Christmas.[62]

---

*A magic square is a square array of numbers arranged in such a way that the sum of the numbers in any horizontal, vertical, or diagonal direction is always the same. The most famous magic square has been long known in China as the Lo-shu—it is the 3 × 3 arrangement of numbers from one to nine such that all the straight lines of three add up to fifteen.

Conway had a strong desire to see four dimensions, which he truly believed was possible (and still does). He walked around wearing his helmet in the Fellows Garden of his college at Cambridge. "I suppose I had a limited amount of success in that quixotic quest," he recalled. "I got to the point where I could see four dimensions, but there was no hope of going beyond, so what's the point?"[63] His discoveries since his helmet days are in dimensions much, much higher—the Conway group is in twenty-four dimensions, and the group he studied and dubbed the Monster group exists in 196,884 dimensions.[64]

△ ⬭ ◈ ⊛ ⬡

Addressing the exercise of thinking in four dimensions in his book *Regular Polytopes*, Coxeter offered three methods: the axiomatic, the algebraic, and the intuitive.[65] Coxeter preferred the third, and seeking some historical enlightenment to bolster his position he quoted French mathematician Henri Poincaré, a staunch advocate for the use of intuition and pictures in mathematics: "A man who devoted his whole life to it, might succeed in visualizing the fourth dimension."[66]

Coxeter would have been slightly more optimistic. Before he began with this pursuit, however, he issued a cautionary disclaimer. It is an "insidious error," he said, to assume that "because the fourth dimension is perpendicular to every direction known through our senses, there must be something mystical about it."[67] He then dashed off a footnote quoting the Platonist philosopher Henry More—"Spirits have four dimensions"—by way of a wrongheaded example.[68]

After those words of warning, he proceeded by acknowledging that visualizing four dimensions is no walk in the park for ordinary mortals who are accustomed to a firm three-dimensional footing. "But a certain facility in that direction may be acquired," he encouraged, "by contemplating the analogy between one and two dimensions, then two and three"—as Abbott did with Flatland—"and so (by a kind of extrapolation) three and four. This intuitive approach is very fruitful in suggesting what results should be expected."[69]

As a teenager, Coxeter had used the very same method while imagining four dimensions in the infirmary, and afterward he wrote a school essay on the subject called "Dimensional Analogy." In the introduction he began by saying,

The number of dimensions possessed by a figure is the number of straight lines each perpendicular to all the others which can be drawn on it. Thus a point has no dimensions, a straight line one, a plane surface two, and a solid three . . .

In space as we now know it only three lines can be imagined perpendicular to each other. A fourth line, perpendicular to all the other three would be quite invisible and unimaginable to us. We ourselves and all the material things around us probably possess a fourth dimension, of which we are quite unaware. If not, from a four-dimensional point of view we are mere geometrical abstractions, like geometrical surfaces, lines and points are to us. But this thickness in the fourth dimension must be exceedingly minute, if it exists at all. That is, we could only draw an excessively small line perpendicular to our three perpendicular lines, length, breadth and thickness, so small that no microscope could ever perceive it.

We can find out something about the conditions of the fourth and higher dimensions if they exist, without being certain that they do exist, by a process which I have termed "Dimensional Analogy."[70]

More specifically, the process of dimensional analogy works by one of two means: section or projection. "According to the first method," instructed Coxeter, the inhabitants of Flatland "would imagine the solid figure gradually penetrating their two-dimensional world, and consider its successive sections."[71] This is like dipping a cube in water—as the corner breaks the surface, and then more and more of the cube slides in, you envision the cross section of the solid as delineated by the waterline. "The sections of a cube, beginning with a vertex," Coxeter said, "would be equilateral triangles of increasing size, then alternate-sided hexagons,* 'truncated triangles,' and finally equilateral triangles of decreasing size, ending with a single point—the opposite vertex."[72] While Flatlanders imagined three-dimensional solids scanned through their two-dimensional reality, Coxeter and his ilk fathomed solids of four dimensions, or more, slicing through our three.

According to the second method—projection—Flatlanders studied the shadow of a solid figure in various positions (as Aristotle did with the Earth's shadow on the Moon, determining that if the shadow is always a circle the Earth itself must be spherical). A cube's shadow projected from a light directly above one face would appear as a square, while a shadow projected from a light directly above one corner would appear as a hexagon.[73] Four-dimensional polytopes, similarly, can be projected down to three dimensions.

Despite the fact that intuition was Coxeter's forte he conceded that intuitive results should be checked by one of the other two procedures, the axiomatic and algebraic methods. "For instance," he said, "seeing that the

---

*The process of sectioning lends itself to another Coxeterian geometric trick: try to cut a cube of cheese so that the cross section (the cut line) is a hexagon.

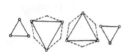

Icosahedron, vertex first

Dodecahedron, vertex first

Icosahedron, edge first

Dodecahedron, edge first

Icosahedron, face first

Dodecahedron, face first

Sequences of "parallel sections," slicing through the icosahedron and the dodecahedron, from Coxeter's *Regular Polytopes*.

circumference of a circle is $2\pi r$, while the surface of a sphere is $4\pi r^2$, we might be tempted to expect the hyper-surface of a hyper-sphere to be $6\pi r^3$ or $8\pi r^3$. It is unlikely that the use of analogy, unaided by computation, would ever lead us to the correct expression, $2\pi^2 r^3$."[74] Using algebraic computation, we can orient ourselves in this new abstract reality by allowing any point in four dimensions to be represented by one of Descartes' inventions—Cartesian coordinates—just like any point on a three-dimensional graph.

In his work *La Géometrie*, French mathematician and philosopher René Descartes (1596–1650) applied the symbols of algebra—a "barbarous" medium, he called it[75]—to the shapes of geometry, thereby inventing his Cartesian coordinates and Cartesian, or analytic, geometry.

As a student Descartes scorned philosophy and ethics; only mathematics gave him satisfaction, he said, "on account of the certitude and evidence of their reasonings."[76] Obtaining a law degree did nothing to convince him otherwise, so he decided to devote his life to learning and applying the methodology of mathematics to master the secrets of the universe. He expected to discover a complete and all-encompassing truth explaining every aspect of existence within his lifetime. He told sick friends to hang on just a little longer, cures for what ailed them, even the secret for eternal life, were on the way.[77]

Although Descartes did not accomplish these grand ambitions, he succeeded in instituting an entire new regime for the study of geometry. He had never liked the Greek approach—he thought it was obscure and fatiguing to the imagination.[78] He may have found, as many have since, that when facing a problem in elementary Euclidean geometry one often has no clue where to begin, the only recourse being to wait helplessly for inspiration.[79] Descartes did away with these uncertainties by introducing lines and shapes to a quantified graphical construct. Shapes could now be investigated with precision, each line represented by an equation, a steam engine that drove a new kind of geometrical proof and discovery.[80]

Cartesian geometry demarcated space between two axes at right angles to each other, x being horizontal and y being vertical, forming a two-dimensional xy plane, with any point on that plane identified by (x,y) coordinates. This graphical domain evolved to include a third dimension of measurement, the z-axis, producing (x,y,z) coordinates, and then to include any number of dimensions. Coordinates of a point in four-space are customarily denoted by (x,y,z,w).[81]

Descartes' ideas about analytical geometry came to him in a dream, revealing, just as he wished, "the magic key which would unlock the treasure house of nature and put him in possession of the true foundation . . . of all the sciences."[82] He published his *Method of Rightly Conducting the Reason and Seeking Truth in the Sciences* in 1637. He outlined his analytic treatment of geometry in one of the book's three appendices, *La Géometrie*.[83]

Within the decade, Descartes' geometry became part of university curriculum. In historian E. T. Bell's estimation, this alliance between geometry and algebra set the stage for classical geometry's near demise three centuries later. "Algebra is easier to see through than a cobweb of lines in the Greek manner of elementary geometry," he argued. The real power of the new method lay in its capacity to reduce geometry in its entirety to algebra. "We start with equations of any desired or suggested degree of complexity and interpret their algebraic and analytic properties geometrically. Thus we have not only dropped geometry as our pilot; we have tied a sackful of bricks to his neck before pitching him overboard. Henceforth algebra and analysis are to be our pilots to the unchartered seas of 'space' and its 'geometry.'" Bell also noted: "Though the idea behind it all is childishly simple . . . the method of analytic geometry is so powerful that very ordinary boys of seventeen can use it to prove results which would have baffled the greatest of the Greek geometers—Euclid, Archimedes, and Apollonius."[84]

△ ▢ ◈ ◎ ⬡

Donald's writings on dimensional analogy won him a school essay prize and grew to five notebooks—filled with both visual diagrams and charts of algebraic computations. Family history, as proudly replayed by Coxeter's three half sisters, his children, and Coxeter himself, records that when his prodigious mathematical talents came clearly into view, he was taken by his father to see an expert in the field: mathematician and logician Bertrand Russell.[85] Author of *The Principles of Mathematics* (1903), Russell was one of England's good and great men. He had been dismissed from his position at Trinity College in 1916, when he was convicted and later imprisoned for antiwar activities.[86] Donald's father, also a pacifist, had made Russell's acquaintance at conscientious objector meetings in London. They became friendly, and when Russell and his second wife opened an experimental school for young children in 1927, it would be on land lent to him by Harold Coxeter. When asked for his opinion of this boy wonder, Russell suggested Donald meet Eric H. Neville, the mathematical scout who had brought self-taught numerical genius Srinivasa Ramanujan from India to study at Cambridge in 1914.[87] Donald was also recommended to Neville by the Fabian socialist and suffragette, Professor Edith Morley,[88] whose letter of endorsement read:

Dear E.H.,

I have taken a liberty which I hope you will forgive! A certain Donald Coxeter, aged 15, who is supposed to be a rather unusual mathematician and musician for his years, has spent his summer holiday writing what I am told is an entirely original treatise on the fourth dimension. The boy is a friend of my friend Mrs. McKillop: I don't know him personally but I have heard a great deal about him and know that he does not get any real sympathy or understanding at school in his mathematical pursuits.

I think you will forgive me for sending him word he may write to you and ask you to help him. Apparently he has read your little book (I think I am right in saying this): at any rate he has heard of it and feels you are the one person who can help him.

If there is no promise in his work, you can easily choke him off: if there is, your advice may be invaluable to him. He is to go to Cambridge later on. He will write to you direct when he plucks up courage to do so and I hope you will not think either of us very presumptuous.

Yours V. Sincerely, Edith Morley[89]

On exactly the same day, September 11, 1923, Coxeter also put pen to paper:

Dear Professor Neville,

Professor Edith Morley says I am to write to you and say she suggested it. I
am going to buy your book on the Fourth Dimension, as I am awfully keen
on that sort of thing. I am writing a book myself on "Dimensional Analogy," of which I enclose an outline . . .

Yours Hopefully, Donald Coxeter[90]

A month later, Donald received a reply and zipped off his return: "I was
thrilled to get your letter," he wrote. "I confess I had given you up almost."[91] Neville arranged to meet Donald at the boarding school and grilled
him with loaded questions.

"Do you know what a limit is?" Neville asked. Donald floundered, giving
various poor definitions.

"What has a limit? What could have a limit?" Neville prodded.

"Well, a function or number," Donald replied.

"You should have said a sequence!" Neville corrected. "You must leave
school at once! They're not teaching you right!"[92]

Neville advised Donald to drop all subjects, save mathematics and German (many of the best mathematicians and texts being German), and fast-
track with private tutelage for Cambridge.[93]

A suitable tutor was found in Alan Robson, a well-known mathematics
teacher of the day, and senior mathematics master at Marlborough College,
two hours or so west of London. Stonehenge became a frequent getaway
spot, a hilltop clearing with vistas in all directions, the perfect setting for
Donald to invite visual confections of space and let them wander through his
head. At Marlborough, Donald rented a room with a family in town and rode
his bicycle every day to the school. Robson coached him during a spare
period—the school would not enroll a boy as old as sixteen, for his mind was
no longer a blank slate, having been sullied by years of teaching elsewhere
(most students enrolled at the school in the primary grades). When Donald
began, his marks ranked at the bottom of all Robson's students (in what
would be his class). His obsession with the fourth dimension caused him to
be dismally behind on some of the basics. To correct the imbalance, Robson
insisted Donald focus on his deficiencies. His tutor forbade him from thinking in the fourth dimension, except on Sundays. Donald did his best to ab-
stain from relations with the polytopes, and as a result, from 1923 to 1925,
his marks skyrocketed and he earned the highest standing among his peers.[94]

Donald wrote the Cambridge entrance exams in 1925 and was accepted
at King's College. Robson felt he could, and should, do better—mathematics

at Cambridge's more illustrious Trinity College was unmatchable. Donald completed another year of study, took the exams a second time, and won a scholarship to Trinity.[95]

He was sent on his way with one final gift from Robson. His tutor suggested he submit some of his work to the *Mathematical Gazette*, a time-honored mathematical periodical, founded in 1894 by the Mathematical Association.[96] Over the years, Russell, Bell, J. E. Littlewood, and G. H. Hardy graced the pages of the *Gazette*.[97] With a push from Robson, Donald sent in his work evaluating the volume of a spherical tetrahedron, which led him to some definite integrals.[98] In volume 13, published in 1926, Coxeter proposed: "Can any reader give an elementary verification of the results which have been suggested by a geometrical consideration and verified graphically?"[99]

With his query dangling in the mathematical ether, Donald marched off to Cambridge for the fall term of 1926, bolstered by a substantial supply of homemade marzipan from his mother (he was careful not to eat too much; he allowed himself only a little each day to make it last as long as possible).[100] His good friend John Petrie went to University College London, but they kept in touch.[101] Petrie made many productive trips to Trinity, and he and Coxeter continued contemplating new geometric shapes, which led to a trio of discoveries.

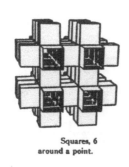

Squares, 6
around a point.

Hexagons, 4
around a point.

Hexagons, 6
around a point.

The skew polyhedra, or regular sponges, discovered by Petrie (top, and bottom left) and Coxeter (bottom right). The black areas are considered holes.

Petrie had previously invented an ingeniously unique way of viewing a polyhedron, tracing its edges in a zigzag pattern until you find you have returned to the vertex from which you embarked (the only rule being that you can trace two consecutive edges belonging to a face, but not three; after tracing two edges you must move your route along, traversing a different face). With a cube, the resulting shape is a "skew hexagon," but in tribute the general term is a "Petrie polygon." In 1926, during one of Petrie's early visits to Trinity, he and Coxeter generalized the concept of a regular skew polygon to that of a regular skew polyhedron. As a result, Petrie discovered two completely new geometric beings, and Coxeter discovered one. Not bad for two nineteen-year-old geometers. These entities are now known as the Coxeter-Petrie polyhedra.[102]

As the crisp autumn weather turned cold in November, Coxeter received in the mail a response to his *Gazette* query—a registered letter from G. H. Hardy, then a professor of geometry at Oxford and recognized as the greatest mathematician in England. "I tried very hard *not* to spend time on your integrals," Hardy scribbled around the perimeter on one of several pages of calculations, "but to me the challenge of a definite integral is irresistible."[103] With that, Donald Coxeter performed a rite of passage. He had entered the mathematical dialectic, striding alongside Hardy no less. He was floating on air for days.

CHAPTER 3

AUNT ALICE, AND THE CAMBRIDGE CLOISTER

The art of doing mathematics consists in finding that special case
which contains all the germs of generality.

—DAVID HILBERT

Students who chose mathematics as their path of higher learning at Cambridge met with an immediate academic hazing—"struck by the shattering blow that for three years they are to do mathematics, all mathematics, and nothing but mathematics!"[1] This warning, issued in a university publication, advocated reform to Cambridge's pure mathematics curriculum to make it less aloof and remote, and more relevant to the real world. The antiquated course of study was limited to "the exceptional being who could live through years and years of nothing but mathematics for its own sake . . . That not many stand up to this impact is shown by the large number of students who after one year change to economics, to physics, to anything but mathematics . . . The normal student soon suffers from an attack of mental indigestion and brings up mathematical wind."[2]

Coxeter's director of studies was John E. Littlewood,[3] an analyst and another of Britain's revered mathematicians. Littlewood expressed his view of geometry in his book *A Mathematician's Miscellany*, wherein he stated that a good measure of mathematical talents was to ask the individual under evaluation: "What did you get out of geometry in school?"[4] Coxeter, having spent two years studying his custom-made curriculum with Robson, and factoring in his private obsession with polytopes, was certainly better versed in geometry than most. On the whole he was prepared as best he could be for the rigors of pure mathematics at Cambridge, the undisputed center of mathematics in Britain. Through the 1920s and '30s, Cambridge came to rival leading universities anywhere in the world. Two noteworthy developments included the creation of the PhD degree in 1924,[5] and the increasing appearance of women—Dame Mary Cartwright, for instance—in the pages of research journals.[6]

Of all the disciplines of study at Cambridge, mathematics was the oldest and most respected, and as such it was one of the last bastions for training in classical geometry.[7] As Coxeter's academic shepherd, Littlewood advised which lectures to attend: analytic geometry, projective geometry, differential geometry, topology, theory of groups, theory of numbers, as well as electricity, celestial mechanics, the theory of relativity, and the geometry of time and space.[8] Littlewood's job also entailed putting Donald through his paces in preparation for the daunting Mathematics Tripos examinations. Although the Tripos amounted to a slick and superficial test of talents, ambitious students strove for Senior Wrangler, the highest achievement—the title "wrangler" deriving from the contentious discussions students in earlier centuries underwent in order to qualify for a degree.[9]

Coxeter welcomed the grind, dashing to the dining hall for breakfast by 7:30 a.m., wheeling across town on his bicycle for lectures starting at 9:00 a.m.[10] Mathematics lectures took place in the old Arts School, tucked away in the center of town where new buildings boxed the old into the middle of the block—finding the Arts School for the first time was as frustrating as navigating to the center of a labyrinth. Upon entering the main hall, a blackboard divvied with a wooden frame into columns and rows indicated which lectures were where. The creaking seminar theater, the main venue, was furnished wall to wall and nearly floor to ceiling in oak, an intellectual tinderbox of mathematical cogitation.[11]

Buried in nothing but mathematics, Coxeter was in his element. He did not suffer the prognosis of intellectual indigestion, but he did develop a duodenal ulcer, perhaps making him the subject of a poem published in the 1926 Trinity yearbook titled "A Simple Story:"

> A Trinity mathematician
> Would not take sufficient nutrition,
> Till his bedder one day
> Threw his text-books away,
> And he's now in the pink of condition.[12]

A strict vegetarian diet cured Coxeter, a regime he would maintain for both digestive and ethical reasons for the rest of his life. He ate nothing but raw vegetables with olive oil, raw fruit, honey, Blake's Vitaveg biscuits, whole-meal bread, and lactic cheese. He lost a lot of weight, making him a thin linear man of a geometer. At one point his condition was so grave that his mother made a fretful trip to Cambridge to check on him, and she stayed the night, which purportedly made Lucy Gee the only woman ever to have slept in residence at Trinity, save Queen Victoria.[13]

△ ⬜ ♦ ⊛ ⊗

Trinity's well-vaunted history also included alumnus Sir Isaac Newton. A portrait of Newton (1642–1727) hung in the dining hall, a constant reminder to students of the man who epitomized Trinity's contribution to mathematics and science.[14]

When Cambridge had closed during the Great Plague in the summer of 1665, Newton, then a student, went home to Woolsthorpe, sixty miles north. He took advantage of the time off for some independent study in mathematics (his aim at Trinity initially was a law degree). Years later Newton told friends that his great insight—containing a fertile germ of generality—came to

Portrait of Sir Isaac Newton in the dining hall at Trinity College (by John Vanderbank, 1725).

him during that respite, first hitting him in his garden: the force that caused the apple to fall from a tree, he realized, might also account for the pull that orbits the Moon around the Earth, and all the planets around the Sun.[15] As James Gleick described in his Newton biography, "The apple was nothing in itself. It was half a couple—the moon's impish twin. As an apple falls toward the earth, so does the moon: falling away from a straight line, falling around the earth. Apple and moon were a coincidence, a generalization, a leap across scales, from close to far and from ordinary to immense."[16] Newton did not produce his theory of universal gravity with this one insightful moment. He continued with his studies at Trinity, and did so well in mathematics that his teacher, Isaac Barrow, resigned as the Lucasian Professor of Mathematics to allow Newton, whom he spotted as an "unparalleled genius," to take his place.[17]

In 1687 Newton published his masterpiece, *Philosophiae Naturalis Principia Mathematica,* hailed by French mathematician and physicist Jean le Rond d'Alembert as "the most extensive, the most admirable, and the happiest application of geometry to physics which has ever been made."[18] Newton did things with geometry that no geometer had done before; geometry was no longer merely about space. Newton married it to motion. As the scientist himself stated in *Principia*: "The description of right lines and circles, upon which geometry is founded, belongs to mechanics. Geometry does not teach us to draw these lines, but requires them to be drawn . . . from the same principles, I now demonstrate the frame of the System of the World."[19] He put forth a set of mathematical laws describing all forms of motion in the Universe. The backdrop for all motion, Newton's notion of space, was classically rigid and inflexible, based on the foundations of Euclidean geometry. Time, also, was absolute, ticking away like a metronome, keeping the universe in sync.[20]

Euclidean geometry was also crucial to *Principia* since it was the parlance Newton chose to popularize his work. Newton claimed he used his new mode of mathematics, his calculus, to get results on his gravitational theory in the first place. But he translated his findings into geometrical terms for publication, believing in geometry as the classic language of mathematics, and the language his elite audience would most readily understand. In contrast to the popularity of Descartes' analytic geometry, the central role Newton gave to Euclidean geometry reestablished its importance.[21]

Newton left Trinity in 1696, accepting a job at the Royal Mint, but before he left he established a formal school of mathematics and mathematical physics (today Cambridge has its high-tech Isaac Newton Institute for Mathematical Sciences). At the end of his life, Newton remarked, "If I have been able to see further, it was only because I stood on the shoulders of giants"— Copernicus and Galileo, Tycho Brahe, and Johannes Kepler.[22]

Kepler (1571–1630) is best known for his three laws of planetary motion, the work Newton extrapolated upon in *Principia*. But by Coxeter's estimation, Kepler's most notable contributions to pure mathematics were his work pertaining to polygons and polyhedra.[23] And in fact, Kepler's at once insightful and quixotic advancements to the extant knowledge of polyhedra were the precursor to his planetary laws.

In 1596, Kepler had published his book *Mystery of the Cosmos*, theorizing that the proportions of the five Platonic Solids governed the paths of the six then-known planets.[24] His polyhedral planetary scheme worked like Russian nesting dolls:

> The Earth's orbit is the measure of all things; circumscribe around it a dodecahedron, and the circle containing this will be Mars; circumscribe around Mars a tetrahedron, and the circle containing this will be Jupiter; circumscribe around Jupiter a cube, and the circle containing this will be Saturn. Now inscribe within the earth an icosahedron, and the circle contained in it will be Venus; inscribe within Venus an octahedron, and the circle contained in it will be Mercury. You now have the reason for the number of planets.[25]

Recounting the description of this scheme by the Hungarian science writer and novelist Arthur Koestler, Coxeter said: "It was a kind of Wonderland

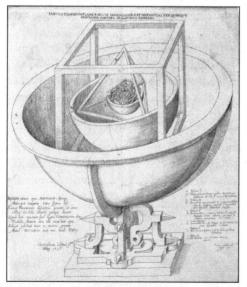

Kepler's polyhedral planetary scheme, from *Mysterium Cosmographicum*, 1596.

croquet through mobile celestial hoops."[26] Kepler presented this whimsical hypothesis to the Duke of Württemberg, and submitted various plans for models. One proposition grabbed the duke's fancy: a planetary punch bowl, the sphere of each planet containing a different beverage dispensed through a network of pipes at the turn of a faucet. The duke commissioned it in silver. The silversmith, however, ran into problems and the project languished.[27]

Eventually, Kepler recognized that his polyhedral theory of the planets, together with the punch bowl, would not hold water. He later became assistant to the foremost astronomer of the day, Tycho Brahe.* After Brahe's death, Kepler inherited his trove of astronomical observations and put them to good use in developing his laws of planetary motion.[28] In 1619, he published *Harmony of the World*, containing a more sophisticated mathematical model than the nesting scheme.[29]

Harmony of the World also included the first systematic treatment of polyhedra, extrapolating on all that was known in the day. Since Euclid's time, geometers had studied one polyhedron or another, but these findings were rather haphazard and scattered. Kepler took a comprehensive approach. He defined classes of polyhedra, identified all their members, and proved his set complete. Kepler redetermined the class of convex uniform polyhedra known as the Archimedean solids (Archimedes' own work on them having been lost), and he discovered that prisms and the antiprisms belonged in the same class. The Archimedean solids are also called the semiregular solids—like the Platonic solids, they have regular polygon faces and the same arrangement of faces at each vertex, but they have more than one type of face per solid.[30]

Kepler also conducted a dig into his own imagination and happened upon two regular star polyhedra, created by "stellation"—the edges or faces of a polyhedron are extended until they meet in such a way that their new faces form stars. Kepler called the resulting critters the small and large dodecahedral hedgehogs, due to their prickly appearance. They are also called the Kepler star polyhedra; earlier renditions of these shapes existed, but Kepler was the first to recognize that they met the criteria for regular polyhedra.[31]

The stellation torch passed to Coxeter at Cambridge, the only Platonic solid whose stellations remained to be investigated being the icosahedron.[32] Coxeter came by this project via Littlewood, who had received a letter about models of stellated icosahedra from a Mr. H. T. Flather, in St. Albans, midway between London and Cambridge. Littlewood sent Coxeter to have a

---

*Brahe (1546–1601) was a Danish astronomer and mathematician who lost part of his nose in a duel with a mathematician (some accounts say they were arguing over a geometry problem).

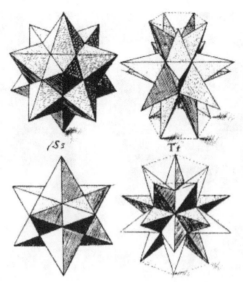

Kepler's star polyhedra, from *Harmonice Mundi*, 1619.

look. Coxeter arrived at the proper address, rang the bell, and when the door opened had a momentary shock. "I was looking straight in front," recalled Coxeter, "and I saw nothing. Then I looked down and saw a tiny dwarf. And that was he." Flather was quite elderly, but invited Coxeter in and exhibited the models. They were remarkably small models, and very intricate. "One could hardly imagine how they could be made," said Coxeter. "Except by his very small, child-sized hands." The series included more than fifty stellations of the icosahedron. Littlewood accepted Flather's models as a gift to Trinity, and Coxeter agreed to write an accompanying enumeration and description, which became *The 59 Icosahedra*.[33] An expert analysis of this work might judge it "nonsense"—it was fun and aesthetically pleasing, quite popular among fledgling geometers, though not at all important in the grand scheme of polyhedra research. But then again, for Coxeter the aesthetics were reward enough, beyond which he did not pretend any great shakes.[34]

Coxeter managed his undergraduate degree without difficulty. He received first class status on part 1 of the Tripos, after his first year in 1927, and with the completion of part 2 the following year he attained the coveted status of Senior Wrangler.[35] With those results, he received a research scholarship and returned the following year. His success was the result of his discipline. "I too

Coxeter striking a cerebral pose at Cambridge.

often feel the need for nine lives, to get done everything that is worth doing,"
he wrote in a letter home to Katie, his stepmother. "There is such a lot of literature, for instance, and I can't spare much time for reading."[36] Though on
one occasion he defended an indulgent day punting up the River Cam, followed by some skinny-dipping in Byron's Pool (whereupon he came to
the conclusion that bathing suits "are the ideal garments to wear *after*
bathing"[37]). That day on the river was not typical, Coxeter insisted. "I work
hard most days, but one must rest sometimes," he said, in response to Katie's
questions about how he found time to laze about in boats—and he added that
"a description of the work-days would make much duller reading."[38] Truth
be told, he said, unfettered leisure and relaxation were the best prescription
to facilitate inspiration. His only regimented extracurricular activity on
record was his membership in the Magpie & Stump debating club. He joined
in his second year with a fellow mathematician, as chronicled in the 1928
Trinity yearbook: ". . . we have two veteran new members, Mr. J. A. Todd,

who is too funny for words, and Mr. H. S. M. Coxeter, who is always very good and unintelligible, but terribly brief."[39]

△ ▭ ◇ ◉ ⬡

In 1928, Coxeter purchased a horse with part of his scholarship money. He named her Trixie and rented a stall at a farm nearby.[40] Being desperately shy, and finding human relationships a bit of a bother, riding Trixie gave Coxeter an excuse to get away from the stresses of social interaction.[41] Coxeter was a hermit within his own head, and, at age twenty-one, he had not had a girlfriend; he was in love with geometry.

Sheltered by his introversion, the carnal pleasures crept up on Coxeter rather unexpectedly. Much to his embarrassment and shame, Coxeter discovered he derived improper pleasure from riding Trixie bareback. He confessed this to his father, who immediately connected his son's repressed sexuality and the shock of the divorce, compounded by his failed crush on Katie.[42]

Harold suggested with minimal moral coercion[43] that his son spend the summer of 1928 undergoing psychoanalysis in Vienna. For June, July, and August, Coxeter placed himself in the useful hands of the great psychoanalyst Dr. Wilhelm Stekel.[44] Stekel was a protégé-cum-dissident of Freud's, who, as Coxeter recalled, disagreed with Freud because the latter would take patients for years and years while Stekel believed that if you couldn't cure a person in a few months then treatment was no use.[45] Harold had made Stekel's acquaintance through Rosalie Gabler at the Royal Psychological Society meetings; Rosalie was the English translator for several of Stekel's books.[46]

Coxeter does not fit precisely any of the anonymous case studies Stekel later published, though a compulsive vegetarian with an obsession for counting comes close. One of Stekel's books points to how the doctor might have treated Coxeter. Entitled *Disguises of Love*, it was published, in Gabler's English translation, in 1922. The pertinent chapter opens with the following tale. "Plutarch tells us a wonderful story of the hereditary prince Antiochus of Syria," wrote Stekel:

> To the grief of his father, Seleucus, he fell sick of a severe disorder, which sapped his strength; the cause of this no physician could discover. Only to the penetrating insight of the celebrated master of the healing art, Erisistratus, was it given to discover that an incurable love for his stepmother, the beautiful queen Stratonice, consumed the prince. Plutarch does not tell us if the prince was aware of his passion. Those who are surprised that we can be in love without knowing it, do not know the enigmatic subterfuges of love and its cravings.

I have repeatedly been able to show that people have fallen sick, under all sorts of disguises of illness, while, in reality, they loved and desired without knowing it.[47]

During treatment Stekel asked leading questions and instructed his patient to keep a dream diary. Coxeter recorded sixty-four dreams over the course of the summer. He dreamed he was an invalid, belittled, and laughed at by a boy. He dreamed about missing trains and boarding buses that drove too fast and took the wrong route. He dreamed he grudgingly shared his raincoat and had to walk long distances (which, in his dream at least, he didn't care to do). In several dreams he was alone and, encountering horses in a field, he tested himself to see, with the analysis, whether he could resist temptation— "(i.e. avoid sexual excitement)." He was pleased to observe that he would not have to pass the farm again in order to get home. And in another dream he found himself walking with John Petrie and his sister, Ann. "I thought, what a pity I am not in love with Ann, and wondered what my feelings would be on meeting the Pritchards next year: I hoped I would love the sister and not the brother (homosexually)."[48]

His nanny, May Henderson, made a cameo appearance, as did a German girl with long dark hair in braids. "Surely she is the ideal," Coxeter said to himself in his dream, noting in the margin of his dairy that this might be Katie. Pleasing his parents was a repeating theme. "My father was about to perform a peculiar manipulation on my body," Coxeter noted. "He explained that I must be tied up so that I couldn't move or make any sound. He would then strike my chest over the heart with his elbow. He said it might kill me, but that would be better than leaving me as I was. I agreed I would prefer even that to the only alternative—suicide."[49]

The only other aspect of his treatment Coxeter remembered was that Stekel advised him to read Oscar Wilde; he thought the poem "The Ballad of Reading Gaol" contained some good psychology. Wilde subsequently became one of Coxeter's favorite authors, and he empathized with the writer over his imprisonment for homosexuality, as Coxeter did (in his own style of generosity and humanitarianism) with anyone he thought victim of an injustice.[50]

Regardless of any progress made with psychoanalysis that summer, Coxeter's time in Vienna proved unexpectedly productive professionally. He loitered in the reading rooms of the University of Vienna Library and there made a meeting that influenced the course of his career—the work of Ludwig Schläfli (1814–95).[51] Once Coxeter was asked which, of any mathematician in all of history, he wished he could meet and converse with. He chose Schläfli.[52]

Coxeter placed Schläfli among the vanguard of nineteenth-century mathematicians who conceived of geometry in more than three dimensions.[53] Schläfli also invented a simple notation that represents all the Platonic solids and all the regular polytopes. Schläfli wrote his notation (now called the Schläfli symbol) as (p / q), which Coxeter later amended to {p, q}—with p representing the shape of each face, and q representing the arrangement (or number) of shapes at each corner or vertex. Take the tetrahedron, represented by the notation {3,3}—p = 3, for the three sides of the equilateral triangle, and q = 3 for the number of triangles at each vertex.[54]

Schläfli is remembered, too, for his proof that in four-dimensional space there are only six regular convex polytopes. The limit of six occurs for the same reason that in three dimensions there are only five regular polyhedra—only a certain number of shapes satisfy the criteria for regularity. In four dimensions the six regular polytopes include: the simplex or 5-cell, each cell being a tetrahedron, and three tetrahedra meeting any an edge; the 8-cell, or tesseract, made of eight cubes, three cubes meeting at every edge; the 16-cell made of sixteen tetrahedra; the 24-cell made of octahedra; the 120-cell made of dodecahedra; and the 600-cell made of tetrahedra.[55]

These four-dimensional regular polytopes are represented by the symbol {p, q, r}—the first two numbers of the notation indicate the type of component polyhedron, and the third number indicates how many polyhedra converge around one edge. The system of notation carries on for higher dimensions. In the fifth dimension, the analog to the tetrahedron is {3,3,3,3}, often called the 5-simplex. It has 6 vertices, 15 edges, 20 triangular faces, 15 tetrahedral cells, and 6 tetrahedral hypercells. The component cell is the 4-simplex—the first {3, 3, 3 in the notation—and to each edge, three of these are joined (in general, in n-dimensional space, the simplex has n + 1 component cells, each being an (n - 1)-simplex, and at each edge three of these are joined).[56]

Schläfli proved that in higher dimensions regular polytopes become a rarer breed. Only three regular polytopes exist in five or more dimensions, continuing to infinite dimensions: these are the simplex (the generalized tetrahedron), the hypercube or "measure polytope" (the generalized cube), and the orthoplex or cross polytope (the generalized octahedron).[57] (See appendix 2 for a chart of the Schläfli symbols for the 3-D and 4-D regular polytopes.)

Unfortunately, Schläfli's work with polytopes was little appreciated while he was alive. His book *Theorie der vielfachen Kontinuität* (Theory of Continuous Manifolds) reached publication only as a memorial volume six years after his death. "The French and English abstracts of this work . . . attracted no attention," lamented Coxeter. "This may have been because their dry-sounding titles tended to hide the geometrical treasures that they contain, or perhaps it was just because they were ahead of their time, like the art of Van Gogh."[58]

⟁ ▢ ⬦ ⊛ ⬡

Inspired and back at Cambridge after his summer in Vienna, Coxeter set about his PhD (in which Schläfli's work figured prominently),[59] with Henry F. Baker as his advisor. Coxeter kept Baker on a high pedestal, never thinking of him as a contemporary. "I thought of him as a god," Coxeter said, adding all the same: "There were things, such as the polytope theory I was doing, that even he didn't understand."[60]

Geometry reached its apex at Trinity with Baker (1866–1956). In 1912, the Fourth International Congress of Mathematics had been held at Cambridge. Baker, then a professor (he was a Trinity alumnus who never left), made an address foreshadowing his ambitions for the field. He praised the distinguished geometers from other lands in attendance, and expressed his hope that their presence would stimulate English geometry to new activity.[61] Baker's sympathies were known to be markedly Italian—admiring the algebraic geometry of Corrado Segre, Annibale Comessatti, Federigo Enriques, and Guido Castelnuovo, with whom he shared a liking for extending what was known about objects in three-dimensional space to higher dimensions.[62]

Two years later, Cambridge's Lowndean chair of Astronomy and Geometry fell vacant. The astronomers were not happy when it was awarded to the geometer Baker.[63] In 1925, Baker published four volumes of his six-volume tome *Principles of Geometry* (the remaining two volumes came out in 1933). And Baker established his geometrical Saturday afternoon "tea parties"—hardcore research seminars softened by the niceties of teacups and biscuits.[64] Baker gathered around him a group of young men infected by his vision and enthusiasm.[65] The substance of their work, however, had a frivolous and trivial element to it—as did classical geometry in that day. Baker's acolytes often chased after superficial and carefree questions, saying, "Oh, here's a nice thing we can do in two dimensions, let's see if we can do it in three, let's see if we can do it in four." These algebraic investigations were clever and tricky, and had a certain charm, but in the end they lacked substance. "It's not where the subject of geometry was going," said Jeremy Gray. "Its take was a bit naïve."[66]

Coxeter managed to avoid this pitfall and carve his own path. "What Donald was able to do," said Gray, "and one or two other people who came to Cambridge at that time, was to tap into a more substantial part of the mathematical river." The frivolous explorations into hyperdimensions were a bit of a backwater. "This was a little cottage industry Baker's people had.

They did very well and they got Smith's prizes out of Cambridge, which is the big thing to do, and sets you up for a research career," said Gray. "Yet they themselves would say they never really in the end got the right generalizations for progress in the subject. Donald somehow did that. He leapt beyond that into really substantial mathematics."[67]

Every Saturday morning, before the tea party, Coxeter made a ten-minute bicycle ride from his residence in Great Court over the River Cam to Baker's house, where he reported progress on his PhD research. Saturday afternoons, all the regulars gathered for the tea party, held in the old Arts School. Students took turns presenting their most recent findings, followed by sparring discussion, debating the cut and thrust of related points. During Coxeter's turn in 1928, he described the sequence of "pure Archimedean" polytopes from three to eight dimensions, having 6, 10, 16, 27, 56, and 240 vertices, respectively.[68] He had begun this line of investigation while studying with Robson, and although he promised his tutor he would refrain from indulging in his polytopes, except on the day of rest, Coxeter was unable to resist entirely. He smuggled in his polytopes, spending a good deal of his spare time adding further volumes to his "Dimensional Analogy" essay. He never forgot the thrill, the frisson of synaptic excitement he felt, when, sitting under a tree in the Savernake Forest, he rediscovered the pure Archimedean polytopes[69]—the analogs of the Archimedean solids in higher dimensions. Even as a rediscovery it was an intellectual coup, and the impact of following such a giant's path stuck with Coxeter for years.

When Coxeter discussed this work at Baker's tea party, the usual banter followed, questioning and connecting his results to other areas of study. "One of the algebraic geometers immediately expressed interest," Coxeter noted, "because 6, 10, 16, 27, are the numbers of lines on the Del Pezzo[70] surfaces in 6, 5, 4, 3 dimensions. Du Val went one step farther by declaring 2×28 to be the number of lines on the 'Del Pezzo surface' in 2 dimensions, which is a repeated plane joined to itself along a quadratic curve of genus 3; the lines are the repeated bitangents."[71] This subject led to Coxeter's first published paper: "The Pure Archimedean Polytopes in Six and Seven Dimensions," printed in the *Proceedings of the Cambridge Philosophical Society*.[72]

Coxeter's research and analysis grew by orders of magnitude, broadening and deepening the reservoir of data on polytopes. One day he escaped from the monastic confines of Trinity for a solitary bike ride to the Gog Magog Hills,[73] in the south Cambridge countryside, a rural oasis inhabited by singing skylarks, aromatic wild marjoram, and a grove of oak, beech, dogwood, and field maple. The fresh air and peaceful landscape did

Patrick Du Val and Coxeter lounging on the lawn at Cambridge.

nothing to calm Coxeter's mathematical thoughts; indeed, what Coxeter "saw" that day was not the bucolic scenery. Rather, he had a flash of insight—witnessed with his geometric mind's eye—into how these Archimedean polytopes could be exhibited as members of a larger family, indexed by means of a clever notation, much like Schläfli's: n_{pq}—for a figure in $n + p + q + 1$ dimensions.[74] This was Coxeter's first original contribution to the domain of polytopes, and led to his second major paper, "The Polytopes with Regular-Prismatic Vertex Figures," published—all ninety-six pages—in the *Philosophical Transactions of the Royal Society*, a journal of even greater repute.[75]

△ ▢ ◈ ⊛ ⬡

When Coxeter's turn came up for another session at Baker's tea party, he invited his "Aunt Alice" to deliver a joint lecture.[76] More widely known as Alicia Boole Stott (1860–1940), she was a housewife geometer and polytope aficionado forty-seven years Coxeter's senior (he twenty-one and she sixty-eight). Stott became one of his dearest friends and professional soul mates. According to Coxeter, Stott had introduced the word "polytope" to the English language with the first publication of her work in 1900.[77]

Alicia Boole Stott

Over the years, Coxeter became Stott's loyal promoter, telling her story at every opportunity. Stott was the middle of five daughters born to George Boole, known for his algebra of logic (the Boolean logic that drives Google searches and computer circuitry), and Mary Everest (the highest mountain in the world was named in honor of Stott's uncle, the surveyor Sir George Everest). Her father died when she was four years old, and she spent her early years, repressed and unhappy, with her maternal grandmother and great-uncle in Cork. "When Alice was about thirteen," wrote Coxeter,

> the five girls were reunited with their mother (whose books reveal her as one of the pioneers of modern pedagogy) in a poor, dark, dirty and uncomfortable lodging in London. There was no possibility of education in the ordinary sense, but Mrs. Boole's friendship with James Hinton attracted to the house a continual stream of social crusaders and cranks. It was during those years that Hinton's son, Howard, brought a lot of small wooden cubes, and set the youngest three girls the task of memorizing the arbitrary list of Latin words by which he named them, and piling them into shapes. To Ethel, and possibly

Lucy too, this was a meaningless bore; but it inspired Alice (at the age of about eighteen) to an extraordinarily intimate grasp of four-dimensional geometry.[78]

Howard Hinton exposed Alice to his mystical interpretation of higher dimensions. But she did not care to follow him along these occult lines of thought, said Coxeter, noting that she "soon surpassed him in geometrical knowledge. Her methods remained purely synthetic, for the simple reason that she had never learnt analytical geometry."[79]

In the 1880s, Stott rediscovered the six polytopes in four dimensions and then, using a ruler and compass, cardboard and paint, she produced complete model sets of their central sections.* During the intervening years, before she met Coxeter, Stott "led a life of drudgery, rearing her two children on a very small income."[80] She returned to her geometric work when her husband, Walter Stott, happened upon the work of Pieter H. Schoute, at the University of Groningen, the Netherlands, who was investigating the central sections of the very same four-dimensional polytopes. Stott wrote to Schoute with the news that her findings corroborated his (Stott's powers of geometrical visualization, Coxeter noted, supplemented Schoute's more orthodox methods). Schoute arranged for the publication of Stott's discoveries, and their partnership continued until Schoute's death in 1913, after which Stott abandoned her polytopes again until she met Coxeter.[81] Coxeter's alliance with Aunt Alice was a great source of joy.[82] "The strength and simplicity of her character," he said, "combined with the diversity of her interests to make her an inspiring friend."[83] They conducted an ongoing conversation about polytopes, by letter and with visits back and forth. Aunt Alice once sent Coxeter on his way after polytopes and tea with a present—two matching lamps with wooden truncated icosahedra for bases, which he carefully carried on the train back to Cambridge.[84]

△ ▢ ◈ ◉ ✪

Coxeter submitted his PhD dissertation in 1931, the same year Godfrey H. Hardy returned to Trinity from his stint at Oxford.[85] As one of only four individuals made a Trinity Fellow that year,[86] Coxeter received a copy of the college ordinances, marked "Private: for Masters and Fellows only," outlining his

---

*When Aunt Alice made her appearance as Coxeter's guest at Baker's tea party, she brought her models and donated them to the department for permanent exhibit. As recently as 2003 they could still be found there, in the office of Professor Raymond Lickorish, director of the Newton Institute of Mathematics, displayed on his desk. Another set is on display at the University of Groningen.

privileges[87]: he could now walk on the grass in the courtyards, shortcutting the ninety-degree paths by the hypotenuse if he wished.*

Hardy, coming to Trinity to occupy the Sadlerian Chair of pure mathematics, was also among the privileged. But he despised anything bourgeois and refused to live in the best rooms allotted to him in residence. A man known for his eccentricities, Hardy had a pronounced phobia of mirrors. When he walked into a room, he covered any reflective surface with a piece of cloth. It wasn't so much a superstitious hatred, but a dislike of his own looks—few photographs of Hardy are therefore known to exist.[88]

It is amusing, in this respect, to imagine the fallout when Hardy bumped into Coxeter on campus, since Coxeter was known to carry a set of mirrors on his person almost everywhere he went. He had the mirrors custom cut by a workshop into triangular wedges which hinged together to form a large kaleidoscope—a tent of mirrors, 24 inches across and 12 inches tall. This was the tool Coxeter used to investigate the symmetrical properties of polyhedra.[89]

When invented by Sir David Brewster in 1814, the kaleidoscope produced quite a stir, selling a reported 200,000 instruments in London and Paris in three months. Sir David was deeply dismayed, however, because his invention gained popularity merely as a toy, though appealing to children and adults alike—he had invented the kaleidoscope as an artistic and scientific instrument. In defending the importance of the instrument, he published a tirade expressing outrage that his patents were being infringed in the kaleidoscope's mass production, and—worse yet—that the knock-offs were so sloppily constructed (in his book he provided illustrated instructions for building them properly). He was greatly chagrined that of the hordes who witnessed his instrument's beautiful effects, not even one thousand experienced a "correct idea of the power of the Kaleidoscope."[90] How relieved and rewarded Sir David would have been to see Coxeter toting his kaleidoscopes for geometric investigation. Using kaleidoscopes, Coxeter generated precise two-dimensional and three-dimensional geometric patterns—he found kaleidoscopes useful in exploring his polytopes, since many of their geometric symmetries are generated by reflections, such as mirror symmetry.

Kaleidoscopes operate by the laws of optics; when a ray of light from an

---

*As a Fellow he also could dine in Hall at High Table and eat the finer food; he could linger after meals in Old Combination Room, sipping tea and coffee and aperitifs in well-worn leather armchairs; and he could order a supply of wine for his private consumption and have it delivered to his room (quantities of wine so supplied could not exceed two dozen bottles a year); he had access to the library key, allowing him to enter and borrow books after hours; and he was given a key that opened the gate to the Fellows' Garden on the outskirts of the college—its winding walking paths punctuated by the occasional bench for thought.

image falls upon a mirror, the angle of incidence (the angle at which a ray hits the mirror) equals the angle of reflection (the angle by which it bounces off the mirror). In this manner, a kaleidoscope generates a repeating sequence of reflections, an effect produced not solely by its real mirrors, but also by its chambers of virtual mirrors, the mirrors reflected in themselves again and again and again. When a sequence of reflections travels through a kaleidoscope's hall of mirrors, one of two things happens. Depending on the angles at which the mirrors meet, the reflected images either multiply endlessly, creating an infinite pattern, or the reflected images coincide, falling back on themselves as they retrace the same path. When the latter occurs, the optics conspire to create the image of a finite geometric pattern or shape in the mirrors. This is what Coxeter (and Brewster) were interested in.[91]

In his investigations, Coxeter deployed two types of kaleidoscopes. The first were simple kaleidoscopes: two hinged mirrors replicating whatever object fell between them, with images of the object ricocheting off the mirrors and

Coxeter assembling one of his Cambridge-era kaleidoscopes some years later.

forming a perfectly symmetric rosette, like a two-dimensional snowflake. To produce this finite effect, the mirrors must be arranged at angles that exactly divide 180° (or pi). Otherwise, the images produced in the kaleidoscope will not match up—the snowflake would be shattered into disjoint pieces.[92]

A square can be generated in a two-mirror kaleidoscope, which can be constructed on a makeshift basis with any two mirrors at your disposal. On a sheet of paper draw a square, and fold it perfectly in half in all the ways possible so that each half is the mirror opposite of the other—this should produce a square divided into eight identical pieces, each a 45°-45°-90° triangle. Cut out one of these triangles, and place mirrors along two adjacent edges, with the mirrors meeting at the vertex that would be the center of the square. If you then look into the mirrors, you'll see the image of an entire square— the fraction (one-eighth) of the square between the mirrors is reflected back and forth and this parade of images configures to form the whole polygon.[93]

Coxeter used kaleidoscopes to produce patterns not only of 2-D polygons, but also images of the 3-D Platonic solids. While two mirrors are needed to generate the 2-D polygons, three mirrors are needed to generate the 3-D solids—images of each of the Platonic solids can be produced by a certain arrangement of three mirrors,[94] and for this reason the symmetries of the Platonic solids are said to be "kaleidoscopic."[95] The German mathematician August Möbius, more famous for his twisted strip, first studied the practice of using mirror arrangements to generate the Platonic solids in 1852.[96]

In order to create images of the Platonic solids, the three mirrors are arranged in a triangular cone, somewhat like the corner of a room, but again the mirrors must be aligned at angles that are specific fractions of 180°.* The mirror planes of symmetry of a cube divide it into forty-eight congruent sectors, and thus when a kaleidoscope generates the cube, it does so by reflecting and reproducing this one sector—with forty-eight reflections bouncing around the kaleidoscope—causing the entire cube to take shape (this composite image of the cube consists of the real object in the chamber of the real mirrors, and the virtual objects or reflections, generated in a multiple array in the virtual mirrored chambers).[97]

And while the image of the square was generated in the kaleidoscope by a piece of paper, geometers use a particular kind of prop to generate the Platonic solids. One prop commonly used is a simple stick (even a pencil or pen will do), which generates only the edges of the solid—for instance, it would

---

*A three-mirror kaleidoscope arranged slightly differently, with the mirrors joined end to end forming a fenced-in triangle, generates an infinite tiling of the plane—images interlocking like the tiling of a floor, and bouncing around endlessly.

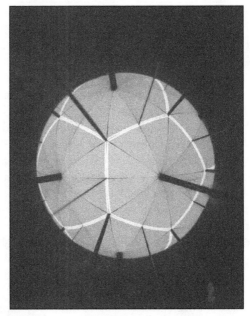

A mapping of a dodecahedron onto a sphere, generated by mirrors of
an icosahedral kaleidoscope.

be a skeletal-looking cube, or the frame of a cube. Another type of prop is a
round ball or blob of any material, which generates the vertices or corners of
the cube—leaving the edges and the faces to the viewer's imagination. This
method of generating the Platonic solids with mirrors and props is called
"Wythoff's construction," invented by the Dutch mathematician Willem
Abraham Wythoff (1865–1935).[98]

Using kaleidoscopes to explore symmetries can also be extended to investigate the polytopes of multiple dimensions. Each dimension simply requires
another mirror. In the physical world, of course, the geometer hits a practical impasse beyond three dimensions, since kaleidoscopes representing four
dimensions and more cannot physically be constructed. But this did not de-
ter Coxeter. The physical limitations motivated him to develop a systematized theory of kaleidoscopes, a generalization of how they would continue
to generate the polytopes of higher dimensions, continuing on to infinity, if
our physical reality would permit.[99]

Coxeter's kaleidoscopes were his prized possessions, and he tended to
them with loving care: "improved the mirrors," he noted in his diary, "fixed
mirrors," "made gadgets for the mirrors."[100] He asked his mother to sew
green felt pockets into which each individual sector of mirror could be

slipped, to minimize breakage and chipping during transport.[101] He pulled out his mirrors for show-and-tell at every opportunity—especially with his father and Aunt Alice. "I can't tell you how thrilled I am at the thought of seeing your magic mirrors!" Stott exclaimed in a letter. "It seems to me too wonderful for words and I am longing for next week to come. *How* you will wake up the stuffy mathematicians on Saturdays!!"[102]

Hardy, with his hatred of mirrors, no doubt agreed to meet Coxeter only if he left his mirrors at home. If he saw Coxeter walking toward him from across Great Court, those wretched mirrors under his arm, Hardy might have given him only a tentative wave, understood to mean "Good day!" but also "Stay the dickens away!"

△ ▢ ◈ ◉ ⊗

Rubbing elbows with Hardy as he neared the end of his PhD, Coxeter was exposed to a fine mentor, the finest example of the caliber of mathematician he hoped to become. Hardy was nearing the end of his career, but it was a noble career and he had earned Trinity as home for the rest of his days. In 1940, Hardy published *A Mathematician's Apology*, reflecting on his career. He asked, "Is it really worth while to make a serious study of mathematics? . . . What is the proper justification of a mathematician's life?" Lines drawn in the margin alongside the text of Coxeter's copy highlighted Hardy's response, which culminated with perhaps the most oft–cited quotation about mathematics: "A mathematician, like a painter or a poet, is a maker of patterns . . . The mathematician's patterns, like the painter's or the poet's must be *beautiful*. The ideas, like the colours or the words, must fit together in a harmonious way. Beauty is the first test: there is no permanent place in the world for ugly mathematics."[103]

Writing his PhD thesis, which he dedicated to John Petrie and titled "Some Contributions to the Study of Regular Polytopes," Coxeter expressed exactly the same sentiment in the introduction, with a bit of self-deprecating humor:

> Although it is unnecessary, from a practical point of view, to consider regular skew polygons of more than five dimensions, the human weakness of a mathematician compels him to examine the general case, although the trigonometry involved is extraordinarily complicated . . . The only excuse for this part of the work must be its intrinsic beauty.[104]

Coxeter's writings on polytopes won him the coveted Smith's prize,[105] indicating he had more than withstood the stultifying labors of haughty pure mathematics at Cambridge. He had not, as a large number of math students were known to do, changed his course of study to something more sensible—anything but mathematics. According to the article advocating reform of the mathematics curriculum, aiming to make it less lofty, more practical and down-to-earth, Coxeter was an exception: the exception who lives through years and years of such mathematical rigors and retains his enthusiasm is "a veritable mathematical specialist . . . someone ideally prepared for a world which does not exist."[106]

CHAPTER 4

COMING OF AGE AT PRINCETON WITH THE GODS OF SYMMETRY

Symmetry, as wide or as narrow as you define its meaning, is one
idea by which man through the ages has tried to comprehend and
create order, beauty and perfection.

—HERMANN WEYL, *SYMMETRY*

Known as the "papa daddy"[1] of Princeton mathematics professors, Solomon
Lefschetz visited Cambridge in 1931, in search of suitable candidates for the
illustrious Rockefeller research fellowships. He met Coxeter and invited him
to apply.[2] As a Rockefeller Fellow, Coxeter stood to benefit from a one-year
research professorship at Princeton with a handsome monthly salary of $150,[3]
a princely sum during the Great Depression, when the average American
family income fell to $1,300 annually and unemployment was at 25 percent.[4]
The prospect no doubt pleased Coxeter immensely, and not for financial or
self-gratifying reasons. Rather, it surely appealed to his humanism. Established in 1913 by John D. Rockefeller, and funded by the Rockefeller Foundation, the fellowships embodied a grand philanthropic enterprise to promote
"the well-being of mankind throughout the world."[5]

Critics argued that Rockefeller's benefactions were set up as a shield
against public censure, considering the industrial fortune he had made in oil,
steel, railroads, and banking. Still, this act of kindness was undeniably far-
reaching in its generosity. With an endowment of nearly $250 million, the
foundation made medical and public health education a priority at home in the
United States, as well as in Africa, India, the Middle East, and Latin America.[6]

The foundation's reach expanded even further after the First World War.
It strove for a grand "advance of knowledge" disseminated by international
research professorships, intellectual missionaries sent out to bring greater
order and civility to society.[7] Wickliffe Rose, the president of the foundation's International Education Board, made a high-minded declaration of his

mandate in a memorandum titled "Scheme for the Promotion of Education on an International Scale:"[8]

> This is an age of science. All important fields of activity from the breeding of bees to the administration of an empire, call for an understanding of the spirit and technique of modern science. . . . Science is the method of knowledge. It is the key to such dominion as man may ever acquire over his physical environment. Appreciation of its spirit and technique, moreover, determines the mental attitude of a people, affects the entire system of education, and carries with it the shaping of a civilization . . . The nations that do not cultivate the sciences cannot hope to hold their own . . .[9]

There was nothing new about this liberal-minded venture: "It was at least as old as the Greeks," stated an historical account of the International Education Board. "Rose was driven to it as a result of the disillusion of the world tragedy of 1914–1918, and the desperate need . . . of some ingredient which would heal the dissensions of nations."[10] Knowledge was that ingredient: "Knowledge is the unifying principle of civilized life—the centripetal force which holds it together. Knowledge is the solvent in which boundaries can be dissolved and barriers burned away. It is the common republic of mankind in which citizenship is denied to no nation and no group. It is the avenue that leads to the ultimate unity of the world."[11]

Rose embarked on a five-month tour of Europe to evaluate the status quo, shadowing scientists in their laboratories, inquiring about their problems and needs. The foundation boasted that never before had a search for superior brains been prosecuted over so wide and diverse an area, and many specimens of inherent brilliance and talent were discovered. Several recipients of a Rockefeller Fellowship later received Nobel prizes, among them Enrico Fermi, a physicist from Italy, and Germany's Werner Heisenberg, the author of the uncertainty principle. Two primary centers of mathematical scholarship judged worthy of support were Göttingen University, in Germany—the home base of Hilbert and his formalism[12]—and the Henri Poincaré Institute in Paris. The meeting of Rockefeller trustees at which Rose presented his proposal for these institutions was recorded in the minute books as a perfect snapshot of his earnest intentions: "[He] reported, with the aid of elaborate charts and diagrams, not on mathematics at Göttingen or Paris alone, but on mathematics in every leading institution around the world. He was reporting on where man had arrived in his mathematical thinking, and where the opportunities for progress seemed brightest." Rose favored Göttingen and Paris because they represented the peaks in mathematical science.[13] Rose's mantra: "Make the peaks higher."[14]

As he systematically surveyed "the mountain ranges of intellectual endeavor,"[15] Rose realized that a parallel course had to be taken in America. American universities received funds to bolster their resources in mathematics, physics, astronomy, chemistry, and biology. And by the mid-1920s, the foundation decided that in addition to funding American scholars abroad, it should also bring the "bright lights" over to study in America. In this spirit, Coxeter took Lefschetz's advice and applied for a fellowship. Upon acceptance, he met with one complication. This was in the middle of his time as a fellow at Trinity College. He hated to give that up, so by a special decree of the Trinity College counsel, his fellowship was extended for an extra year to compensate for the year away. Several other fellows later took advantage of this precedent and were said to be "doing a Coxeter."[16]

In August 1932, Coxeter chose to forgo a family holiday in Vienna, allowing him to sail early for America and spend some time in New York before he was due at Princeton. On the eve of his departure, his father pondered what Donald's future might hold:

> So Donald is off tomorrow after all . . . If only he could really fall in love—even a hopeless passion would be better than none—to make him give himself away. I had such a nice talk with Mrs. Stott—alone—about him. She is a fine woman, and sees the danger of his becoming nothing but an intellectual, and also sees in him the possibility of a fine soul as well as a fine intellect.[17]

Donald's transformation began as soon as he set foot aboard the RMS *Aquitania*, abetted by his traveling companion, Frank O'Connor, a family friend who was a doctor in New York (Harold thought his son was in good hands; O'Connor was "a man of the world").[18] To Donald's astonishment, days on the ship were filled with play—a horse-racing gambling game and dancing every night. The dancing made him feel inferior; he didn't know how to dance "American style." To make matters worse for a stiff fellow who could barely screw up courage to dance at all, on certain nights partners were assigned by pure chance in a "Cinderella dance"—women threw off a shoe and each man danced with the girl whose slipper he grabbed. Coxeter was not quick enough to get the one he wanted, but he did manage to draw the attention of one young lady. She was curious about geometry and he obligingly taught her to draw a four-dimensional polytope. She was also the center of attention among a crowd of rowdy young men. In his first letter home, Coxeter signed off saying, "But that is as far as I have got, and the voyage ends tomorrow, when I prepare to enter a whole continent of such people."[19]

The approach of the New York skyline was "just as in the flicks, except that now one no longer had the sneaking suspicion that the skyscrapers were faked." Almost immediately after landing, owing to a misunderstanding entirely of his own fault, Coxeter became separated from O'Connor, with whom he was staying. The Rockefeller representative sent to meet Coxeter sought him out at all the speakeasies in the vicinity of the dock, clearly ignorant of the character he was looking for.

Coxeter experienced a good sampling of America before making his way to Princeton. O'Connor introduced him to his secretary, and she and a friend took Coxeter up the Empire State Building. "A fine view from the top," he said in his second letter home. "A thunderstorm approached, and came right over us, the cloud obscuring everything. After one great clap of thunder the girls said they could feel the building swaying in the wind (it actually sways only 2 inches, so I expect in reality they were trembling)." The next day a rich patient of O'Connor's sent a chauffeur forty miles to fetch them for a visit to his Mount Kisco estate. "Swimming pool on the lawn; special dressing places with hot and cold water and a telephone. Everything else on the same scale. (When he heard I was a vegetarian, he got one of his six chauffeurs to take us in one of the six cars to see the vegetable garden.)"

On August 31, Coxeter and O'Connor drove to New Hampshire to watch a total solar eclipse (this, in fact, was the reason he wanted to be in America early, to catch the eclipse that Europe would miss). "We selected an open hillside, ten miles in from 'the belt of totality.' At 4.25 ½ there was no sun left but a tiny crescent . . . At 4.29 ½ we got a very good view of 'Bailey's beads.' The darkness over the earth was very impressive, and we saw the corona well. It was all over at 4.30, and we hurried southward to get ahead of the crowds." Back in New York, Coxeter recounted all this and more—"each day holding sufficient events for a month"—in an eight-page letter home, writing it "wearing nothing but a towel, Samoan fashion, on account of the heat."[20]

After a time, Coxeter set out for Princeton, an hour's train ride southwest from New York City. Before getting down to work, he bought a bicycle and explored the town. While descending a grassy slope at considerable speed he was "pulled violently off my bicycle . . . by a man who apologized profusely, declaring that he was 'very drunk.'"[21] He also traveled back to New York for a World Series baseball game between Babe Ruth's New York Yankees and the Chicago Cubs, and he attended his first football game, between Princeton and Amherst. Reflecting on these American pastimes, he offered a caustic verdict: "I do not want to be bored by a repetition of either."[22]

△ ◻ ◈ ◉ ❀

Princeton, by the early 1930s, was poised to overtake Göttingen as the mathematical center of the world. In 1932, after the Nazis took power, Hitler purged German universities of all Jewish academics, propelling many to

America. Einstein arrived at Princeton in 1933. He called the town "a wondrous little spot, a quaint ceremonious village of puny demigods on stilts."[23] And Einstein's disciple, the Polish physicist Leopold Infeld, noted, "It is difficult to learn anything about America in Princeton—much more so than to learn about England in Cambridge." Infeld observed that in the mathematics building, Fine Hall, "English is spoken with so many different accents that the resultant mixture is termed 'Fine Hall English.' "[24] The architecture of the university, similarly, was an "amazing tutti-frutti of all possible and impossible styles." Mostly it was neo-Gothic, an amalgam of Cambridge and Oxford, including an imitation (a grotesque one, Infeld thought) of Trinity's Great Gate.[25]

Fine Hall, built in 1931, was the most decadent mathematics facility ever designed—and designed deliberately so mathematicians would be loath to leave.[26] The individual working quarters were sanctuaries, more living rooms than offices, each well-appointed with a big desk, plump upholstered chairs, oriental rugs, and a clothes closet. The walls were paneled with dark oak, shipped from England at the cost of one-fifth of the building's budget, and select wall panels cleaved open like cupboards to reveal hidden blackboards, inlaid both into the interior wall and the opening flaps. Even the lead-paned windows were part of the design scheme—etched with mathematical formulae and figures. The air, Infeld said, was suffused with mathematical ideas and formulae: "You have only to stretch out your hand, close it quickly and you feel that you have caught mathematical air and that a few formulae are stuck to your palm . . . Even the sun rays must remember, when passing through the windows, the law to which they are subject according to the will of God, Newton, Einstein and Heisenberg."[27] The windows in the professors' lounge (decorated by the famed New York architectural firm McKim, Mead, and White, and overseen by a departmental wife) had Einstein's gravitational law and general relativity theory indelibly recorded on the glass, and one of the studies refracted light through a stained glass emblem of the Platonic solids. To top it all off, there were reading lights in the bathrooms—Fine Hall was in tune to the unpredictable timing of inspiration—and an onsite shower and locker room allowed mathematicians to return to their research, refreshed with adrenaline pumping, after nipping out for a tennis match on the backyard courts. In the insouciant lyrics of a faculty song, Fine Hall was a "country club for math, where you can even take a bath."[28]

Soon after Coxeter's arrival at Princeton, he met with his recruiter, Professor Lefschetz, to chart a course of study that outlined sufficient work to occupy about thirty hours each day.[29] Loud and obstreperous, Lefschetz both awed and frightened graduate students. Russian-born and trained as an engineer in France, he came to the United States at age twenty-one to find work. In 1907, he lost both hands in a transformer explosion at Westinghouse

Electric Company, in Pittsburgh, the tragedy pushing him into the more philosophical field of mathematics. He obtained his doctorate, and ultimately landed at Princeton in 1924. He became known for his profound geometrical intuition, in the algebraic hyper-dimensional tradition of the Italian school. But Lefschetz's area of specialty became topology—or "rubber sheet" geometry*—a modern genre of geometry that contained only select morsels of interest to Coxeter.[30]

Lefschetz could smell a theorem where many mathematicians wouldn't even suspect one, but seldom could he be bothered to work out the details of a proof; the running joke was that he never wrote a correct proof nor stated an incorrect theorem.[31] If something was clearly true, he considered producing a proof just for the sake of verification to be a waste of time—when a student proudly showed him a clever result for one of his theorems, Lefschetz barked back: "Don't come to me with your pretty proofs. We don't bother with that baby stuff around here."[32] In reality, however, proofs were necessary for publication, and Lefschetz wrangled his grad students into finishing his work. And with his two hooks for hands, over which he usually wore shiny black gloves reaching his elbows, he also depended on students for practical matters—every morning a graduate student pushed a piece of chalk into his glove and removed it at the end of the day. A film from his early days at Princeton shows him giving a lecture, gesticulating wildly with his shiny-gloved appendages. The faculty song composed about Lefschetz was a toast: "Here's to Lefschetz, Solomon L./Irrepressible as hell/When he's at last beneath the sod/He'll then begin to heckle God."[33]

Officially, as Coxeter's Rockefeller personal record card stated, he had come to Princeton to study with Oswald Veblen. But Veblen, another topologist, and Hilbert's counterpart in America seeking to modernize geometry, was not much more up Coxeter's alley than Lefschetz. Coxeter and Veblen discussed mathematics during long walks in the woods (as Veblen was known to conduct his work). But on the whole, Veblen left Coxeter free to do his own thing. Informality was the custom at Princeton. Mathematicians, especially visiting mathematicians on research fellowships, conducted their business by sidling up to pertinent parties in the common room, testing the waters for collaboration, and by drifting in and out of alluring lectures.

---

*Topologists study properties of shapes that are preserved when the shapes are deformed through stretching, twisting, or compressing, though tearing is not permitted by the rules of topology. A circle is topologically equivalent, with some stretching, to an ellipse; even a cube and a sphere can be deformed into one another, and are thus are said to be "homeomorphic." A topological characteristic of polyhedra was discovered by Swiss mathematician Leonhard Euler (1707–83), and is now known as the Euler characteristic of polyhedra, or Euler's formula: for any polyhedron, the number of its vertices (V) minus its edges (E) plus its faces (F) equals two, or $V - E + F = 2$.

Coxeter attended Veblen's lectures, during which Veblen posed an extended question: "What is geometry?"[34] The shifting grounds of geometry in the early twentieth century had made it a nebulous concept. Veblen focused his lectures for weeks on that question, trying in vain to hit upon a satisfactory answer. The problem was that everything the geometers came up with as a definition for the purview of their field could, semantically, be twisted to include all of mathematics. In the end, Veblen settled on a deliberately and amusingly vague definition. Geometry, he reckoned, was "that part of mathematics which a sufficient number of people of acknowledged competence in the matter thought it appropriate so to designate, guided both by their inclinations and intuitive feelings, and by tradition."[35]

Coxeter also attended lectures by the Hungarian mathematician John von Neumann, another Rockefeller Fellow, who always wore a suit one size too small. Von Neumann was a magician of a lecturer, able to take what was given and with mathematical sleight of hand unveil logical conclusions with sweeping and illusive dexterity. He was so fast with his delivery that students asked him questions for the sole purpose of slowing him down. Von Neumann playfully engaged in an ongoing game with Swiss mathematician Henri Frederic Bohnenblust (known as Boni), whereby either man tried to catch the other working. The rules stated they could burst into one another's office in Fine Hall at any time without knocking, in an effort to catch their opponent in the act. If caught, the loser doled out ten dollars. Von Neumann was never caught, since he did his work late into the night, and spent the daylight hours apparently doing nothing (he was thinking).[36]

Coxeter also attended lectures by two mathematical physicists, Eugene Paul Wigner and George Pólya,[37] both Hungarians, and both at Princeton as Rockefeller Fellows (Wigner split his fellowship with von Neumann, each taking half a year). Pólya considered himself more a mathematician with a physicist's inclinations. "I am not good enough for physics and I am too good for philosophy," he once said wryly. "Mathematics is in between."[38] Coxeter's encounters with Pólya and Wigner (he attended Wigner's lecture on "how to make a single crystal of copper as big as a human head") were facilitated by the fact that Fine Hall was connected to the physics building, Palmer Laboratory, by a second-floor corridor—an architectural detail nicely symbolic of the relations between mathematics, queen of the sciences, in service by analogy and abstraction to the kingly physics.[39]

These encounters engaged Coxeter with the scientific applications of mathematics. He was a pure mathematician at heart, studying and developing the art of mathematics for its own sake. Pure mathematicians are propelled by the internal logic of mathematics as an abstract and symbolic structure, rather than by any insights about the world. In contrast, insights about the world propel applied mathematicians to orient their work toward an immediate usefulness, in the physics, biology, sociology, or elsewhere. However, these seemingly black-and-white realms are not, or at least seldom remain, entirely isolated. Pure mathematics discovered and investigated in one era— and considered in that era to have no practical value whatsoever—is often found, at a later date, to hold startling and unexpected practical applications (this was the case with Coxeter's work, especially his Coxeter diagrams and Coxeter groups, which will be discussed later).

Coxeter was often praised as the purest of the pure, a mathematician who sequestered himself in his study and reveled only in the pearls of the intellect. But a glance at his bibliography shows otherwise—with papers, "On Wigner's Problem of Reflected Light Signals in de Sitter Space," "The Space-Time Continuum," and "Virus Macromolecules and Geodesic Domes."[40] He welcomed the chance to converse with applied mathematicians and scientists, sometimes seeking them out with a spontaneous phone call in a flash of insight. He delighted in exploring geometry's appearances in the sciences, even if he wasn't actively engaged in uncovering these connections himself. Coxeter was not a pedantic purist who strictly observed the barrier, and certainly not the hostility, between pure and applied.[41]

To that end, at Princeton Coxeter exposed himself to topics slightly beyond his ken. He went to lectures on neutrons, on cosmic rays (with a Geiger counter exhibited so that the audience could observe the rays coming in at about one per minute), on various kinds of expanding universes, and on the primeval atom (created 10^{10} years ago), which, as he recounted in a letter home, "while breaking up under a kind of super-radioactivity, emitted hard rays analogous to the x-rays from radium, and these 'birth cries of the universe,' have gone rushing around and round space ever since, to be observed today as 'cosmic rays.' "[42] After a lecture by Governor Holt on emotion and arithmetic, Coxeter talked with him about physicists and mystics. And he argued with Irving Robertson at supper about telepathy.[43]

In his residence at the Graduate College, Coxeter and his new acquaintances—"all delightfully at ease in the American manner"—listened to President Franklin Roosevelt's fireside chats on the radio, outlining the details of his New Deal.[44] They took turns reading poetry aloud—Edgar Allan Poe's "The Raven"—until they were all almost asleep.[45] Coxeter bragged in his continuous flow of letters that he was entertained at grand dinner parties with distinguished professors and their wives, much older than himself.[46] And he noted in the postscript to one letter that his new friends observed America had already made quite an impression on him; he appeared to have matured noticeably since his arrival, and now actually looked his twenty-five years.[47]

Keen to fend off his parents' concern that their only son was doomed to become "a fusty old bachelor,"[48] Coxeter resolved to do two things while in America: to shun all Englishmen, and to find a suitable woman to bring home to England and make his wife. He filed this progress report to his father:

> I have met a really perfect girl—a fortnight ago, her first and my second time at the square dancing class—very graceful, with straight black hair, good features and pale complexion. I thought, "surely, she cannot still be unmarried, as she looks at least 22"; but her beautiful hands were ringless.

> The mere touch of those hands seemed unlike that of anyone else's. The next time, a week ago, I contrived to have her for my partner almost all the time, but it was impossible to talk. Afterward, I asked who she is, and was merely told that she is one-quarter squaw. At that I was enormously thrilled, and wove the most marvelous fantasies about her. I thought I would begin by inviting her to the Faculty Dance. And then today I learnt that her husband is assistant lecturer in geology. So you see fate has been kind in every matter save the most important of all . . . I must relinquish all my lovely fantasies, which culminated in a triumphant return to England with my bride and the prospect of one-eighth Red-Indian children. (The shock was pretty bad, and I consider myself very unfortunate: Burwell says 99.9% of American married women wear a ring.)

> Would you say that if I am as keen as all that, her being already married ought to mean nothing to me? I can only reply that her husband must be a better man than I, else she would not have accepted him. And I think they are happy. I am writing this in bed in the middle of the night. Being too bowled over to sleep, I get some comfort by relating this sorrowful story to you, my dearest friend. Now I will try to find solace in [Felix] Klein's "Lectures on the Icosahedron."[49]

Resigned to leaving romantic matters in arrears, Coxeter had a productive year at Princeton, his polytopes research proliferating to such an extent that Professor Lefschetz nicknamed him Mr. Polytope—"because I had long been specially interested in the figures which [Lefschetz] insisted were simply 'polyhedra' in space of any number of dimensions."[50] Coxeter understandably found this remark a tad dismissive, though Lefschetz's was a perfectly proper definition of polytopes. On the whole, Coxeter said in a letter to his father polytopes are an "anathema"[51] in America. Coxeter was somewhat consoled to receive a letter from a Cambridge compatriot, Gilbert de Beauregard Robinson, who by then was at the University of Toronto. Robinson wrote inviting

Coxeter to deliver one or two lectures on polytopes at Toronto, an offer which Coxeter gamely accepted.[52] But on the whole, Coxeter was working upstream, against the flow of the more fashionable areas of geometry, such as topology. The less trodden path, for him, was more seductive, and, perchance, would lead toward unexpected and more rewarding destinations.

Coxeter pressed ahead with polytopes that year and made a break-through, indeed one of his greatest career achievements, though the extent and range of his discovery would not be fully realized for some years to come. Coxeter discovered what he at the time called "graphical symbols"[53] for kaleidoscopes and the polytopes they generated.

The kaleidoscopes Coxeter carried around at Cambridge—his custom-made contraptions of mirrors and hinges—were the tools he used to investigate the symmetrical properties of polytopes generated in up to three dimensions, with one mirror per dimension. Beyond three dimensions it was impossible to physically construct kaleidoscopes. But mathematicians do not dwell in three dimensions—not intellectually, anyway. They live in hypothetical n dimensions, exploring patterns and polytopes that stretch to infinity. The physical impossibility of building n-dimensional kaleidoscopes did not hold Coxeter back. The challenge of gaining a glimpse of higher-dimensional shapes—and an understanding of their symmetrical properties—turned in his mind, rotated and reflected, until he realized he could simply devise another sort of contraption for climbing the dimensions. He discovered a way of exploring n-dimensional kaleidoscopes with a different tool, a mental crutch, formulated from imagination and ingenuity alone. This tool mimicked his physical kaleidoscopes and took the shape of a simple symbolic diagram constructed with pencil and paper. His graphical symbol became known as the Coxeter diagram. Once it caught on, with the contagion of an indispensable high-tech gadget, it became widely used by mathematician and scientists alike as they investigated symmetries—whether symmetries of shapes, numbers, equations, or symmetries in the fabric of the universe and all its contents.

"A Coxeter diagram is a code,"[54] said Neil Sloane, a mathematician and telephony scientist at the AT&T Shannon Lab, in New Jersey. (Sloane is best known for creating the On-Line Encyclopedia of Integer Sequences; one sequence, discovered by Bernardo Recamán Santos, contains a pattern of numbers so difficult to decipher that those who have tried dubbed it "How to Recamán's Life.") "A code is a way of converting data from one format to another, encoding it in such a way that you've concealed information from prying eyes or protected it against distortion, or you've taken out redundancy, or just put it in a nice clean format. Certainly, Coxeter diagrams do that. They convey information"—about polytopes and their kaleidoscopes and the groups of symmetries they generate—"in a precise and elegant format. A Coxeter diagram is a good vocabulary for talking about groups," said Sloane. "In that sense, it's a bit like Morse code. It's a language."[55]

As it happens, the analogy between Morse code and Coxeter diagrams works on two levels. Both codes convert essential information into a concise and visual short form. And superficially, the basic components of Morse code, the dots and the dashes, are also the basic components of a Coxeter diagram. In a Coxeter diagram, which imitates the physical contraption of the kaleidoscope, each dot, or node, represents a mirror. This, for example, is the Coxeter diagram for the kaleidoscope that generates an icosahedron in three dimensions:

The numbers under the segments connecting the nodes indicate the angles at which the mirrors meet. Here, 180° degrees divided by 3 = 60°, and 180° divided by 5 = 36°. In this way, the numbers dictate what kind of kaleidoscope it is—tetrahedral, octahedral, or icosahedral.

Being a proponent of elegance and economy, Coxeter devised abbreviations to simplify his diagram. For instance, when nodes are not joined, such as the nodes to the far left and far right in the above diagram, their respective mirrors in reality meet at an angle of 90° (or in a manner such that if they were joined by a line there would be a 2 beneath). For another abbreviation, Coxeter decreed that when mirrors are joined at an angle of 60°—since they frequently are, such as in the icosahedral kaleidoscope above—the number 3 should be left out; so when you come across two dots joined by a line with no number beneath, you assume the angle between those mirrors to be 60°. Thus, the proper Coxeter diagram for the icosahedral kaleidoscope is:

Such abbreviations may seem needless, but in practice anything that can go unstated serves as a simplification and increases a tool's utility and efficiency, much the same way you might prefer to use an Internet hookup with a minimum of connections and passwords to quickly access the desired information. Coxeter, being a master of concision, wanted to make his diagrams as spare as Shaker furniture, with a minimum of lines, nodes, and numbers cluttering the design.

Another indispensable feature of his diagrams was that they not only conveyed the kind of kaleidoscope used for exploration, a Coxeter diagram also indicated what shapes could be generated within. Following Wythoff's method of positioning a prop—a blob or a stick—in the kaleidoscope to generate the polytope, Coxeter added one more symbol to his diagrams. He circled the appropriate node to indicate where the prop should be placed on the mirrors. So this is the Coxeter diagram not for the icosahedral kaleidoscope (as above), but for the icosahedron itself:

Likewise, the following Coxeter diagram indicates the dual of the icosahedron—the dodecahedron—the only difference being that the prop is placed in a different position:

The dimensions can be easily increased, and hyperdimensional polytopes easily investigated, by simply adding nodes, or mirrors, to the diagram—one node per dimension. Here is the Coxeter diagram symbolic of the icosahedron in four dimensions, also known as the 600-cell:

Icosahedra and dodecahedra do not exist in dimensions higher than four, which suited Coxeter fine. "Four is my favorite dimension," he once said. "The things that happen in four dimensions are extra special and agreeable."[56] To his eye, four dimensions produced the most exquisite collection of regular polytopes. Coxeter probed up to 8 dimensions, but beyond that he left it to other people.[57] But even though Coxeter did not often ascend into these altitudinous domains, he nonetheless enumerated all possible kaleidoscopes that produce polytopes continuing to infinite, or n-dimensions. Coxeter diagrams are like a rope ladder by which mathematicians climb, notch by notch, node by node, dimension by dimension, through the eternal exosphere of infinity. (See appendix 3 for all the kaleidoscopes Coxeter enumerated, organized by dimension.)

Coxeter diagrams may appear rather quaint and arcane, even esoteric and useless explorations into a very narrow corner of symmetry, but deceptively so. To mathematicians exploring the symmetries and patterns of numbers, and scientists exploring the symmetries of the universe, they became invaluable tools. Coxeter diagrams provide a quick summary of the symmetrical properties being studied, exactly like shorthand. It is much easier to hold the whole message in the mind's eye—get the whole picture—about symmetries when they are laid out with a diagram, rather than spreading the message through a disjointed list of algebraic identities and equations. The group of symmetries produced by one set of mirrors, or generators, can be quite different from those of another set, and the Coxeter diagrams help in the differentiation, providing an identification tag that twigs a mathematician's memory to the complete characterization of the group.[58]

Take for example an 8-dimensional polytope, which has 711,244,800 symmetries, 240 vertices, 6,720 edges, finishing with its 7-dimensional cells, namely 17,280 simplexes and 2,160 orthoplexes.[59] The Coxeter diagram representing that polytope helpfully compresses such a stream of data:

Gaining a glimmering of a polytope's appearance in the intangible higher dimensions through projection—projecting the shape like a shadow down to three or two dimensions—also efficiently compresses the complexities. Here is a three-dimensional projection, a model, of the four-dimensional icosahedron, the 600-cell, followed by a projection of the same polytope down to two dimensions[60]:

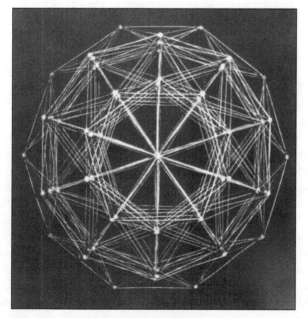

Paul Donchian's model of the four-dimensional icosahedron, or 600-cell, projected down to three dimensions.

The 600-cell projected down to two dimensions, drawn by the German
mathematician Salomon van Oss, circa 1906.

Clearly, Coxeter's graphic representation of the 600-cell is trim by comparison:

"It is really remarkable," said John Conway, "because you have a higher-
dimensional and tremendously symmetrical object and Coxeter's way of
conveying it is with this astonishingly compact notation."[61] Marjorie Senechal
is similarly astonished at Coxeter's pithy diagrams: "He tugs us upward
through the thickets and wickets of higher dimensions so deftly and delightfully that we (almost) feel at home there."[62]

As this invention took shape at Princeton, Coxeter delivered the inaugural lecture introducing his graphical symbols. Lefschetz sniffed at his
accomplishment: "It's good to talk about trivial things occasionally,"[63] he
said. Little did Lefschetz know where these trivialities would take Mr. Polytope. Unfazed and confident in the intrinsic value of his work, Coxeter produced a long paper outlining his research results. Oddly enough, he left
it with Lefschetz,[64] a daring, or foolish, choice for a shepherd. Coxeter
hoped the paper could be published in the Princeton journal *Annals of
Mathematics*.

His year at Princeton coming to a close, Coxeter planned to spend the next two years in England, to complete his tenure at Trinity, and then reconsider his options. His friends at Princeton organized a farewell pub crawl around town[65] but tried to convince him to stay, on the assumption that a job offer was sure to materialize. And, they pointed out, it was a pity to leave just as Einstein arrived. "Not that I should really appreciate Einstein's work," Coxeter said in a letter to Katie. "Veblen's relativity reaches practically the same conclusions, in a more geometrical manner."[66] As if he'd heard the compliment, Veblen told Coxeter that he had liked having him to an unusual extent and wished he was staying another year. Coxeter felt inclined to reply, "If you had said that earlier, I might have arranged to stay."[67]

Little did Coxeter know, and never in his wildest dreams would he have guessed, that his discoveries, his Coxeter diagrams and his enumeration of the symmetry groups of polytopes, would three-quarters of a century hence make unexpected appearances in applications of the very same work by Einstein that had left Coxeter so nonplussed—in string theory. Although Einstein's relativity may not have been sufficiently geometric for Coxeter's liking, it was rooted in geometry nonetheless—the revolution of non-Euclidean geometry. This was one of those times when a piece of pure mathematics, dismissed in the moment as having no practical connection to the world in which we live, later fell into the lap of the right person and came alive again in application.

△ ◻ ◈ ◉ ◉

Before heading home, Coxeter cruised across America on a road trip with his father, stopping at the Chicago's World Fair. The theme was "A Century of Progress," with exhibits featuring "Live Babies in Incubators," "A House of the Future," and the leggy fan dancer Sally Rand. A Hall of Science exhibit caught Coxeter's attention—as touted in a newspaper headline: PAUL S. DONCHIAN OPENS DOOR TO A FAIRYLAND OF PURE SCIENCE, HIS WIRE AND CARDBOARD MODELS EXPLAIN HIGHEST MATHEMATICS.[68] Donchian, a rug salesman and amateur geometric model maker[69] from Hartford, Connecticut, manned a display table covered by a gaggle of his wire polytope models, big ones the size of beach balls, little ones bobbing alongside like ducklings after their mother.* The subhead to the newspaper report, which appeared in Donchian's hometown paper, the *Hartford Courant*, chronicled how his

---

*As a result of Donchian's distraction with his models, the family rug business struggled financially; he had cultivated a serious hobby that evaporated weeks at a time as he painstakingly constructed his elaborate wire sculptures. When a hurricane hit New England, causing massive flooding, an apocryphal tale has it that Donchian saved his models first, his children second, and his livelihood, his sopping wet rugs, last.

appearance at the fair had ENTHRALLED THOUSANDS—the models drew a gaping public that stood in befuddlement at the four- and five-dimensional entities, and they drew "wizards of higher mathematics perking up their ears"—Coxeter was one of them and he and Donchian became fast friends. The father of relativity showed up to take a look as well, but as the newspaper headline reported, EINSTEIN WAS BARRED FROM EXHIBIT LEST CROWDS CRUSH BOTH HIM AND MODELS. Being Einstein, he was granted a private viewing after the hall closed for the day.[70]

After the road trip, Coxeter returned to England, but his Princeton year lingered as the happiest time of his life.[71] Even the idyllic Cambridge did not seem so perfect by comparison. When Princeton's *Annals of Mathematics* published his kaleidoscopes paper—"Discrete Groups Generated by Reflections"—Coxeter came down with a bout of nostalgia that even his tightly calibrated rational disposition could not suppress. On the recommendation of a friend, the Indian astrophysicist Subrahmanyan Chandrasekhar,[72] a fellow at Trinity with whom he often dined at High Table, Coxeter applied for and received an Eliza Procter fellowship, allowing his return to Princeton.[73]

△ ⬜ ◇ ◉ ⬡

And so for the academic year of 1934–35, Coxeter found himself back across the pond, in time to witness the unparalleled stir caused by Einstein's presence; the gawking photographers hadn't lost interest even after his eminence had been about town for over a year. Reporters from small-town newspapers and the *New York Times* camped out in the swath of evergreen forest in front of Fine Hall, hoping to catch a glimpse of Einstein's bouffant-haired silhouette in the window of the professors' lounge; or they loitered on his walking route, hoping to bump into him sauntering to work, eating an ice-cream cone in his sloppy clothes and grubby tennis shoes (once a dapper dresser, by the time he got to Princeton, Einstein was completely uninterested in his appearance).[74]

Coxeter's second stint at Princeton was busier and headier than the first.* His Trinity coeval Patrick Du Val came as a Rockefeller Fellow, the two of them basking in captivating discussions with members of the newly formed Institute for Advanced Study.[75] The institute did not yet have a building of its own. It shared Fine Hall with the university's math department, having

---

*By one measure, as a Rockefeller Fellow he had been appointed treasurer of the Fine Hall tea club and was charged with the task of "collecting one dollar from each of about forty unwilling professors and students," but this time around he was chairman of the tea club, since that duty always went to the Procter Fellow.

the effect that this year there was no plush office space for Coxeter. He worked from a carrel in the mathematics library, which took up Fine Hall's entire third floor and was open twenty-four hours a day, every day of the year including Christmas. Coxeter's working quarters in the library may have been slightly second-rate, but the company he kept that year was out of this world.

By some accounts, Einstein made an effort to fit in with the Fine Hall culture. He played ping-pong in the common room; he was uncoordinated and the ball often landed in his hair. He attended at least some of the social events, such as the buffet luncheon held every few months, catered with home cooking by the faculty wives. The dish prepared by Mrs. Einstein was "always very exotic and something to be avoided." Certain mathematicians were known to huddle with dread as the luncheon began. They drew straws and the person with the shortest was sent down the buffet table on a surveillance mission, giving the "high sign" when he found the unsavory Einstein salad, and flagging it for those who came after.[76]

By other accounts, Einstein isolated himself at Princeton, perhaps fatigued by his fame. He had moved on from relativity and was now laboring over his unified field theory, seeking to unite all the laws of physics. But his initial discoveries still captured public and scholarly attention. In particular, his relativity work rankled the mathematicians, providing fodder for gossip, skepticism,* and resentment—much more vehement opposition than Coxeter's mild disinterest. Being a physicist and not a mathematician, Einstein faced criticism because he hadn't thoroughly learned the mathematics he needed for his results. His friends allegedly alerted him to some results they thought he should look at, including the non-Euclidean geometry espoused by Germany's Bernhard Riemann (1826–1866),[77] which subsequently produced his theories. This wasn't uncommon. It would be an imposition for a physicist to study mathematics intensively just to glean the crucial bits and pieces he needed.[78]

Nonetheless, mathematicians and mathematical physicists were disinclined to forget or forgive that Einstein had achieved his work by thinking in physics rather than mathematics. One Fine Hall mathematician suggested that Einstein's general theory "made natural the surmise that all physics might be looked at as a kind of extended geometry." Veblen opined in a letter that, "though the great physicist used mathematics as a tool, [Einstein]

---

*Einstein had won the Nobel in 1921, but not for relativity, since it was still somewhat in dispute; instead he received it for his 1905 essay on photoelectric effect.

probably could not have discovered the general theory of relativity without the four-dimensional geometry earlier worked out at Göttingen." David Hilbert at Göttingen noted: "Every boy in the streets of our mathematical Göttingen understands more about four-dimensional geometry than Einstein. Yet, despite that, Einstein did the work, and not the mathematicians." And speaking to an audience of mathematicians, Hilbert remarked, "Do you know why Einstein said the most original and profound things about space and time that have been said in our generation? Because he had learned nothing about all the philosophy and mathematics of time and space."[79]

Einstein himself didn't begrudge credit to mathematics and geometry. In 1921 he gave an address, "Geometry and Experience," to the Prussian Academy of Sciences. "We may in fact regard geometry as the most ancient branch of physics," he said. "Without it I would have been unable to formulate the theory of relativity."[80] And the next year, during a lecture in Kyoto, he stated: "Describing the physical laws without reference to geometry is similar to describing our thoughts without words."[81]

△ ▢ ⬦ ⊛ ⊛

Einstein's theories set our universe to a new score, orchestrating a booming crescendo between symmetry and physics. Prior to the twentieth century, the laws of nature were believed to operate like gears and pulleys. Physicists considered the concept of symmetry inconsequential, a sideshow of pleasing eye candy; occasionally symmetry simplified a problem, but it certainly had no fundamental role to play in the core dynamics of the physical world. Though Einstein isn't usually thought of in these terms, he cast symmetry as the underlying foundation of space and time. A unique aspect of his thinking on special relativity was the assumption of a "symmetry principle."[82]

"Symmetry doesn't so much control as it does describe or account for nature," explained Leon Lederman, Nobel laureate in physics, and director emeritus at Fermilab, in Illinois. "As we go deeper, deeper into our understanding of the physical universe, even the biological universe, it gets more complicated—seemingly, the equations get more complicated, the things you have to describe are more varied. And you'd like some unifying principle. Then, out pops symmetry. I give Einstein credit for introducing symmetry into modern physics. He did that with his special theory of relativity—$E = mc^2$. Wow! The big increase in knowledge is the statement that the laws of physics apply to any system that you want; the laws are invariant to a change in the velocity of the system—that's relativity. And then it became very clear to even grubby plumber-like experimental physicists—which is what I am—that symmetry in fact makes things much more simple, that it is the overriding basis of the mathematics of physics, or all sciences, that symmetry produces an elegance and a beauty to the description of nature."[83]

Another mathematician known for revolutionary work in symmetry was Emmy Noether, also around and about on the Princeton campus during Coxeter's second year there. An academic refugee in one of Hitler's early sweeps (being Jewish, and a woman), Noether accepted an invitation to the all-women Bryn Mawr College, in Pennsylvania, arriving in 1934. She made weekly trips to lecture at Princeton's institute, and to visit her friends Einstein and Hermann Weyl. She could be spotted walking up from the train station, always wearing the same outfit: a shiny shapeless jumper that only accentuated her proportions, being almost as big around as she was high. One day a mathematician standing in the common room window of Fine Hall saw her approaching. "How can you tell a penguin from Emmy Noether?" he asked. "A penguin doesn't carry a briefcase." (Weyl gave a more flattering portrait of Noether, saying, "The graces did not preside at her cradle.")[84]

During her seminars, Noether stood at the blackboard with a wet sponge in hand, ready to erase as she worked. Intellectually too impatient to wait for the water to dry, she wrote on the blackboard when it was wet. This brought two unfortunate results: her chalkings initially were invisible, emerging only upon drying; and then, once the writing had finally dried, it was nearly impossible to erase. Her multitasking complicated matters further. She'd write something of import below while erasing with her wet sponge above, producing a trickle of water rolling, ominously, downward. She didn't know quite what to do to stop the drop of water that was threatening to destroy her work. So she blew on it to direct it off course.[85]

In 1915, Noether had made one of the most remarkable contributions to human knowledge in a theorem pertaining to symmetry and derived from Einstein's insights. Known as Noether's theorem, it states: For every symmetry in the laws of physics, there must exist a conservation law (if there is symmetry, something is conserved).[86]

On a metaphysical level there are symmetries with respect to time, as Einstein proved—if scientists such as Lederman, in Illinois, and his friendly rivals at CERN, in Geneva, do the same experiment, either years apart or separated only by nanoseconds, both parties will achieve results governed by the exact same laws of physics, since those laws do not depend on any absolute time. Similarly, there is symmetry in the law of conservation of energy—energy can neither be created nor destroyed. "Why?" asked Lederman. "I don't know. That's the way nature is."[87]

Noether's theorem was extraordinary due to its capacity to unite the mindset of physicists and mathematicians, catalyzing their interaction.* Symmetry, studied in the confines of an icosahedron's corner, or in the widest frontier of the ever-expanding universe, is omnipotent and omnipresent. And thus, it can be applied by analogy from Coxeter's geometrical niche to Einstein and Noether's macroscopic expanses of physics. "All of mathematics is the study of symmetry, or how to change a thing without really changing it," Coxeter said in a 1972 radio documentary. "It is symmetry, then, in its various forms, which underlies the orderliness, laws, and rationality of the universe, and thereby also the language of mathematics."[88]

Coxeter, studying the symmetries of polytopes—how they can be transformed and conserved in appearance by reflections and rotations—had no purpose in mind for his work other than its stunning aesthetics. And in the 1930s, his work not only lacked connection to the symmetries of physics, but even such mathematicians as Lefschetz were unable to see the inherent value in his pure mathematical pursuit—trivial topics like polytopes are fine for the occasional intellectual romp, Lefschetz had said, pooh-poohing Coxeter's lecture, a comment which stabbed the geometer in the moment (enough to quote it in his diary), and surely echoed in his head. Coxeter, it seemed, preoccupied himself with a rusty relic of classical geometry, a futile endeavor as far as most modern mathematicians and scientists were concerned. Then again, as is said of pure mathematics, if it is beautiful and elegant, if it is good and profound, there is always the latent promise that it will open something up. Eventually, almost inevitably—often inadvertently and unbeknownst to its inventor—a beautiful piece of pure mathematics will fall into the pattern of crystallizing with an application in science.

During his second stint at Princeton, Coxeter sat in on a few lectures where Einstein either spoke or was present, but they had no concrete interaction.[89] Coxeter's most relevant intersection with Einstein, therefore, was vicarious, through Hermann Weyl, known for his Weyl gauge theory (Gauge theories are a class of physical theories rooted in the notion that symmetry transformations can be executed locally as well as globally).

Before coming to Princeton, Weyl and Einstein had collaborated on the concept of a unified theory.[90] Weyl's earlier unification attempts had failed,

---

*For this, Lederman praised Noether's contribution to the skies: "Noether's theorem provides a natural centerpiece for any discussion that unifies physics and mathematics . . . in a way that enlivens them both . . . The worlds inhabited by theoretical physicists and mathematicians are often quite separate and independent. It is during the rare moments when the two worlds converge that the bugles blow, the drums roll, and science moves forward!"

and Einstein had put his finger on one of the serious flaws. From then on, the unification plot consumed Einstein, who worked on this problem for the rest of his life. "The physics community thought that he was wasting his time," recalled John Moffat, a physicist at the Perimeter Institute, in Waterloo, Ontario, who corresponded with Einstein in the early 1950s. "But as usual, Einstein was ahead of the game. Because he was doing something that subsequently became one of the main activities of fundamental physics today—unified theory, or string theory."[91]

At Princeton, Weyl continued working with Einstein on the unification theory, but the emphasis of Weyl's work was elsewhere. Serendipitously, it was along the lines of what Coxeter was doing, and Coxeter's published paper on kaleidoscopic reflection groups caught Weyl's eye.[92]

Weyl looked more the part of a proprietor of a German delicatessen than the world-renowned mathematician that he was. By the estimation of some, he was the closest the twentieth century had to a truly universal mathematician. He boasted a perspective that was both deep and broad, spanning geometry, algebra, mathematical physics, topology, and analysis. His masterful results, however, did not come easily. He had a harsh voice and, toiling in his study, he let out groan after groan, his exasperation reverberating through Fine Hall. One colleague said Weyl sounded like the perpetual recipient of the worst possible news. Another colleague compared Weyl's delivery of theorems to a woman giving birth to a child. Weyl's lecturing style also was unnerving, as his faculty verse attested: "Here we have a punning Aryan/Who likes to make groups Unitarian/He is that most saintly German/The One, The Great, The Holy Hermann." Lefschetz invented the nickname Holy Hermann for Weyl due to his ponderous way of lecturing with protracted sentences, stuffed with long multisyllabic words, and the verb as far toward the end of the sentence as was linguistically possible.[93]

Weyl delivered a course during the 1934-5 academic year called "The Structure and Representation of Continuous Groups." While Coxeter worked with a family of groups known as discrete or finite groups, symmetry groups that were generated by discrete or finite objects such as polyhedra, Weyl worked with the mysterious and amorphous family of continuous groups, such as the infinite symmetries of the sphere. The sky, similarly, is continuously symmetrical as we gaze into the well of outer space, and the Earth, rotating around its axis, with the Sun rising in the east and setting in the west, gives us the same sense of continuous spherical motion. Within the continuous groups there are five particularly "exceptional groups," as they are called, whose character has been likened to Uranus and Pluto—in a way these planets are familiar to us, but really they remain intractable to investigation and unknowable.[94]

Coxeter was a regular at Weyl's series of seminars on continuous groups, and the unexpected connections between their fields soon became apparent. Coxeter's discrete symmetries, it turned out, were special cases of the larger family of continuous symmetries, and thus informed, by analogy, some of the more sporadic and exceptional infinite groups. One way to think of it is that the infinite symmetries of a sphere's continuous rotation are hard to get a handle on; their unwieldy nature makes them overwhelming. But it is an elementary fact of geometry that a rotation can be broken down into smaller parts, smaller movements, or reflections—reflections upon reflections add up to a full-turn rotation. The relevance of Coxeter's work for Weyl, then, was that Weyl's continuous groups now had a skeleton, a basic underpinning framework, a root system, reducing their infinite and intangible properties to abstract and more manageable finite pieces. Coxeter's work—his Coxeter diagrams and his enumeration of the finite reflection groups—were just the tools Weyl needed. Weyl's enigmatic infinite symmetries were rendered slightly less amorphous and mysterious when treated with the Coxeterian tools. This amounted to a tremendous simplification and is now part of the standard treatment of continuous groups.[95] "The overall classification of these symmetry groups is probably one of the most important handful of ideas in human history in the last century," said Ravi Vakil, a younger-generation algebraic geometer at Stanford. "Because they underlie so much."[96] Coxeter had hit upon a primordial and indispensable tool that permeates the field of mathematics.

Weyl, duly impressed, enlisted Coxeter to take the official notes for the seminars. And when Weyl's lessons reached the more general realm of "various topics in group theory," he invited Coxeter to present his research with discrete symmetries. Coxeter took over five of Weyl's seminars in total.[97] After the course concluded, Weyl immortalized Coxeter's contribution in the official course notes,[98] adding it as an appendix. This was a coup of exponential proportions for Coxeter, both locally and globally. With Princeton's reputation, the course notes were mimeographed and distributed worldwide. International orders occupied two secretaries who were occupied solely with the task of duplication and mailing. Reaching such a captive audience raised Coxeter's profile and catapulted him into an elite tier of mathematicians.

Over the following years, as Coxeter continued with his polytope opus, amassing, piece by piece, chapters for his first book, *Regular Polytopes,* he consulted occasionally with Weyl. He wanted to devote a number of pages to an idea of Weyl's that added to the narrative on polytopes. He wrote asking Weyl's permission, and sent along the relevant chapters in rough draft. "Of course you are welcome to use 'my' formula for the order of the special sub-group of an infinite group generated by reflections," Weyl wrote back.

"Thank you very much for your kindness in letting me see these two chapters of your book . . . I have no criticisms to offer . . . But I look forward to reading the whole book when it comes out."[99]

△ ⬭ ⬧ ⊛ ⊗

At the end of his second turn at Princeton, Coxeter again wasn't hearing of any job offers, either with the math department or the Institute for Advanced Study. Between the Depression and the turmoil in Europe, such venerated figures as Weyl and Einstein were available for the taking; no one but the brightest minds in the firmament had prospects.[100]

Coxeter's only job opportunity in America came from Carmelita Hinton—the wife of Alicia Boole Stott's nephew, Sebastian Hinton. In 1935, Carmelita, a progressive educator and reputedly a friend of Chairman Mao, was in the midst of establishing the Putney School, at Elm Lea Farm, in Vermont—America's first coeducational boarding school, which based its curriculum on the ideas of farm work, academics, travel, and the arts (to this day the school observes this tradition). Carmelita was a great believer in John Dewey's philosophy of "learning by doing." She offered Coxeter a position on the teaching staff—she was trying to get "real people" to teach and Coxeter, with his great gift for mathematics, as well as his talent in music and love of the outdoors, was a perfect fit. Coxeter considered himself well suited for the job, especially since while at Princeton he had taken on the task of editing W. W. Rouse Ball's *Mathematical Recreations and Essays*, a book full of good teaching material. Coxeter, clearly delighted to be chosen as Ball's successor, wrote to his father: "Did I tell you that I have undertaken to edit . . . Rouse Ball's Recreations and Essays? It was Hardy who proposed me as the man to do it. I think it will be fun, and familiarity with the sort of stuff that it contains cannot fail to be an advantage to a teacher (if such I am to be)."[101] Charged by the prospect of "pioneer" work at a new school, his first instinct was to jump at the chance, even going so far as to suggest to his father that his eldest half sister, Joan, be sent overseas for a year at the new school.[102]

Coxeter's mathematical peers, however, cautioned him against such a move, the strongest warning coming from Oswald Veblen. His antiteaching bias ran in the family. Veblen was the nephew of the sardonic iconoclast and economist Thorstein Veblen, whose five-chapter memorandum, "The Higher Learning in America," argued that the whole American apparatus of degrees and undergraduate teaching was a sham. In contrast to the goal of practicality, Thorstein believed "the sole end of the truly inquiring mind should be irresponsible scholarship, idle curiosity, and useless knowledge."[103] In the same vein, Oswald Veblen bedeviled the Princeton administration, urging visitors to the university, graduate students, and department professors alike not to waste time doing any more teaching than they had to. Veblen stressed to Coxeter the importance of having time to write down ideas during the brief creative years of one's life—"of which one year is a considerable fraction." And he cautioned Coxeter that he would miss, more than he realized, "the companionship of mathematical minds." The sanctum for research that Trinity provided was a precious thing. "I could have created so much more," Veblen said, "if time and energy had not been taken up with teaching."[104]

Coxeter finally lost his battle of indecision when the Trinity council refused to grant him an additional leave of absence. He had written the bursar to test the lay of the land, financially and otherwise, should he want to defer his fellowship another year. Coxeter was prepared to give up one of two remaining years at Trinity, but he knew that to sever the connection entirely would be damaging to his career. He turned down Carmelita Hinton's offer—"I don't feel a bit happy about it, and I hate having to confess this victory of reason over emotion. But everyone assures me I will see the advantages of remaining in the academic world, as soon as the pangs of remorse have softened."[105]

He made the most of his remaining days at Princeton. His diary entry one day was, "Drawing circles," and on another, "Drew some more circles," and on another, "Got up early to draw (4,6) triangles." He exclaimed, "I have overworked this week!"—and stayed in bed the next day reading W. Somerset Maugham's *Cakes and Ale*.[106] He made numerous jaunts into New York—for an exhibit at the Metropolitan Museum of Art; for a magicians' meeting; for the annual gathering of the American Mathematical Society, after which Pat Du Val took him to see some burlesque.[107] On June 14, 1935, Veblen told him that he was "the best-liked Englishman who has come to Princeton," and six weeks later Coxeter was once again on his way back to England.[108]

CHAPTER 5

LOVE, LOSS, AND LUDWIG WITTGENSTEIN

For pairs of lips to kiss maybe
Involves no trigonometry.

—FREDERICK SODDY, "THE KISS PRECISE"

Back at Trinity, Coxeter unpacked and settled into his suite of rooms in Great Court. He obediently explained his work to his father, showed Aunt Alice his "group pictures," and dined in the college hall with Pat Du Val, who "got drunk and tried to show how he could sing."[1] And then, out of the blue, a job offer arrived: an assistant professorship at the University of Toronto. Coxeter's visit there while at Princeton had been a vetting of sorts. "My lecture seems to have been well liked," Coxeter said of the job offer it brought.[2] But he was not sure whether to accept, and discussed his uncertainty with his father. When would he have to start? How long would he have to stay? What minimum period would not seem too shabby to them? Would he lose forever this last year of his Trinity Fellowship? Could he refuse Toronto now and change his mind at a later date? What were his chances for a job in England?[3]

Coxeter had just been appointed a lecturer at Trinity, and his first few lectures received good reviews from the college chairman, although Coxeter noted in his diary, "Palmer, my pupil, showed that I know no geometry!"[4] A college lectureship was a lesser position than a university appointment—not nearly the status of a professorship[5]—but, buoyed by this small success, Coxeter refused the University of Toronto offer,[6] much preferring to remain in England if he could find a comparable position. His chances were slim since only one professorship was coming open in his field—the Lowndean Chair of Astronomy and Geometry, at Cambridge, from which his PhD advisor, Henry F. Baker, was due to retire. On January 8, 1936, he composed his letter of application.[7]

△ ⬚ ⬦ ⊗ ⬡

While awaiting news of the Lowndean chair appointment, Coxeter kept himself preoccupied. He took in a bad Laurel and Hardy film, and attended a lecture by Godfrey Hardy: "How to Write and How Not to Write a Mathematical Paper."[8] And he resumed his often "painful" discussions with Austrian philosopher, Ludwig Wittgenstein then at Cambridge, who was among the twentieth century's most influential philosophers, alongside Bertrand Russell and Jean-Paul Sartre.

The reclusive Wittgenstein had taken a liking to Coxeter when he was a student, and they kept in touch. "I had tea with Wittgenstein yesterday," Coxeter recorded in his diary. "He talked very interestingly about blindness and deafness, and why you get seasick on a camel but not on a horse. He doesn't seem any more abnormal than before."[9]

Coxeter had enrolled in Wittgenstein's "Philosophy for Mathematicians" lecture for the 1933–4 year. To Wittgenstein's horror, so did a total of forty students, far too many for the intimate lecture he was willing to deliver. "There are too many of you," the philosopher protested. "Will three or four please leave?" After only a few weeks, Wittgenstein informed his still too numerous students that the class would continue no longer. He deigned to lecture for only a chosen few. He would dictate his thoughts, and his select students were instructed to copy the notes and distribute them to the rest of the class in what became known as his Blue Books. The select group included Wittgenstein's five favorite students: Francis Skinner (a promising mathematics student who became Wittgenstein's constant companion, confidante, and collaborator); mathematician Louis Goodstein; philosopher Margaret Masterman (a pioneer in the field of computational linguistics, her beliefs about language processing by computers were ahead of their time and are now fundamental to the field of artificial intelligence); philosopher Alice Ambrose (of the analytic school, who also wrote papers on pi, mathematics, and the mind); and Coxeter.[10]

For Coxeter, Wittgenstein was largely unintelligible and intellectually precious. Wittgenstein refused to lecture for the customary 50 minutes, but required 150 minutes, partly because he needed an hour to warm up and partly due to his habit of stopping midsentence as he worked through his next point internally. Coxeter timed a pause at more than twenty minutes, after which Wittgenstein carried on where he left off, as if nothing was amiss. On another occasion, Wittgenstein complained the lecture hall was too formal. Coxeter offered the sitting room in his suite in Great Court, and Wittgenstein agreed to the new location. Of one lecture, Coxeter noted: "Found Wittgenstein really interesting for the first time (Locality of Thought)," but shortly thereafter he ceased attending, judging his time better spent on his own mathematical research (nonetheless Wittgenstein continued using Coxeter's room).[11] Coxeter later remarked of his Wittgenstein excursion: "I

Ludwig Wittgenstein

couldn't understand that kind of philosophy. I thought it was nonsense. It didn't appeal to me at all. The only thing I remember of his work is that his book *Tractatus-Logico-Philosophicus* began in chapter one with the proposition, 'The world is everything that is the case,' and ended in the final chapter with the proposition [and the only sentence in that chapter], 'Whereof one cannot speak, thereof one must be silent.' "[12]

The Cambridge professorial elections were held on February 28, and disappointingly Coxeter lost the Lowndean chair to William Hodge, winner of Cambridge's Adams Prize. Never before had the biannual Adams Prize competition, since its inception in 1848, been on a geometrical topic. Usually it addressed natural philosophy or physics. The Adams question this time called for an advance in existing geometrical theory.[13] Coxeter's groups generated by reflections would have qualified (there is no record as to whether he made a submission), but Hodge, backed by none other than Weyl and Lefschetz,[14] won for work developing the relationship between geometry, analysis, and topology—a contribution described as ". . . one of the great landmarks in the history of science in the present century."[15]

Initially, Coxeter's reaction was calm disappointment; he commiserated with his friend Patrick Du Val, also an unsuccessful candidate (as was Eric H. Neville).[16] By June, however, panic set in, forcing Coxeter to reconsider the offer from Toronto. His father admitted that with the change in circumstances, the balance seemed to favor that he go. "I see that it would add to your prestige later on when a suitable position in England turns up," said Harold. "I get the sense that in some ways England is asleep or sitting on the top of a shaky pedestal. But the whole world seems to be shaky too and without firm foundation . . . Probably your mother will be very disappointed . . . but I do not think you should consider your mother—still less us—in anything which vitally concerns your career."[17] Coxeter discussed the matter with Hardy and Littlewood, and a few days later Hardy concurred with a wire: "Reluctantly agree you better go this year."[18] And after talking it through with Baker over dinner, Coxeter made his final decision to go. "I could not have dissuaded you from going to Toronto," wrote Baker. "Many good men have begun away from England; Europe seems now to be mad; and anyway, Toronto is an inspiring place."[19] Coxeter wired Samuel Beatty, head of Toronto's mathematics department: "May I accept after all?" Beatty's response came in the affirmative and Coxeter was slated to start in the fall.[20]

△ ⬠ ◈ ◎ ⬡

Squeezed into Coxeter's busy schedule of research and teaching duties and preparations for the move, were the emotionally fraught demands of his mother. All of a sudden Lucy seemed old and slowing down, almost winding down on life. Her son took her shopping and dropped in for tea, reading aloud from Oscar Wilde's *De Profundis* while Lucy painted colored faces on an icosahedron model.[21] Still, she always wanted more of his time, and resented his legion of friends and commitments that kept him away. Coxeter's father had been visiting Lucy more regularly as well. "I went to visit Lucy a few days ago and found her really ill," Harold said in a letter to Rosalie. "She has greatly altered recently and is much kinder and considerate . . . I should not be surprised if she has not very long to live, and I think she feels this herself. Donald also senses the same, I think, as Lucy says he has been sweeter to her than he has ever been."[22]

On a Saturday afternoon in March, during tea at a friend of his mother's, Coxeter made the chance acquaintance of "an attractive Dutch girl,"[23] Hendrina (Rien) Brouwer.* Her parents had both died recently and, minimally

---

*Their first conversation revealed, disappointingly, that Rien was of no relation to the Dutch mathematician Luitzen Brouwer, who devoted much of his career to defending intuitionist and visual mathematics, for which he considered logic only a helpmate, as discussed in his paper, "On the Untrustworthiness of Logical Principles."

Donald with his mother, Lucy, and father, Harold, at Much Hadham circa 1933.

educated, she came to England to find work as an au pair. She had wide-set blue eyes and an entrancingly open and symmetrical face framed by fine blond hair, always pulled back.[24] Coxeter arranged their first date by post two days beforehand:

Dear Miss Brouwer,

I will call for you at 10 o'clock. It takes about an hour to reach Cambridge. Then we can look at colleges for a while, have lunch in my rooms, meet my best friend Pat Du Val for tea, and drive to the station . . . I hope you will forgive my not driving you back, but I shall have to stay in Cambridge for a few days to finish off some work.[25]

For their second date, they visited the Fitzwilliam Museum, the art museum at the University of Cambridge, and afterward Donald gave Rien a demonstration of his kaleidoscopes. And from there, the courtship proceeded at a clip. Coxeter's father, knowing his son's track record, had no compunction about offering advice, sending this missive:

May I be so rash as to say it might be worthwhile to be careful that Rien does not fall for one or other of your friends or acquaintances or any other attractive or attracted young man among the men at Cambridge while you are "getting to know each other." She will be susceptible in a new country, and a different life, and springtime, and unless I am a bad judge, many of the said young men would be susceptible also to her; or might feel less inclined to caution than HSMC! Having said this much, I now say take no notice of what I have said, but go your own way and make your own decisions for your own life.[26]

Coxeter took notice. Less than a week later, on May 24, he proposed to Rien in a Cambridge cemetery, recording in his diary, "R to supper. Asked her, and she did not say No."[27] Rien also documented the day, noting that they "chatted a long time" and that Donald asked "whether or not I wanted to be his wife . . . It was like a dream. Is it really to become a reality?" Rien was smitten. She thought Donald was a "darling," and he provided the stable future she sought—"Yes, I have everything."[28]

Donald's father was ecstatic about the engagement and immediately gave his blessing.[29] His mother was "tiresomely hurt" that Harold and Katie received word of the engagement before she did. "How easily misunderstandings arise and bitterness and jealousy and hurt pride is aroused," Harold wrote to his son. "Don't be too worried or depressed about Lucy," he consoled. "She will be all right. It is difficult for her to feel she is losing you, when in many ways you have taken my place in her life. We have definitely decided that neither Katie nor the children will appear at your wedding. Perhaps we could have a second celebration here after the wedding sometime . . . It is a great joy to me to feel that you will have so loveable and suitable a wife and I believe you will bless the day you met her. There will be much to discuss, but there is time."[30]

△ ◻ ◈ ◉ ⬟

Prior to the wedding Coxeter and his father planned a vacation in Norway for the entire month of July, timed to coincide with the International Congress of Mathematicians in Oslo. The close relationship between Donald and his father was more that of brothers, though in preparation for the trip the domineering Harold told the full-grown Donald what to pack and how to behave. "Take a nice pair of leather shoes with rubber soles for walking. No sneakers please. And several tie pins and suspenders for your socks . . . Don't invite anyone else to tour with us, without my approval . . . What fun it will be."[31]

Donald and his father on their trip to Scandinavia, shortly before Harold's death, 1936.

Landing at Oslo, they enjoyed two weeks of touring before the congress began—climbing mountains, bathing in the snow and swimming in idyllic lakes, Coxeter driving through the hills in their rented convertible and Harold, a freer and wilder spirit than his son, standing naked on the passenger seat in the wind. Expeditions like these were old hat to Harold and filled many family photo albums—pictures of men scaling iced mountainsides with walking sticks and safety lines stringing the party together (a cigarette company financed one trip; Harold was a chain-smoker and won the prize for the customer who sent in the largest collection of emptied packages).[32] Coxeter, on the other hand, wasn't so accustomed to the long days of trekking—"very footsore," he moaned in a letter to Rien.[33]

The conference started on July 14. Coxeter and a few Cambridge peers attended. One told Coxeter he found himself sitting between two mathematical behemoths—"I felt like a very small hyphen." Another confided that he sat through only a single session and then skipped out to spend the rest of his time hiking and swimming in the mountains. Three days into the conference, Coxeter ducked out to visit the National Gallery, for the second time.[34]

At the congress, Coxeter met Irish mathematician J. L. Synge (nephew of the playwright J. M. Synge), who had founded the department of applied mathematics at the University of Toronto. J. L. Synge became a lifelong friend, and he and Coxeter would correspond with each other into old age. Synge made wide-ranging contributions to mathematics, in classical mechanics, geometrical mechanics and geometrical optics, gas dynamics, hydrodynamics, electrical networks, mathematical methods, differential geometry, though his main achievement was to apply geometrical terms to Einstein's theory of general relativity. He restated many problems in simple geometrical language, applying this method to the stability of the bicycle in a paper titled "Steering Gear," and launched into elasticity with one on "The Tightness of the Teeth, Considered as a Problem Concerning the Equilibrium of a Thin, Incompressible Elastic Membrane."[35] Synge also wrote a fantastical mathematical novel, *Kandelman's Krim*. Coxeter loved it and plundered its pages, excerpting twelve passages in his book *Introduction to Geometry*, such as this one prefacing a section on rational numbers:

> "The northern ocean is beautiful," said the Orc, "and beautiful the delicate intricacy of the snowflake before it melts and perishes, but such beauties are as nothing to him who delights in numbers, spurning alike the wild irrationality of life and the baffling complexity of nature's laws."[36]

The congress was that year distinguished as being the first time a Fields Medal was awarded, and Synge had been instrumental in planning the award with another Toronto mathematician, John Charles Fields. It was formally called the International Medal for Outstanding Discoveries in Mathematics, though soon dubbed the Nobel Prize for Mathematics, since Alfred Nobel had neglected the category of mathematics. A Swedish industrialist and inventor of dynamite, Nobel has stipulated in his will that the interest from his $9 million endowment be used to establish prizes for inventions or discoveries of the utmost practical benefit to humankind. Why mathematics was excluded is unknown, though a long-circulating rumor at mathematics conferences suggests that Nobel was angry about the attentions another mathematician had showered upon Nobel's mistress, and this was his revenge.[37]

The honor of a Fields Medal recognized outstanding mathematical achievement of a mathematician no older than forty: ". . . while it was in recognition of work already done, it was at the same time intended to be an encouragement for further achievement on the part of the recipients and a stimulus to renewed effort on the part of others." Embossed with the head of Archimedes, the gold medal also bore a dictum from the Roman poet Manilius: "to rise beyond your understanding and make yourself master of the universe." In its inaugural year in Oslo, Fields medals were awarded to Lars Valerian Ahlfors, of Harvard University, and Jesse Douglas, at the Massachusetts Institute of Technology.[38]

△ ◻ ◈ ⊛ ⊗

Returning from Norway at the end of July, Coxeter faced the wedding and the move to a new country and felt overpowered with things to do.[39] He and Rien set their wedding date for the first of September, mere days before they were due to leave for Toronto. Trinity College Chapel was already booked, so they settled on the Round Church in the heart of Cambridge—the main rotunda, fittingly, was a perfect circle.

Then, on August 15, a wire from Katie brought tragic news. Having observed his mother's surrender to life, Donald was shocked to learn it was his father who had died.[40] Harold had been holidaying in a coast guard cottage at the seashore near Brighton. "I am managing to get the whole family to bathe [swim] before breakfast every day so far, and 2 or 3 times during the day,"[41] he wrote in his last letter to his son. Harold suffered a heart attack while teaching his youngest two daughters to swim underwater in the shallows of the English Channel. The girls noticed that he failed to surface and, by the time they fetched help, their father had drowned.[42]

The following days went by in a numb and telescoped rush: to the undertaker, to the inquest, to the burial at Woking, and "in face of this terrible calamity," as one friend said, "all your plans have to be made anew." Donald and Rien decided to marry early, to cut short the painful waiting. Two days after the burial—"two of the most ghastly days ever spent"—they wed with only a few family members present. Many of the intended wedding guests were surprised to open Cambridge's evening newspaper and see a photograph of Donald with his bride on the porch of the Round Church, two weeks ahead of the date on their invitation.[43] In the wedding photographs, Donald's fists were clenched tight, knuckles white with stress.

Congratulations and condolences arrived in tandem. One friend said, "Fate is very cruel, why had he to die—as you ask—when all the family was happy and content?" Another commiserated: "Why Harold? I can hardly believe it yet but to you who were nearest the loss must be doubly bitter in its reality . . . One's philosophy falters when Death appears so indiscriminate."[44]

A wedding present in the form of a mathematical poem came from Thorold Gosset—"The Kiss Precise," by Oxford's Frederick Soddy, a Nobel

Donald and Rien on the steps of the Round Church on their wedding day, Cambridge 1936.

laureate in chemistry. The poem revealed Soddy's formula for the relationship of the radii of four mutually tangent circles. "Although Professor Soddy has put his discovery in such a frivolous form," wrote Gosset, "it is really rather an interesting geometrical proposition. He says it took him three years to work it out . . ."[45]

> For pairs of lips to kiss maybe
> Involves no trigonometry.
> 'Tis not so when four circles kiss
> Each one the other three.
> To bring this off the four must be
> As three in one or one in three.
> If one in three, beyond a doubt
> Each gets three kisses from without.

If three in one, then is that one
Thrice kissed internally.

Four circles to the kissing come.
The smaller are the benter.
The bend is just the inverse of
The distance from the centre.
Though their intrigue left Euclid dumb
There's now no need for rule of thumb.
Since zero bend's a dead straight line
And concave bends have minus sign,
The sum of the squares of all four bends
Is half the square of their sum.

To spy out spherical affairs
An oscular surveyor
Might find the task laborious,
The sphere is much the gayer,
And now besides the pair of pairs
A fifth sphere in the kissing shares.
Yet, signs and zero as before,
For each to kiss the other four
The square of the sum of all five bends
Is thrice the sum of their squares.

Also known as the "Kissing Circles Theorem," it had been published by Soddy the previous June in *Nature* magazine.[46] Gosset closed his letter of congratulations with a suggested addition, to describe the even more general case, "in N dimensions for $N+2$ hyperspheres of the Nth dimension," which he published in *Nature* the following year.

And let us not confine our cares
To simple circles, planes and spheres,
But rise to hyper flats and bends
Where kissing multiple appears,
In n-ic space the kissing pairs
Are hyperspheres, and Truth declares
That n plus two will osculate
Each with an n plus one fold mate
The square of the sum of all the bends
Is n times the sum of their squares.[47]

Friends and family sent notes wishing Coxeter bon voyage, "not only for the Atlantic, but for the wider sea of life." He underscored a passage saying, "You must now be looking forward to the future, a new life and new relations in a new world rather than to the past . . ."[48] From Aunt Alice, Coxeter received an antique stained-glass Archimedean solid lampshade. Her condolences were full of hope: "My dear! I don't know how to write to you— words seem so futile beside so great a separation! But indeed one can rejoice, for his sake, that it happened so . . . While I have been writing my mind has gone back to the lovely world we have visited together, and which you have made so much your own. I wonder where you will get to in it! How I wish I could follow."[49]

Coxeter said his good-byes "without sentimentality."[50] His mother, as well as Katie and his three half sisters, could not quite believe he was holding to his plan to leave.[51] The newlyweds set out for Canada on September 3, the trauma of the last weeks gradually sinking in. As Coxeter noted in his diary, Rien "nearly lost her breath in the night," which he found quite disturbing in light of recent events. And he noted: "R told me I must stop mourning for Harold."[52]

CHAPTER 6

"DEATH TO TRIANGLES!"

He who despises Euclidean Geometry is like a man who,
returning from foreign parts, disparages his home.

—H. G. FORDER

Logic teaches us that on such and such a road we are sure of
not meeting an obstacle. It does not tell us which is the road that
leads to the desired end. For this, it is necessary to see the end
from afar, and the faculty which teaches us to see is intuition.
Without it, the geometrician would be like a writer well up in
grammar but destitute of ideas.

—HENRI POINCARÉ, *OEUVRES*

The Coxeters sailed into Quebec City, endured a long wait for the immigration inspector, and then continued by train to Toronto,[1] somewhat of a backwater in that day, second in status to Montreal. By Toronto standards, this English couple were more cosmopolitan, avant garde, and their arrival created a bit of a stir—they painted rather than wallpapered the interior of their house, and wore stylish clothing.[2] For the Coxeters, the weather took some getting used to. The first winter was a shock—Donald constantly had colds (he assaulted them with ascorbic acid and shoulder stands), and Rien began lobbying for a fur.[3] But before long, Coxeter was ensconced at the University of Toronto. He noted in his diary: "showed mirrors at seminar" and "1/4 hour after my graduate lecture students were still taking notes."[4]

The job came with an annual salary of $1,500. Coxeter was a penny-pincher, need be or not, writing on the overside of each page so as not to waste an iota of space (for this reason, no working copies of his books survived). He recycled stamps left clean, complaining that Canadian postage was absurdly expensive. And the couple saved on hot water bills by sharing bathwater, Rien bathing first.[5] They were the perfect picture of a "Jack Sprat" couple[6]—Rien, a lover of sweets, was forever trying to reduce, while Coxeter, skinny and vegetarian, was still plagued by the duodenal troubles that had first beset him at Trinity. The course of treatment prescribed by a Toronto doctor stipulated that he pump his stomach at closely timed intervals after eating, leaving food in his stomach long enough for nutrients to be absorbed but not long enough for it to reach his faulty digestive tract[7] (this continued on and off for ten years until Coxeter underwent a gastorectomy).[8]

Coxeter and Rien, the mathematician and the housewife, established a nice economy of existence. Rien prepared their separate diets (she made a failed attempt at vegetarianism), with mealtime scheduled like clockwork at eight, twelve, and six, the table set, the radio tuned to the news. Coxeter always dressed for breakfast, even on weekends and holidays, and always wore a tie (Rien insisted, finding his neck unattractive).[9] All the workings of the household were similarly calibrated to facilitate Coxeter's profession. This dynamic was etched in the mind of one of his early students, John Coleman, now professor emeritus at Queen's University, in Kingston, Ontario. He needed to consult with Coxeter outside office hours, and telephoned his professor at home to make plans to drop by. Rien greeted Coleman when he knocked: "WHAT—DO—YOU—WANT?" barked this beautiful, blunt woman, her accent a thick Dutch-English. "She was like a bulldog at the door," remembered Coleman.[10] Rien treated her husband as a precious object, protecting his time and his space, optimizing the parameters of his life for a career of mathematical productivity—early in 1937 he recorded in his diary: "I made myself sick working on $(3,3,5; 5)$."[11] The only obstacle hindering Coxeter's ambitions as a classical geometer was bad timing; he set out on his research career exactly when this antiquated field was firmly entrenched as passé. Coxeter's kind of geometry was fun as a hobby, but few mathematicians in their right-angled mind were staking careers on it.

△ ⬭ ⬙ ⊛ ✦

At the International Congress of Mathematicians in Paris in 1900, Hilbert, the father of formalism, had delivered a rousing address at the Sorbonne in which he posed twenty-three unsolved questions that he felt would shape mathematics in the coming century. He asked, "Who of us would not be glad to lift the veil behind which the future lies hidden; to cast a glance at the next advances of our science and at the secrets of its development during future centuries? . . . We hear within ourselves the constant cry: There is the problem, seek the solution. You can find it through pure thought."[12]

Only three of Hilbert's problems would have been even remotely tempting to Coxeter (and there is no evidence he worked on any of them very much).[13] Fewer and fewer questions were being asked in classical geometry. And even if Coxeter had discovered a surprising new theorem about Euclidean geometry—for example, Frank Morley's 1899 "miracle" theorem that Coxeter liked so much, trisecting the angles of a triangle (see appendix 5)—this would hardly garner him a plum job at the likes of Princeton or Cambridge.[14] With all the interesting questions in classical geometry seemingly answered and its theorems discovered, mathematicians turned elsewhere to fuel their ambitions, to churn out papers for publication and scurry up the ranks of academia. "As long as a branch of science offers an abundance of problems, so long is it alive," Hilbert had said in his address—"a lack of problems foreshadows extinction."[15]

By that measure, classical geometry, like an old-growth forest, was an endangered domain. And in general, the pendulum of scientific methodology had swung far away from the intuitive visual approach. "In the sciences in the last century and a half, the pictorial and the logical have stood unstably perched, each forever suspended over the abyss of the other,"[16] observed Peter Galison, professor of the history of science and physics at Harvard. "It goes back and forth, and not in an accidental way," Galison said. "Pushing hard on the visual methods ends up pushing toward the anti-visual. Beliefs swing between an almost theological dogma that images are stepping stones to higher knowledge, or that they are deceptive idols that keep us from higher understanding . . . Ultimately we need both sides."[17] For the time being, however, geometry was subsumed and shrouded by the algebraization of each and every branch of mathematics—shapes were expressed in terms of algebraic equations.

△ ▢ ◈ ⊕ ⊗

A dramatic example of this shift was reflected in the widespread acceptance of Nicolas Bourbaki's approach to mathematics.[18] Mathematical folklore allows it that Bourbaki, an enfant terrible of French mathematics, first made his mark in the mid-1930s. Bourbaki was known to frequent the Café Capoulade, in Paris, near the Luxembourg Gardens, where he worked on an ambitious tome, an encyclopedia of all mathematics. He called his treatise *Eléménts de mathématique*—the singular "mathematic" hinting at the unifying intentions (though it was translated in English to *Elements of Mathematics*)[19]—and ultimately it was to include six books, a few chapters of each book published every couple of years.* Bourbaki's first installment, chapter 1 of Book I, on set theory, was published in 1939. He wrote various books of his treatise simultaneously, the next installment, published in 1940, being chapters 1 and 2 of Book III, followed in 1942 by chapters 3 and 4 of Book II, and so on.[20]

In a retrospective on Bourbaki's impact, a *Scientific American* article reported that this mathematician had quickly attracted international attention. "His works are read and extensively quoted all over the world. There are young men in Rio de Janeiro almost all of whose mathematical education was obtained from his works, and there are famous mathematicians in Berkeley and in Göttingen who think that his influence is pernicious," wrote the mathematician Paul R. Halmos. "He has emotional partisans and vociferous detractors wherever groups of mathematicians congregate . . . The legends about him are many, and they are growing every day. Almost every mathematician knows a few stories about him and is likely to have made up a couple more."[21]

But there was something odd about Bourbaki, something more than a little suspicious. The great French professors, whose reputations were pegged to producing great students, were all in a dither because none of them could claim this outstanding Bourbaki as their own. "*Qui est ça Bourbaki?*" they muttered to themselves. The illusive Bourbaki accepted invitations to conferences, agreed to give lectures, but then he never showed up, sending word that he was sick or had missed his plane.[22] As Halmos climactically concluded in the article: "The strangest thing about him . . . is that he does not exist."[23]

In reality, Bourbaki was a pseudonym for a secret society comprised of the crème de la crème of French mathematicians.[24] The founding members, or "prima donnas" as they've been called, included Henri Cartan[25] (son of Elie Cartan), Claude Chevalley, Jean Delsarte, Szolem Mandelbrojt (uncle to Benoît Mandelbrot), René de Possel, and Jean Dieudonné. The brains of the movement was André Weil, brother of the philosopher Simone Weil, who, by contrast, considered algebra, as well as money and mechanism, a "monster of contemporary civilization"; she believed algebra in particular should be deemed "an error concerning the human spirit."[26]

Why the members chose Bourbaki as their nom de plume was cloaked in a number of running jokes, perpetually recycled in the pages of mathematics journals. One story holds that the surname referred to a French army officer in the Franco-Prussian War—General Charles Denis Sauter Bourbaki—who fled from France to Switzerland with a small remnant of his army in 1871,

---

*Book I. Set Theory, Book II. Algebra, Book III. Topology, Book IV. Functions of One Real Variable, Book V. Topological Vector Spaces, Book VI. Integration.

Cartoon from *Scientific American*, May 1957, representing Bourbaki as a
"milling throng of French mathematicians."

was arrested and interned, then botched a suicide attempt, and lived to the
venerable age of eighty-three. A statue of General Bourbaki in Nancy,
France, establishes a connection with the mathematicians who appropriated
his name, since several were young faculty at the University of Nancy.[27] A
second ancestral line dates to the 1920s, when André Weil, then a student at
the École Normale Supérieure, was exposed to an initiation of sorts, a guest
lecture by a distinguished visitor name Bourbaki. At the end of class students
realized that their visitor was in fact a senior student in disguise—with a
beard and a fake foreign accent—and that his lecture was a scripted piece of
mathematical double-talk, devolving into "sheer nonsense" which included a
"theorem of Bourbaki."[28]

In 1980, *The Mathematical Intelligencer* published perhaps the most
fabulous musing on Bourbaki's origins. The article translated a story by
Frenchman Octave Mirbeau, an anarchist and art and literature critic who
promoted the careers of August Rodin and Claude Monet. In Mirbeau's
story, "Journal d'une Femme du Chambre," the character of one Captain
Mauger takes Celestine, his neighbor's chambermaid, for a tour of his garden. Celestine narrates, reflecting on her outing:

The Captain recounted to me how, last week, he caught a hedgehog under a
woodpile. He is training it . . . He calls it Bourbaki . . . What an idea! . . . An
intelligent animal, a joker, extraordinary, which eats everything!

"My word, yes!" the Captain exclaimed, "In a single day this confounded hedgehog ate a beefsteak, a mutton stew, some salted bacon, gruyere cheese, jam . . . he's marvelous . . . there's no restraining him . . . he's like me . . . he eats everything!"

This tale was an apt metaphor for the mathematician Bourbaki—"Nicolas has pledged to produce a comprehensive treatise of modern mathematics; the staggering scope of his enterprise makes plain his kinship to Captain Mauger's omnivorous pet."[29]

Bourbaki, the group—some members of the collective to this day hold to the conceit and refer to Bourbaki in the third person singular[30]—advocated a unified restructuring of mathematics into an architecture of "mother-structures."[31] The Bourbaki style epitomized the dry and formalist trend, algebraic and axiomatic, and in this respect Bourbaki members considered themselves Hilbert's heir.[32] The Bourbaki method, as the *Scientific American* article reported, was based on a dogmatic belief in the right order for learning, a gratuitously invented terminology, and an economical organization of ideas that was "so bent on saying everything that it leaves nothing to the imagination and has, consequently, a watery, lukewarm effect."[33] In an omen of Bourbaki's insulting impenetrability (for some mathematicians, at least), the first publication was armed with a users' manual in the form of "Instructions to Readers."

Building on the *Elements* treatise, with new chapters published periodically, the Bourbakists became notorious among a certain set of mathematicians for being antigeometry. Mythology held that "Down with Euclid! Death to Triangles!" was the Bourbaki battle cry.[34] Geometry was slighted predominantly by the conspicuous absence of pictures or diagrams in the Bourbaki publication. This was perhaps the most distinctive characteristic of the Bourbaki method (and the most offensive characteristic, to Coxeter and like-minded mathematicians)[35]: it banned the use of diagrams in mathematics. The only exception was a symbol, a backward S-curve printed in the margin to caution readers, like a traffic sign on a treacherous mountain road, when a slippery or "dangerous turn" in the argument was ahead.[36]

"Bourbaki made a *point* of *no* pictures," said Pierre Cartier, a retired second-generation Bourbaki member (all members must retire at age fifty[37]), now director emeritus of the Institut des Hautes Études Scientifique, in France. "Rather, it was based on pure logical reasoning, as little visual insight as possible. Visual insight was considered a concession to human weakness."[38]

The impetus behind both the invention of Bourbaki in the first place, and Bourbaki's distrust of diagrams, can be traced to the First World War. "[T]his war, we can very well say, was extremely tragic for the French mathematicians," recounted Dieudonné, the Bourbaki scribe. "In the great conflict of 1914–1918, the German and French governments did not see things in the same way where science was concerned. The Germans put their scholars to work, to raise the potential of the army by their discoveries and by the improvement of inventions or processes, which in turn served to augment the German fighting power. The French . . . felt that everybody should go to the front . . . This showed a spirit of democracy and patriotism that we can only respect, but the result was a dreadful hecatomb of young French scientists." This had dire repercussions for French mathematics. After the war there was a dearth of young talent. While the older mathematicians who had made Paris an international center were still alive (being too old to serve), they knew only the mathematics of their youth. The teaching at École Normale stagnated. There were no liaisons with cutting-edge mathematics in the international community. Bourbaki was born as a means of reinvigorating the teaching of mathematics in France and reestablishing its world-class standing.[39]

As for the antivisual bent, Bourbakists' opposition to figures had the noblest of intentions. They didn't consider figures evil, nor did they target triangles with an irrational vendetta. Rather, in seeking mathematical truth, they rationally distrusted the subjective visual sense, and thus felt it should not be employed; the Platonic intellectual powers were trustworthier for penetrating the perfect and pure world of mathematical forms. The Bourbakis invoked, in religious terms, the "third eye of the soul" to compensate for the fallibility of the two physical eyes.[40]

The move away from the hegemony of the eye was a far-reaching trend in early-twentieth-century France. One thesis holds that it was instigated in part by the First World War and the failure of sight to protect soldiers. The extreme conditions of trench warfare created "a bewildering landscape of indistinguishable, shadowy shapes, illuminated by lighting flashes of blinding intensity, and then obscured by phantasmagoric, often gas-induced haze . . . When all that the soldier could see was the sky above and the mud below, the traditional reliance on visual evidence for survival could no longer be easily maintained."[41]

The antivisual trend was also a reaction against France's long-standing "hypertrophia of the eye." The eye had grown into an enlarged organ with swollen importance. Louis XIV's Versailles demonstrated this fetish for visual appearance, as did the camera, early cinema, and Paris as a city of spectacle and light—its image of parade, phantasmagoria, dream, dumbshow, mirage and masquerade, all "metaphors of visual untruth." Unlike Germany, where music rose in importance, and England, where words dominated, in France visual appearance remained paramount for an extended period, reaching its peak in the late nineteenth century, when decoration, decadence, artifice, and art nouveau[42] held sway.[43]

Reacting against this state of affairs, French intellectuals from a broad spectrum of disciplines suffered a loss of confidence in the eye and abandoned the value of visual evidence.[44] Bourbaki was part of this phenomenon. Visual reasoning in mathematics and science was supplanted by the power of equations and abstract methods to conceptualize and explain reality, delivering a hard blow to the classical, and very visual, tradition of geometry.

According to Imre Toth, a Hungarian philosopher of mathematics and emeritus professor at the International College of Philosophy in Paris, "Bourbaki was an enemy of everything that is geometry." Coxeter, by contrast, stood strong. He was the rock, the stone, which the Bourbakis could not destroy. "He remained a geometer and represented a high fidelity to geometry," said Toth. "He was the preserver of the classical geometrical spirit; Coxeter was the citadel of geometry, the unconquerable fortress of geometry, against this huge deluge of Bourbaki."[45]

Pitting Coxeter against Bourbaki is suspect for some, since Coxeter and the Bourbakis never so much as shot derisive glances at each other across conference halls—that was not Coxeter's style.[46] The relevance of making opponents out of Coxeter and Bourbaki is that the two embodied vastly different approaches to geometry as practiced in the last century, a period during which classical geometry was in desperate straits.

△ ▢ ◈ ◉ ⬢

Bucking the antivisual Bourbaki movement, Coxeter, who had begun spinning his polytope opus at Cambridge and Princeton, stoically continued with this pursuit at the University of Toronto.

His polytopes made their way into his updated version of W. W. Rouse Ball's book *Mathematical Recreations and Essays*. In February 1939, the page proofs of the final manuscript arrived, shortly after the birth of Coxeter's first child, Edgar (his daughter, Susan, was born two years later in 1941). He read the proofs while taking breaks from painting the nursery yellow, and he marked a common baby milestone in his diary with an extra geometric dimension of enthusiasm: "E succeeded in rolling over 93 degrees onto his tummy."[47] Coxeter's revised edition came out later that year.

This classic book was as much about magic as it was about math—the mind-bending trickery of math, the games and mystery, with no application but fun. Among the first items in the table of contents was "To find a number selected by someone." There was a section on mazes, their history and rules for traversing.[48] In Ball's edition there had been an entire chapter on string figures, which Coxeter deleted. Putting his subtle stamp on the book, he also deleted the pages on "Mechanical Recreations" to make room for a new chapter on—what else?—"Polyhedra." Therein, in the subsection on "Ball-piling and Close-packing," he gave a solution to the problem of the sand on the seashore and why it gets dry around your foot and wet underneath:

> If you stand on wet sand, near the seashore, it is very noticeable that the sand gets comparatively dry around your feet, whereas the footprints that you leave contain free water . . . The grains of sand, rolled into approximately spherical shape by the motion of the sea, have been deposited in something like random piling. The pressure of your feet disturbs this piling, increasing the interstices between the grains. Water is sucked in from around about, to fill up these enlarged interstices. When you remove your feet, the random piling is partially restored, and the water is left above.[49]

As a testament to the book's appeal, it went through eight reprints in the following three decades, and elicited the letter to Coxeter from an aspiring John Conway.

In addition to these recreational polytopal preoccupations, Coxeter ramped up his rate of publication in scholarly mathematics journals, producing articles with an evocative litany of titles: "Regular Skew Polyhedra in Three and Four Dimensions and Their Topological Analogues"; "An Easy Method for Constructing Polyhedral Group-Pictures"; "The Regular Sponges, or Skew Polyhedra"; "Regular and Semi-regular Polytopes"; "The Polytope 2 2 1, whose 27 Vertices Correspond to the Lines on the General Cubic Surface"; "The Nine Regular Solids"; "The Product of Three Reflections"; "A Problem of Collinear Points"; "Quaternions and Reflections."[50] His papers attracted attention for their beauty, not only in terms of their mathematical content, but also their stylistic merit. Coxeter had an eye for detail and paid close attention to how he marshaled his facts, presenting his argument in the most orderly, logical, and eloquent manner.* He crafted flowing segues from one point to the next and carefully constructed an overall symmetry throughout his arguments, nicely tying-off all the references.[51] Coxeter

---

*He tickled his readers with unexpected turns of phrase such as: ". . . dividing the product of the first three expressions by the product of the last two, and indulging in a veritable orgy of cancellation, we obtain . . ." And a pet word of Coxeter's was "perspicuous"—from the Latin *perspicuus*, as in "perspective," and meaning plain to the understanding, or conveyed with clarity and precision of presentation. "Perspicuous" is a very Coxeterian word both because he used it at least once per book, and because, as an expositor, in the written word or in the classroom or at the conference lectern, he embodied perspicuousness.

routinely wrote papers displaying such mastery. John Conway once tried to match the feat of a Coxeterian-caliber paper and found it required head-aching effort.[52]

As Coxeter refined his research, at once expanding and focusing its scope, he parsed the properties of polytopes in hope of finding something new. Coxeter worked with his polytopes the way a sculptor approaches yet another block of marble. As Coxeter's student, the CBC broadcaster Lister Sinclair likened: "When he got a new block of marble, Michelangelo stared at it, listened to it, touched it, and softly walked 'round and around it. He was asking Pandora's block, 'Who's there? Who's in there?' Only then would Michelangelo begin to let the unseen prisoner loose. I've actually watched that happen many times. I've seen a great artist pace 'round and around a block of new material asking, 'Who's there? Who's in there?' That great artist is Donald Coxeter."[53]

Day to day, pencil and paper always at hand, Coxeter sifted and resifted one polytope and then another through his mind. He looked for interconnections, extrapolations, hybrids, and analogies, these subtleties accumulating and leading toward discoveries that sometimes came to naught—as he noted in his diary one day: "I considered possible new polytopes (useless idea)."[54] He kept Aunt Alice apprised of his work by letter, until her death in 1940.

Coxeter developed another fruitful collaboration with Gilbert de Beauregard Robinson, a colleague at the University of Toronto, who suggested an extremely rich rethinking of the interactions between polytopes and group theory, the algebraic study of symmetry.[55] This link between Coxeter's elementary Euclidean polytopes and the more modern algebraic geometry was tantamount to striking gold. The hybrid attracted the attention of highbrow academics, those who might otherwise have viewed Coxeter's classical inclinations with dismissive disdain. His work in this area was top-rate, eliciting invitations to speak at all the best universities, leading Coxeter to nurture the beginnings of a loyal fan base that would come to circle the globe.[56]

△ ⬡ ◈ ◎ ◈

Benoit Mandelbrot was one such fan. In 1947, Mandelbrot was at Caltech as a graduate student when Coxeter visited to give a lecture. "He was a great reassurance," recalled Mandelbrot, professor emeritus at Yale and fellow emeritus at IBM.[57] Mandelbrot had begun his university education in Paris when Bourbaki began imposing its influence. "Bourbaki had a very destructive aspect which they deny today," he said. "They say it was benign, but I can testify it was not."[58]

Mandelbrot's boyhood education had been put in the hands of his uncle, Szolem Mandelbrojt, a professor of mathematics at the Collège de France and a founding member of Bourbaki. Nephew and uncle, however, had opposing tastes in mathematics. The young Mandelbrot called himself a geometer; he had a special gift for shapes and did geometry in his head. His uncle, an analyst, also had a visual gift, but he kept geometry for Sundays and vacations; geometry, his uncle told him, was exhausted, and one must outgrow it in order to make a genuine scholarly contribution. If he pursued geometry, young Mandelbrot faced ruin and unemployment.[59]

In 1945, after a splendid performance in mathematics on the Grande Écoles entrance exams—Mandelbrot credited some cheating with his photographic memory of an "army of shapes" that allowed him to find a geometric counterpart for any analytic problem—he decided to attend the École Normale Supérieure, but with the intention of avoiding his uncle's type of mathematics. Mandelbrot soon ascertained he had no alternative but to follow Bourbaki. "[T]hey were a militant bunch, with strong biases against geometry and against every science, and ready to scorn and even to humiliate those who did not follow their lead," he once recalled. "It was presented to us students as the best there was. And if we didn't like it, we were advised to move out of math."[60] Mandelbrot was disheartened and after only a few days he moved to a different school. Eventually he drifted out of mathematics, dabbling in economics, engineering, physics, and physiology. Intermittently, geometry coaxed him back, as when he encountered Coxeter at Caltech.[61]

"He was viewed as a throwback," said Mandelbrot, remembering Coxeter in that day. "He was a bit marginal. He could not have been a professor at Princeton or Harvard, but he was at Toronto, which was very good but not quite so central. I remember feeling the strength of his style. The enjoyment Coxeter always had handling shapes, models, and letting models help him dream, is something I find very attractive and very important—the spirit of loving shapes and the role of the eye and the hand, that's what I found so marvelous in Coxeter."

"Most people are not strong enough to have a well-defined personal style," Mandelbrot said. "They would bend according to fashion or circumstances and he clearly did not bend. He kept with his classical tradition of geometry, which had been totally flattened—pulverized would be even closer—by Bourbaki. To learn mathematics without pictures is criminal, a ridiculous enterprise."[62]

Coxeter's unlikely success, flouting the mathematical fates, did not occur by happenstance. He was diligent, doggedly hanging on to his polytopes, and dodging distractions that threatened to lure him away.[63] His wife pushed him to take on departmental duties, hoping he might one day become head of the math department, but Coxeter had no interest. He was well aware that the nuisance of administrative commitments would only keep him from his mathematical objectives.[64]

Coxeter willingly gave his time as founding editor in chief of the *Canadian Journal of Mathematics* the late 1940s, continuing in the position for nine years. That sort of extracurricular activity he considered worthwhile because it provided an ongoing and cutting-edge education. And he maintained a voluminous correspondence with all ranks of mathematicians who, more and more as the years passed and the ingenuity and delights of his work spread, wrote to him from around the world.[65]

Time consumed by teaching he also considered well spent—classrooms being full of fresh incubating minds. Coxeter began each class by asking for questions from the floor. When they were answered he seamlessly segued into his lecture, so smoothly that the last question seemed to have been carefully planted. The exception to this contemporaneous style was when Coxeter planned his lectures around books in progress. He came to class with a manuscript and spread piles of pages on tables in front of him. He gave the manuscript a test run over the course of the term, fielding questions and comments from students. Occasionally he profited when his students produced ideas or solutions to problems that he hadn't thought of himself—he would interrupt the class, run over to the manuscript, and make a note at the edge of the page. And of course, students' contributions were credited in the published text. His mathematical ambitions were almost selfless. In preserving and bolstering the oeuvre of the classical tradition, Coxeter's primary interest was the progress of knowledge and encouraging younger generations.[66]

Coxeter, for the most part, was the master of his domain—he was unimpeachable. He had a remarkable geometric eye, and his powers of intuition were hard to match. "Donald would look at a picture," said Barry Monson, one of Coxeter's PhD students, "of the tiling of a two-dimensional plane—think simply of square tiles on the floor, going out forever and ever and ever. Donald would look at that sort of thing and say, 'Well, it is quite clear that if we magically jump from one tile to another a considerable distance away, then along the route we must pass over the intermediary tiles in such and such a fashion.' This is all fine and easily visualized," said Monson. "Nevertheless there are some statements in there that one can, and perhaps even should, prove quite rigorously, in part because Donald would make such statements not only about the ordinary Euclidean space we live in, but he would go on to say it is obvious that exactly the same thing must happen in non-Euclidean space, and then by further extension, in higher dimensional

Coxeter and genus three.

Coxeter's famed intuition, lampooned.

spaces. Most of us can't readily visualize these things. We need to rely on algebraic arguments to bolster what our intuition tells us. But Donald would not rely on those watertight algebraic arguments, even when writing his papers and books. He would just see things and that is how they were. And he was usually right."[67]

If ever Coxeter was questioned by a student on the veracity of a point that he knew was beyond reproach, he didn't respond in a wishy-washy conciliatory manner, modestly asserting his knowledge. Nor did he snap back in a brusque or rude manner common among many socially clueless mathematicians (a stereotype, but for good reason). In his avuncular way, he simply made it known that he had been around a while, and that he was right. On a personal level, there was not an ounce of egomania about him, but intellectually he had an ego and wielded it, politely when necessary. "He was almost courtly," said Monson. "He was very gentle, even when he managed to show you that you were thinking like an idiot." It was thus quite gratifying for students to catch Coxeter in a lapse and point out he was wrong. "That did not happen very often," said Monson, "and you didn't crow about it when it did, but it was fun to do." Asia Ivić Weiss, now a professor of mathematics at York University, was lucky enough to have such an experience. She was Coxeter's seventeenth and final PhD student (and the only woman). Working on a problem, she couldn't see Coxeter's extrapolation from two dimensions to three, even after reams of calculations. She persevered and, after days and days, successfully proved that the three-dimensional result was not a spiral on a cone, as Coxeter had said was so patently clear, but rather a spiral on a sphere. "I was afraid to go and tell him," said Weiss. When she did, Coxeter's response was pure delight:

"Haaaaaaaa! Look at that!" he let out with glee. Weiss married the following year and Coxeter gave her a beautiful glass ball with spirals winding around it as a wedding gift.[68]

Coxeter's selective attention span, and disinterest in the tedious minutiae of departmental affairs, made promotions at the university somewhat hard to come by. He had been hired as an assistant professor in 1936. Seven years passed before he rose to associate, another five before he received tenure as a full-fledged member of faculty: "I felt like the patriarch Jacob," he joked about his time served, "working seven years for Leah and seven years for Rachel."[69] When the professorship finally came—and soon after a coveted office in the university college tower—he was already a fellow of the Royal Society of Canada, and shortly after a fellow of the Royal Society of London[70] (he signed his name in the Royal Society book, which, a few flips of the page backward in time, also bore the signatures of Newton and Einstein).

More important, in 1948, the year he became a professor, he at long last published his treatise dedicated entirely to polytopes, giving it the singular title *Regular Polytopes*. A postcard of congratulations arrived in the mail from his Marlborough College tutor, Alan Robson: "I am glad to see your Polytopes actually printed; and I like it very much. The pictures and tables are very pleasing. What a long time it is since you made that resolution (do you remember it?) when you were working for the Trinity exam, not to work in 4 dimensions except on Sundays."[71] The book was the consummation of twenty-four years' work.

Regular Polytopes earned a reputation as a modern-day addendum to Euclid's work, and Coxeter had intended it to be exactly that ("In fact," he noted in the preface, "this book might have been subtitled 'A Sequel to Euclid's Elements'"). It picked up where Euclid left off, extending the study and classification of the Platonic solids and a larger family of polytopes to the *n*th dimension. "As for the analogous figures in four or more dimensions, we can never fully comprehend them by direct observation," Coxeter said. "In attempting to do so, however, we seem to peep through a chink in the wall of our physical limitations, into a new world of dazzling beauty."[72]

The book's "chief novelty,"[73] by Coxeter's estimation, was his graphical notation representing kaleidoscopes and the multidimensional shapes they generated—here was one strategy for overcoming the human inability to experience hyperspace. Another strategy he presented was to convert Coxeter diagrams and the information they encoded into their algebraic equivalent, into Coxeter groups. Coxeter groups—a comprehensive and systematic enumeration of kaleidoscopes and the symmetries they generate—elucidated mathematical symmetries in a transcendent way, translating geometric entities into algebraic ones like an English-French dictionary, and giving mathematicians another ladder by which to climb the dimensions.[74]

Coxeter, of course, was not so vain as to give these graphical diagrams and their algebraic equivalents the Coxeter moniker. These tools—and Coxeter himself—became more and more popular as an increasing number of mathematicians picked up his methods. Only gradually did his tools gain their proper-noun nomenclature, working their way into the mathematical lexicon as Coxeter namesakes.

By inventing such versatile tools, Coxeter firmly melded his classical geometry with the more modern algebraic approach. This was not the first time a geometry-algebra interface had been achieved. In the seventeenth century, Descartes made the first crossing with analytic geometry. And in the eighteenth century, Joseph-Louis Lagrange, one of the fathers of group theory, declared: "As long as algebra and geometry traveled separate paths their advance was slow and their applications limited. But when these two sciences joined company, they drew from each other fresh vitality and thenceforward marched on at a rapid pace towards perfection."[75]

Coxeter made his crossing by the bridge of group theory, the mathematical language of symmetry. With his envoys, he injected into classical geometry a dose of modern mathematics. He breathed new life into the subject and made it sparkle with sophisticated symmetry groups. "Coxeter groups are an unexpectedly nice application of group theory," said Jeremy Gray. "Donald's work makes the connection between geometry and groups that much richer. Which is a two-way street, because you then get interesting families of groups that you care more about because you can carry them back to geometrical questions . . . That is a very big trick in mathematics. That's what Donald showed us how to do."[76]

Coxeter's big trick transported mathematicians into towering and daunting dimensions using visual images by analogy. And again, the microcosm-macrocosm metaphor applies. While Coxeter didn't stray too far from his home base of the fourth dimension, 24 or 256 dimensions are of interest to other mathematicians. With Coxeter groups, a polytope—say, a square—in the tangible two or three dimensions is translated into symbolic algebra. The algebraic generalization then informs the symmetries of the square in 256 dimensions, or an arbitrary number of dimensions.[77] Coxeter groups describe the symmetries of a shape—the geometric existential essence of what it means to be a square or a dodecahedron—in any dimension.[78]

Before wading any further into how Coxeter groups work, some exposition on group theory is helpful. The commands issued to a soldier—Attention! Right face! Left face! About face!—might loosely be considered a group. These commands form a group of order four, since there are four in the set. To be more specific, however, a symmetry group comprises the set of commands, or transformations, performed on an object that preserve its initial appearance.

The symmetry group of a square contains eight motions that preserve its appearance (the squares below have been marked to assist in tracking the motions).* These are:

(1) Do nothing. This is called the identity element.

(2) Rotate 90° about the center, counterclockwise.

(3) Rotate 180° about the center, counterclockwise.

(4) Rotate 270° about the center, counterclockwise.

(5) Reflect across the horizontal side bisector.

(6) Reflect across the diagonal through the lower left-hand and upper right-hand vertices.

(7) Reflect across the vertical side bisector.

(8) Reflect across the diagonal through the lower right-hand and upper left-hand vertices.

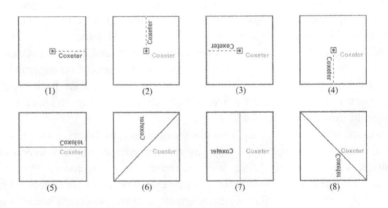

*Other motions or transformations may have the same effect, but they are equivalent to the eight listed. If you reflect the square twice in the same mirror line, for example, that would effectively be the same as doing nothing. Mathematicians are only concerned with the end result of a transformation—the final effect rather than the actual motion that obtained it. Two symmetries are equivalent if they produce the same final effect.

Since there are eight symmetries of the square, the square is said to have a symmetry group of order eight. These transformations together gain the status of a group because they satisfy the four laws of group theory. These are:

1. The "do nothing" identity element is in the set.

2. The "associative" law dictates that when three symmetry transformations are performed one after the other, as long as the order in which they are applied is the same, they can be grouped in two different ways and achieve the same result. For example, multiplication of integers is associative: $(2 \times 3) \times 5 = 2 \times (3 \times 5)$. And with rotations of the square: $(90° + 270°) + 180° = 90° + (270° + 180°) = 540°$ (which is equivalent to a 180° rotation).

3. The "inverse" law dictates that each symmetry must be reversible by another symmetry (its inverse) in the group; that is, performing one symmetry and then its inverse gives the same result as the identity, doing nothing. For example, the inverse of a 90° rotation is a 270° rotation, since $90° + 270° = 360°$, a rotation equivalent to the identity.

4. The magical "closure" law requires that when any two symmetries in the group are performed one after the other, the result of that combination is also a symmetry in the group. As shown above, rotation (2) followed by rotation (3) results in rotation (4), which is also in the symmetry group.[79]

Given these persnickety laws, or axioms, group theory is described by mathematicians who work with it not only as "magical," but also "thorny."[80] To dwell too much on the gnarly complexities of group theory, however, would hardly be in keeping with the Coxeterian spirit of simplicity. "It's a mistake to assume that what mathematicians do is esoteric, deep and difficult," said John Conway. "All the great discoveries are very simple—Einstein's for example. Coxeter's books explain things in elegant and simple terms. And what Coxeter did with his Coxeter groups was simple."[81]

Coxeter always kept the discussion of his groups concrete by referring to the mirrors he used to generate the Platonic solids. Coxeter groups are symmetry groups that can be generated by reflections in mirrors—or as he described them, "the algebraic expression of how many images of an object may be seen in a kaleidoscope."[82] Since the square's eight symmetries can be generated in a kaleidoscope, it is defined by a Coxeter group of order eight.[83]

For a crash course on Coxeter groups, any mirror will do. Imagine standing at a bathroom mirror, and there before you is your mirror image—so there are two of you. Coxeter described this phenomenon by referencing *The Adventures of Alice in Wonderland*: "If Alice could take us through the looking-glass," Coxeter said, "we would still see the same two things, for the image of the image is just the original object."[84] The mathematical description of either scenario is a "Coxeter group of order 2," because there are two images: the original and the virtual opposite twin reflected in the mirror.[85]

In the alphabet of algebra, this Coxeter group of order 2 is expressed as $aa = 1$ or $a^2 = 1$, where you can think of a as the mirror, and 1 as you, or the identity image. So when an object—you—is reflected into mirror a and back out from mirror a producing a second image, the result is you, or the identity image. This is the simplest Coxeter group, and it is given the designation: A_1[86]

Next, imagine mirrors in an elevator, on two adjacent sides of the compartment—this amounts to a simple two-mirrored kaleidoscope. When you look at two mirrors that meet at a perpendicular corner, four images are present: your immediate image in one mirror; your immediate image in the second mirror; yourself outside the mirrors looking in (the real or "original" image); and then there is a fourth image behind the seam between the two mirrors. These mirror reflections generate a Coxeter group of order 4.[87] The algebraic alphabet for this Coxeter group has two characters, since there are two mirrors—a and b. The algebraic statement in this case would be: $a^2 = b^2 = (ab)^2 = 1$.

If yet another mirror was present on the ceiling or the floor, various combinations of the three algebraic symbols a, b, c would algebraically represent this Coxeter group, since it is generated by three mirrors. In this fashion, the algebraic alphabet accumulates. A dictionary of algebraic words accumulates, forming a vocabulary* that facilitates discussions and investigations of geometric entities in an algebraic language—a meeting of geometric and algebraic minds.[88] (See appendix 4 for further exploration into Coxeter groups.)

Coxeter groups proceed in the same manner. When more precisely aligned mirrors are used, and a prop is placed inside to generate a Platonic solid, the mirrors do not behave, in terms of Coxeter groups, exactly like the simple mirrors just described. There isn't a direct correlation, such as with a Coxeter group of order two producing two images—suffice it to say the mirrors of the kaleidoscopes interact to form a more complex pattern. The three-dimensional icosahedron is defined by a Coxeter group of order 120—the reflection of an appropriately placed prop, a blob, will bounce off the

---

*The language metaphor in explaining how groups work continues to be applicable because, like in any language, the algebraic words can be strung together, or interact, according to the equivalent grammatical rules. The grammatical rules dictate how reflections combine in the kaleidoscopes—as the reflections bounce from mirror to mirror like billiard balls ricocheting off the bounds of the pool table. The rules work like multiplication, since the images reflect off mirrors successively, or in "multiple" fashion, and combine to generate a complete image.

Charles Addams's 1957 *New Yorker* cartoon inadvertently illustrates a
Coxeter group of infinite order.

three mirrors, producing twelve images of the blob, which together form the
vertices of a finite icosahedron. Using the algebraic symbols to mimic more
and more mirrors, the study of polytopes ascends the dimensions, adding
new words to the algebraic alphabet and expanding the study of group theory. The icosahedron in four dimensions, or the hypericosahedron, is defined
by the Coxeter group of order 14,400 and has 120 vertices. Mathematicians,
if they have the time, can calculate this data for themselves, or they can grab
their copy of *Regular Polytopes* and look up the information, calculated by
Coxeter for all kaleidoscopes and organized nicely in tables at the back of his
book.

In deepening the union between geometry and algebra, Coxeter acknowledged the value and power of both methods for obtaining results. But he
was a master at using his geometrically rooted tools—his Coxeter groups
and Coxeter diagrams—to establish results in group theory. He plugged in
his symmetry groups and diagrams, bypassed laborious calculations, and

covered leaps and bounds with a few simple and swift steps. "Most people do not have his fantastic geometric insight," said Roe Goodman, a professor of mathematics at Rutgers University. "Coxeter was a real pioneer, who, through dint of great insight and concentration, imagined higher dimensional objects. But for most people it is very hard to fasten onto them. And for those people, like myself, the algebraic description comes in handy. But in a certain sense the algebraic calculations are always a little disappointing. It's sort of like bookkeeping—you see that the account balances but you'd like to know where the fun was in spending the money."[89]

Coxeter groups are forever crawling out of the mathematical ocean, and the fact that they pop to the surface so much as a useful gadget signals the omnipotence and omnipresence of their symmetries, underlying everything from geometry to topology to number theory to algebra to physics, even chemistry, cosmology, biology, sociology, catastrophe theory, economics, and so on. Coxeter groups forge links to all sorts of fields; they are a way of thinking that can be applied to all sorts of problems. The basic concept of a Coxeter group is a template for symmetry, a universal building block for investigation. There are hundreds of mathematicians doing research pertaining to Coxeter groups, and related concepts of symmetry.[90]

Fields of study that investigate patterns often apply group theory, and thus Coxeter groups, as probative tools, because symmetries are invoked to simplify or completely solve complicated problems. This association works, in part, because equations behave according to symmetries, and hence, they behave analogously like shapes. The square, for example, has four vertices and eight symmetries. The equation $x^4 = 3$ has four roots, and in this case the same eight symmetries, or permutations. "The symmetries of the square and the symmetries of the equation $x^4 = 3$ are the same, from a certain point of view. That's an analogy," said Simon Kochen, the Henry Burchard Fine Professor of Mathematics, at Princeton. "Mathematicians don't talk a lot about analogy in mathematics. Not because it isn't there, but just the opposite. It permeates all mathematics." Analogy in mathematics and science—and in bridging the two—is pervasive, and a vehicle to jump-start new advances. Coxeter groups are one way of crafting mathematical analogies.[91]

"It's always amazing when Coxeter groups turn up," said Vakil. "Someone gives a lecture, on something seemingly unrelated, and then the name 'Coxeter' comes up and there is sort of a shiver through the audience. 'Ah! Here they are again!' And why are they there? I have no meta-reason, no quasi-philosophical or religious reason as to why they come up. But there's got to be some reason why they underlie so many different structures. This powerful idea of symmetry, this aesthetically beautiful and extremely simple structure, for some reason underlies the world and so much of mathematics."[92]

While Vakil doesn't offer any philosophical or meta-existential reason for the omnipresence of Coxeter groups, Sir Michael Atiyah is willing to try. "These surprising connections in mathematics are always the most interesting things. I've seen them enough to have a general philosophical view," said Sir Michael, knighted for his work as a geometer, winner of the Fields Medal and currently an honorary professor of mathematics at Edinburgh University. His curiosity often leads him to these unpredictable interfaces, where dissolving boundaries shed new light on both realms as techniques are transported from one side to the other. "And usually what it means," he said, "if you examine underneath, is that you find deep important truths that aren't obvious. These connections are an indication of something really exciting that you've got to explore—like a sign, a warning sign: 'GOSH, LOOK, DIG HERE, HIDDEN TREASURE!'" Some people think of these connections as nice fortuitous accidents, but Sir Michael does not take this view. "These things are not accidents. They are somehow fundamental. Even if you didn't know they were there before, once you see them you have to investigate and by investigating you discover lots and lots of things. They are a very important part of mathematics in terms of directing the search of mathematicians into new areas. This is where new mathematics is forged," he said. "In a lot of mathematics, you build up big theory in a rather straightforward way. But every now and again there are these things that connect up to different parts. They are showing you that you missed something by building this big structure, going up in one direction. You realize that you should have turned off 'right' at one stage and explored something else."[93]

The ubiquity of Coxeter groups in modern mathematics is fodder for some mathematicians in refuting the lament that geometry has suffered any decline in the twentieth century. It's just that geometry has changed; it is a different sort of geometry. Coxeter's ideas have not become less important, they've become more important, they have transcended their origins. In the earlier stages of his career, Coxeter's followers studied precisely what he studied and asked precisely the same questions. Then those questions were answered, and the answers have become the terrain over which the next generation of acolytes pass on their way to another neck of the woods, where new questions await answers, and then more questions again.[94]

"Coxeter's perspective and ideas are in the air we breathe," said Vakil. "It's not that his ideas are used to solve problems, it's that the fundamental problems grow out of his ideas. He's the soil, part of the substrate, part of the building in which we work, in which we live. Coxeter's name gets stated, but in some senses people don't even think of him as a person—he is an adjective that gets applied to so many things. Towards the end of his life, many people I met were amazed that he was still alive. He had become a name. A famous name—these famous names become more a concept than a person. It's like hearing that Beethoven was walking down the street."[95]

CHAPTER 7

TANGENTS ON POLITICS AND FAMILY VALUES

[B]ut let us not confine our cares
To simple circles, planes and spheres . . .

—THOROLD GOSSET, ADDENDUM TO "THE KISS PRECISE"

Regular Polytopes made Coxeter's reputation. He fielded offers for visiting professorships internationally, and for full-time jobs in bigger cities, more in tune with his cosmopolitan sensibilities. He faced a difficult choice when an opportunity arose that would take him back to England, to Sheffield University. It was the most agonizing decision of his life. As was his habit, he made a list of pros and cons, which balanced perfectly, sixty-five on each side. A handsome raise persuaded him to stay in Toronto. But no sooner had he made the decision than he regretted it and wished he could change his mind. Around the same time, there were rumblings of interest from Princeton, which never materialized, and he turned down Notre Dame. In 1951, the American University in Washington, D.C., wanted him to do operations research in linear programming—the use of mathematical models to aid decision-making problems involving a multitude of variables.[1] American also tempted him with a high salary. This time the head of the Toronto math department alerted the university president that Coxeter was considering leaving. Not wanting to lose the star that was putting Toronto—and Canada—on the mathematical map, the president persuaded Coxeter to stay, again with a sizeable salary increase.[2]

Regular Polytopes attracted two distinct crowds: practitioners in Coxeter's classical corner, who appreciated the symmetrical and shapely polytopes for their own sake, and those who wanted to harness his findings for more modern application. "Every reader will find some parts of the book more palatable than others," Coxeter noted in the preface, "but different readers will prefer different parts: one man's meat is another man's poison."[3] Coxeter also made clear his purely quixotic intentions. "The chief reason for studying regular polyhedra," he said, "is still the same as in the time of the Pythagoreans, namely, that their symmetrical shapes appeal to one's artistic sense . . . Such an escape from the turbulence of ordinary life will perhaps help to keep us sane."[4]

△ ▢ ◈ ⬡ ⬢

While Coxeter had put his final years of work into *Regular Polytopes*, the real world was in the throes of the Second World War. "Real mathematics has no effects on war," G. H. Hardy stated in 1940. "No one has yet discovered any warlike purpose to be served by the theory of numbers."[5] Hardy, there, made an uncharacteristic miscalculation. During the Second World War, the Allies utilized mathematical logic in deciphering Germany's Enigma code. Coxeter received his call for service in 1941, when the Canadian government decided it needed to set up a cryptanalysis bureau the likes of Britain's Bletchley Park. "There is grave possibility that every message transmitted is available to . . . enemy sympathizers," warned a Department of National Defense circular. "It is safe to assume that the signals emanating from Ottawa, Winnipeg, Halifax, etc. are just as strong in enemy countries as they are in Canada."[6] The Canadian government knew its signals were being listened to. Canadian receivers "manned by army, navy, and Department of Transport personnel crackled with a constant stream of messages, day and night, on all practical bands. Dot-dot, dash-dot—almost all were in Morse code and the listeners knew that they came from U-boats announcing sinkings, enemy raiders, spies calling home, diplomats reporting negotiations, or German air and army units organizing attacks. Some of the messages were from enemies—Germany and Italy—and some from potential enemies—Japan and Russia."[7]

Samuel Beatty, the head of the math department at the university, recommended Coxeter and Gilbert de Beauregard Robinson when the National Research Council came looking for cryptographer candidates. These two mathematicians knew nothing about codes and ciphers, but they were acquainted with mathematician Abraham Sinkov,[8] then with the U.S. Army's Signal Intelligence Service (Coxeter had met him at Princeton). They acquired the necessary accreditation—"fingerprinted for my visa,"[9] Coxeter noted in his diary—and undertook two reconnaissance missions to Washington to get a briefing on what the work involved. Robinson decided to engage with the Canadian cryptology team, cover name "Examination Unit." Another friend and colleague of Coxeter's, the graph theorist William Tutte, worked as a code breaker at Bletchley Park—along with Alan Turing*—and deciphered a series of German military encryption codes, an accomplishment cited when he was inducted into the Order of Canada as "one of greatest intellectual feats of World War II."[10]

In the end, Coxeter decided against the cryptology work. "I managed to wriggle out of it," he admitted. "I wanted to just go on with my mathematical research."[11] On August 14, 1945, he wrote in his diary "THE WAR IS OVER," and two weeks later he sent in the final draft of *Regular Polytopes* to his editor.[12] Also, he declined because he was a pacifist. Years later, a colleague mentioned to him that Bertrand Russell had been a pacifist in the First World War but not in the Second. "That's one of the things that makes him likeable," Coxeter said, adding that the situation during the First World War was quite different.[13] Nevertheless, Coxeter remained firmly committed to his views against war; he was "disgusted" when a maid in his employ enlisted in the Women's Air Auxiliary Force.[14]

Pacifism was not a popular position during the Second World War, throwing the Coxeters conspicuously out of kilter with others in their social circle. Rien went to a luncheon with departmental wives and reported back that Mrs. Beatty was "unsympathetic about her pacifism, and said: 'How funny the English are!' "[15] And Elizabeth Synge seemed to have run away from Coxeter and Rien at a concert and they wondered: "Is she mad because we still drive a car in war-time?"[16] The Coxeters' wartime contributions included mailing food parcels to England[17] and cabling Coxeter's stepmother, Katie, with an invitation to send one of his half sisters to Toronto for safekeeping.[18]

Coxeter's staunch pacifism was only one facet of his larger sense of social justice. His presence and support were always felt, even if mostly behind the scenes. He was outspoken in his quiet fashion, attending meetings and contributing money, if not always on the barricades.[19] On July 17, 1942, he attended a Canadian Civil Liberties Union rally with the physicist Leopold Infeld,[20] who by then was at the University of Toronto. The Toronto rally, which had been organized jointly with the American Civil Liberties Union, drew a crowd of five thousand, demanding repeal of the ban on the Communist Party of Canada. Coxeter was classified in the *Canadian Who's Who* as a liberal, but by all indications his leanings were rather more to the left of center.[21]

Coxeter supported Infeld when he was forbidden from leaving Canada to spend his sabbatical year in his native Poland, then behind the Iron Curtain. In the Canadian Parliament, the Progressive Conservative leader George

---

*Coxeter and Turing overlapped at Trinity, and later corresponded about their mutual interest in phyllotaxis.

Drew denounced Infeld as a traitor who would provide the communists with atomic secrets. In 1950, after Infeld resigned from his position at the university and remained in Poland, he and his wife and their Canadian-born children were stripped of Canadian citizenship. Coxeter kept in touch with Infeld, visited him in Europe in 1954, and later wrote an introduction to his posthumous autobiography, *Why I Left Canada*. However, Infeld's wife, Helen, decided against including his contribution. Several friends and colleagues had offered introductions, and she found Coxeter's humdrum anecdotes about their children playing together in Toronto's ravines rather beside the point. Mrs. Infeld appreciated Coxeter's private support all the same and kept in touch with the occasional letter, such as one in 1976 in which she expressed her gratitude: "Do you know, my life has been such that I have come to highly evaluate some human qualities and feel that it is good to tell people who have them so. I'd like to tell you that I do admire you as a person of principle, not swayed by general prejudice, emotional blindness or temporary hysteria of others in important matters. Would that people as a whole had such rational understanding, everywhere!"[22]

Chandler Davis, a mathematician at the University of Michigan in the 1950s, also benefited from Coxeter's civil libertarian sympathies. Davis had been called before the House Un-American Activities Committee (HUAC). He refused to testify and was indicted for contempt of Congress. It was a particularly disheartening and deflating experience for Davis, as HUAC investigators had dug up fewer than a dozen indiscretions. He had been an active leftist since high school, and he felt more of his efforts were worth noting[23] (at least Davis was called to testify; subpoenas were prized as acknowledgment of one's stance, and failure to attract the attention of HUAC was said to elicit "subpoenas envy"[24]).

Some of Davis's fellow professors deemed him guilty of "deviousness, artfulness and indirection hardly to be expected of a University colleague." He was fired and blacklisted from university jobs in the United States, spending eight years in limbo, "jobless and under indictment." His political activities were limited to preparing for his court case, which he lost, committing him to a six-month jail sentence. A welcome-home party awaited him upon release, but no employment, until he met the pacifist Coxeter at a mathematics meeting. Coxeter invited him to consider a job at the University of Toronto, and wrote letters of recommendation to the math department and (Davis suspects) to the Canadian government. Initially, the government refused Davis entry but, ultimately, after a letter-writing campaign, they relented.[25]

In subsequent years, Coxeter was active in the cause for nuclear disarmament, sitting on the university's Nuclear Disarmament Committee, organizing speakers, and engaging in a long talk with Nobel chemist John Polanyi about the origin of the Hiroshima bomb.[26] In 1959, he signed his name to a nuclear disarmament petition, making the front page of Canada's national newspaper, the *Globe and Mail*, with the headline: U OF T HEADS SEEKING END TO NUCLEAR TESTING.[27]

And in 1967, he expressed his outrage at American involvement in Vietnam, discussing the issue with his friend and former PhD student Seymour Schuster, then a professor of mathematics at Carleton College, in Minnesota. Coxeter asked Schuster to send him a copy of *Macbird,* by Barbara Garson—a play, based on *Macbeth*, mercilessly attacking President Lyndon Baines Johnson and his war policies. Schuster obliged, but worried Coxeter might be offended by Garson's crass intimation that President Johnson was responsible for the assassination of President John F. Kennedy. "I feel this intimation was, at the very least, in bad taste, and might also be considered unethical," Schuster wrote. "I should like to think further about the question and look forward to hearing your opinion on it." Coxeter's response surprised Schuster: "In reply to your question about the book, I fully appreciate your hesitation, but I find myself able to enjoy it without any qualm. The monstrous insinuation seems to me justified by the consideration that, even though technically false, it is 'in character.' What he is doing now, she could argue, is so atrocious that, if he had been guilty also of the assassination, it would make no difference, just as we would think no worse of Hitler (than we do) if someone proved him responsible for the death of Gandhi."[28]

△ ▢ ◈ ◉ ⊗

On the domestic front, Coxeter did not display quite the same propensity for peaceful relations. Of her husband's dedication to social justice and his all-consuming and unwavering devotion to geometry—his polytics—Rien seldom had much to say (one couple who socialized with the Coxeters observed that Rien seemed not to have any intellectual interests, other than being overawed by her husband's intellect).[29] On the surface, Rien accepted her husband's commitments mostly without quibble, trying not to feel any imposition or jealousy. She humored herself by saying that at least her husband would not stray—math was his mistress.[30]

"I was not able to love Rien as fully and completely as one should his wife,"[31] Coxeter conceded. He devoted most of his time and energy to his geometry. One weekend, he mischievously procured an invitation for his wife and their two children to visit a colleague's cottage without him, so he could stay home and be alone to work on his book. He saw nobody all weekend, working hard to get a chapter done. But his finagling brought regret— "leaving me alone and miserable."[32] And it plummeted Rien into misery. "R thinks I am trying to drive her crazy. I must take more responsibility for the children."[33] Coxeter made token efforts to help with the parenting. He recorded in his diary that he sat writing the final copy of *Regular Polytopes* in the bedroom of his daughter, then age three, who had the measles.[34]

Neither nature nor nurture were enough to make mathematics a fascinating subject for either of Coxeter's children; they hated math. By age nine, as Coxeter noted hopefully in his diary, Edgar was "getting quite good with multiplication."[35] But in his last year of high school, Edgar brought his father to meet with his mathematics teacher, and they "spent an hour explaining his neurotic horror of trigonometry."[36] Edgar commandeered his father into pleading his case so he could drop the course without penalty (but in the end, he did "pretty well" on geometry[37]). Susan fared somewhat better. She chose to take geometry and trig instead of Latin. But by grade thirteen she managed a score of only 50 percent on the standardized mathematics exam (she maintains the marker saw the name "Coxeter" and didn't have the heart to fail her, as she deserved).[38]

Edgar, Donald, Rien, and Susan, circa 1952.

Their unimpressive performances were perhaps less an indication of smarts and more a symptom or side effect of having such a legendary but emotionally absent father. Indeed, Susan and Edgar remember his presence not so much as a father, but as another sibling who needed taking care of. Rien told him what to wear (dressing him in spiffy windowpane-checked suits with diagonal lattice ties), whether to wear his galoshes on a rainy day, and when to wipe his chronically dripping nose.

In 1959, Susan and Edgar lodged a formal complaint with their parents about the tension and lack of affection in their household. And from there followed a traditional pattern of rebellion. Edgar, who had an interest in theology, announced he wanted to become a Jew (he later studied to become an Anglican minister). Susan dated older men, whom her parents tried their darnedest to scare away. They repeatedly accused one of burglarizing their house. Coxeter dispensed with another boyfriend after he caught him in a compromising position with Susan on the living room floor in the wee hours of the morning after a date; Coxeter accused the young man of taking advantage of his daughter (or was convinced to do so by Rien) and had the dean of the University of Toronto faculty of medicine expel him from medical school (the fellow transferred to the University of Ottawa).* Coxeter was a mathematical pioneer, progressive in his politics, but a stern and chilly conservative as far as child rearing was concerned. He seemed to subscribe to a view advocated by Harold H. Punke, at Auburn University, in Alabama: "intellectually competent persons" should breed, Punke argued, but not be bothered with the charges of child rearing in order that they "use their time and energy for other purposes."[39]

In 1957, Coxeter reflected on his creative process in a survey sent to him by Punke, a professor of education. On the topic of creativity, Coxeter observed, "It seems to emerge from the sub-conscious . . . Fresh air, exercise and restful sleep are better than any artificial stimulants." His best times of the day were very early in the morning. His moments of contemplation left him "never lonely or afraid, but surely thrilled." He relied on "mental images" in fostering ideas; "imagination should be allowed unlimited scope." And he said his stream of ideas came to him "not increasingly; but I have no fear of running dry, because the supply of unfinished projects would suffice to occupy me for many years to come, even if no completely new ideas appear from now on."[40]

---

*Forty years later, having forgiven Coxeter, this man had the courage to come courting Susan again; she had been recently widowed by her husband, Alf (eleven years her senior). Coxeter by then was a widower himself. This time Coxeter decided he liked the fellow and issued him a written apology. Susan nonetheless declined the overtures of her suitor.

He appended his survey answers with a story of one "eureka" moment of creativity that occurred at Trinity College, Cambridge. Some of his bright ideas had hit him while resting under a tree in a forest, while riding a bicycle, and at that intermittent stage between dreaming and waking (he usually kept paper and pencil waiting on his bedside table). He chose to give a detailed description of his discovery of a four-dimensional figure, the "snub-24-cell," having 96 vertices, 432 edges, 480 triangular faces, and 144 solid cells—a creative insight that came to him in the middle of a peaceful night's rest in his suite of rooms in Great Court. He relayed the discovery with intricate technicality, indifferent to the limited, if not nonexistent, mathematical expertise of the man he was writing to:

> I had long been trying to extend to four dimensions the familiar construction for the snub cube (one of the 13 Archimedean solids) by taking, as vertices, suitable points inside all the white triangles covering a sphere . . . I knew that the four-dimensional analogue of the network of spherical triangles is an arrangement of black and white tetrahedra covering a hypersphere, the shape of such a tetrahedron being usually "quadri-rectangular."

> The problem was to locate a point inside a white tetrahedron, in such a position that it would be equidistant from the corresponding points in the nearest other white tetrahedra. The snag was that, since the number of "nearest other white tetrahedra" was nine, the equality of their distances would impose eight conditions on the point to be selected: five more conditions than such a point could generally be expected to satisfy.

> So I went to bed and soon slept soundly. About 3a.m. I awoke with the idea of using a symmetrical "isosceles" tetrahedron: a right pyramid based on an equilateral triangle. Such a tetrahedron still has nine neighbours of the same color, but they consist of three of one type and six of another; I could thus choose a point on the axis of symmetry and adjust its height so as to equate the distances of the two types of neighbouring point. I switched on the light and went into my living room to write it down, lest I might find the next morning that it had passed away like any ordinary dream. When morning came, there it was, ready for all the details to be filled in.

Coxeter emphasized there is no use trying to force creativity. At the crucial moment, effort is only a hindrance, but this may follow months of painstaking preparation. "My advice to others who wish to develop creativity," he said, "is to choose a problem so absorbingly fascinating that they are really happy to think about it at every available moment, especially at such times of relaxation as in a bath or in bed, or while out for a pleasant walk."[41]

△ ▢ ◈ ◉ ◉

From outward and material appearances the Coxeters were doing nicely by the 1950s—far better off than their days of sharing bathwater in the dirty thirties. In 1949, Coxeter inherited money after the death of his mother (whom he'd only been back to visit once since he'd left for Canada). He received the bad news while sailing with his family to England, the children expecting to meet their grandmother for the first time—it was Edgar who spotted the name Coxeter in the *Times* obituaries.[42] With the inheritance, the Coxeters purchased a three-story house for $37,500 on Roxborough Drive, in Toronto's tony Rosedale neighborhood.[43] And they could now afford a live-in maid. Finding and keeping decent help was another matter. Coxeter returned home from work nearly every day to be briefed on the never-ending saga of the revolving door of maids and their shortcomings. As Coxeter noted, they were "too ladylike to work properly, and very stupid," "had a weak heart," "like a witch (never again)," or were caught stealing white shoe polish from the study and had to be dismissed. More often than not they left of their own accord. Coxeter chronicled the frequent turnover in his diary.

Dec 8	—	Mrs. Dickinson said she couldn't bear vegetarian cooking, wouldn't come anymore (day 4).
Jan 5	—	Engaged new maid, Ethel. Worked well.
Jan 6	—	Ethel said she couldn't come any more.
Jan 8	—	Interviewed Mrs. Peacock.
Jan 9	—	Mrs. Peacock failed to come.
May 15	—	Maid Violet dismissed.
Nov 8	—	We caught Viola with her young man, gave her notice.
Jan 1	—	Verna failed to show up. No more of her.
Feb 8	—	Irene Dupuis came and worked well.
Feb 12	—	Irene is irritating, goes to the washroom too often at night.
Feb 13	—	Irene didn't turn up.
Feb 19	—	Joan drove R to the slums in search of a maid (no use!)
Feb 28	—	Mrs. Foster works well.
Mar 7	—	Mrs. Foster can't come any more, not well.
Mar 1	—	Maid Nora Kean failed to come.
Mar 28	—	Fired Nora.
Apr 1	—	We engaged May.
Apr 2	—	May works well but eats a lot secretly.

APR 6   — May stayed out til midnight; we were furious.

APR 12   — May was abducted by her friends, who backed their car into the Brown's (neighbouring) driveway and took her luggage thru the window. R rang Mrs. Jeffery, who took steps to try to locate her.

APR 13   — No church, having no maid.

APR 14   — Detectives on May's trail.

APR 15   — I called on Miss Parsons to assure her we did all we could to make May happy. Miss Parsons admitted all her girls are morons.[44]

There was absolutely no way Rien could manage the household without help. The Roxborough house placed them smack in the neighborhood of ladies who lunched. Soon after the Coxeters moved in, the University of Toronto purchased a mansion up the street for its president, adding even more social cachet to their address. They often had university colleagues and their wives in for high tea or dinner, sometimes a gathering of Coxeter's graduate students—"Erdös and Dirac to dinner . . . 20 staff and students for sherry at 9."[45] The hour of a party approaching, Coxeter invariably got himself into trouble. As the doorbell rang, he fumbled hopelessly with the tight zipper of his wife's dress. He lowered the drinks tray down the dumbwaiter such that the sherry and glasses crashed and shattered. He lit a fire in the fireplace but forgot to open the flue. Rien's customary response was to screech at him, in Dutch:[46] "You *ezel*! You *rotzak*! You *lammeling*!!"* Coxeter's evaluation of these evenings: "All my 6 grad students for the evening. Frightful,"[47] or "Fiasco!"[48] or "Infelds came for the evening. We felt dreadfully inferior."[49]

Donald and Rien went out frequently to social gatherings as well. Between all the socializing, sometimes with numerous events each week, one wonders when Coxeter managed to be so prolific. He said he drank too much only once in his lifetime. On New Year's Day 1946, after he and Rien spent the evening quietly at home, he observed with incredulity that they ". . . both felt just as rotten as those who had celebrated with strong drink all night!"[50] Rien loved the ivory tower social scene. But even their busy social itinerary she judged as lacking—such as having no plans on New Year's— and she aired her grievances: "R says nobody asks us to parties; they all find us dull. How depressing." "R very worried because I was too quiet and when I did say something it was the wrong thing." "R says I must talk to

---

*Translation: *ezel*=ass; *rotzak*=rat, stinker, scoundrel; *lammeling*=dead loss, rotter, pain in the neck.

people and stop being so dull." "R depressed since everyone dislikes me." "I made a faux pas."[51]

Coxeter enjoyed the social scene well enough, and by all accounts he was the perfect gentleman, if not effusively convivial. Chitchat and schmoozing were not his cup of tea, but he took great pleasure in finding other people who had interests similar to his own. Rien was fun and gay at parties, but she put people off with exuberant and often inappropriate outbursts.[52] At home she wondered when her miseries would end. She was plagued by eternal discontent over the invitations she failed to receive and she suffered chronic malaise about her rotund waistline. When she went in for a checkup, Coxeter noted, "Dr. Owen found R physically healthy. Prescribed a holiday, a maid, a hobby, and a full life. She would have preferred him to find a curable disease."[53] Rien felt listless and longed for something exciting to happen.[54]

It probably wasn't quite what Rien had in mind but, a few years after *Regular Polytopes* was published, her husband was on the verge of becoming a household name—in mathematicians' households, anyway. Mathematicians were reading his book in such numbers as to warrant a second printing in 1962—some called it their bible, a reference they kept in multiple copies, in their offices at home and at work. "The mathematical community Coxeter gathered round himself with *Regular Polytopes* is many ringed, like ever widening ripples on a pond," said Marjorie Senechal (adding that Coxeter was "no falling stone").[55] The ripples would eventually encircle Bourbaki—even Bourbaki would come to acknowledge the usefulness of Coxeter's tools, his muse of classical geometry notwithstanding—forgiven, or at least overlooked.

CHAPTER 8

BOURBAKI PRINTS A DIAGRAM

A soul never thinks without a mental image.

—ARISTOTLE

As Coxeter's research bridging polytopes with modern group theory gained esteem,[1] he also set out as a missionary to raise the profile of plain, old, classical geometry, popularizing its gems at the grassroots level with grade school and high school teachers and students.

From 1955 through 1957, Coxeter dedicated several weeks of his summer holidays to cultivating the seeds of his beloved geometry.[2] He had been summoned by the Mathematical Association of America (MAA) to be a "roving lecturer," as he described it, touring American universities and organizations as part of the National Science Foundation's "Summer Institutes" for high school teachers.[3]

Over the course of these tours, he stopped in Ann Arbor, Michigan; Chicago, Illinois; and Arkansas City, Kansas. He gave several lectures to the Friends of *Scripta Mathematica*, in New York City, upon invitation from Jekuthiel Ginsburg, the journal's founder, at Yeshiva University. He went as far west as Stanford, California, and as far north as Fairbanks, Alaska. He took advantage of his time spent on of the lecture circuit to test material for his next book, *Introduction to Geometry*, another of the most popular mathematics books of the century.

Willy Moser, one of Coxeter's PhD students at the time, was lucky enough to tag along for the last leg of one of Coxeter's tours. "Donald made many great contributions to mathematics. I made one great contribution," recounted Moser. Moser's opportunity came at the end of Coxeter's 1955 summer of roving lectures, after his session in Stillwater, at Oklahoma State University. Moser drove down to meet Coxeter and serve as his assistant, taking detailed notes of the well-polished lectures. "At the end of

"In Oklahoma," Coxeter noted on the back of this photo, "where my host introduced me to an American-Indian ('Native American') couple."

the summer we drove north, to civilization," said Moser, wryly. "We were in my car and Donald asked me if he could drive. It was a new car. Indeed it was the first car I had ever purchased, a green 1955 Plymouth 2-door. I paid $2,000 for it and drove it to Oklahoma. But I agreed. I was surprised to see that he was an aggressive driver. At one point he was trying to pass a car while driving up a hill on a 2-lane highway. I immediately perceived that this was not a prudent thing to do.[4] He tried to coax the car to go faster but it wouldn't respond. At the last possible moment I shrieked at him, 'Pull back, pull back.' I was probably his only student to shriek at him. He began to pull back and at that moment a truck came over the hill. He managed to get back into the right lane just in time. I HAD SAVED HIS LIFE! And mine. But saving Coxeter's life was my greatest contribution to mathematics."[5]

Published in 1961, *Introduction to Geometry* was Coxeter's second masterpiece. He opened the book, and no doubt his lectures, with one of his pointed comments: "For the last thirty or forty years, most Americans have somehow lost interest in geometry. The present book constitutes an attempt to revitalize this sadly neglected subject."[6] *Introduction to Geometry* became widely used as a university textbook.[7] As testament to its popularity, for a time it was the most frequently stolen item in the University of Toronto mathematics library. And it was one of the first textbooks to be built around the concept of symmetry,[8] and of course, it was full of pictures. "I agree with Alice in Wonderland," Coxeter once remarked. "Wasn't it Alice in Wonderland who said, 'What's the good of a book that doesn't have pictures?'"[9]

Coxeter's book received a rave review from the preeminent critic Martin Gardner in his column in *Scientific American*: "Most professional mathematicians enjoy an occasional romp in the playground of mathematics in much the same way that they enjoy an occasional game of chess; it is a form of relaxation that they avoid taking too seriously. On the other hand, many creative, well-informed puzzlists have only the most elementary knowledge of mathematics. H. S. M. Coxeter . . . is one of those rare individuals who are eminent as mathematicians and as authorities on the not-so-serious side of their profession . . . There are many ways in which Coxeter's book is remarkable. Above all, it has an extraordinary range."[10] And while *Introduction to Geometry* is encyclopedic in its scope, like the Bourbaki treatise, it is at once as engaging and awe-inspiring as a curiosity cabinet. In the chapter on hyperbolic geometry, Coxeter prefaced his section on "The Finiteness of Triangles" with a Shakespearean epigram borrowed from *Hamlet*: "I could be bounded in a nutshell and count myself a king of infinite space."[11]

Coxeter's career, considering his two masterpieces—*Regular Polytopes* and *Introduction to Geometry*—can be viewed as two intersecting circles: they overlap, but their circumferences delineate two distinct realms. One realm encompasses Coxeter's role as popularizer and connoisseur of the beauty and fun of classical geometry (symbolized by *Introduction to Geometry*), whereas the other comprises his contribution as a pioneer, an innovator melding classical with modern geometry (as demonstrated in *Regular Polytopes*). The former achievement won him a wide and varied fan base, and the latter cemented his reputation among mathematicians. As observed by Sir Michael Atiyah, "If his fate was just to be a connoisseur of beautiful pictures, he wouldn't have been so widely recognized, he would have been more of a sideline. But you add this extra dimension of symmetries (finite or continuous), and that lifted him up and made him well known and in touch with other aspects of mathematics."[12]

Fields Medal winner David Mumford, who teaches pattern theory and the mathematics of perception at Brown University, felt Coxeter's impact in both realms. He stumbled upon Coxeter's book *Regular Polytopes* as a high school student in the early 1950s. "It was like I had discovered how math was really done," he recalled. "High school mathematics didn't show how deep the subject was. It was a revelation. It made me realize what mathematics was all about."[13] As Mumford continued his studies in mathematics at a scholarly level, Coxeter's work influenced his interests, specifically with the "compact-ification of modulized spaces"—just as they suggest, these are atlases, of sorts, for algebraic objects—and Coxeter's work fit nicely into the story he and his coauthors told in the book *Smooth Compactification of Locally Symmetric Varieties*. Mumford later met Coxeter in the 1970s, in a replay of the typical scenario. "I had assumed he was dead, and then, 'Oh my god, Coxeter—he's here.' " At the time Mumford was at Harvard, teaching the undergraduate geometry course. He often shook up the syllabus using *Introduction to Geometry* as a text, and he invited Coxeter down to give a lecture.[14]

△ ⬜ ◈ ◉ ⊛

With the twentieth-century stampede toward modern mathematics, toward all things abstract, algebraic and austere, the Bourbaki enterprise thrived be-yond the borders of France—through publication of volumes of its treatise, and via Bourbaki members in the flesh who were installed at various univer-sities on secondments, exposing new guinea pigs to the Bourbaki approach firsthand. Claude Chevalley went to Princeton's Institute for Advanced Study and later to Columbia.[15] Dieudonné spent time in São Paulo, Brazil, as well as at the University of Michigan and Northwestern University.[16]

Over the years the group continued to disguise itself in "mock mystery," and rumors continued to spread about Bourbaki, "the mathematician."[17] In one ruse, Bourbaki applied for an individual rather than group membership in the American Mathematical Society (AMS); the request was refused, twice. The AMS secretary suggested that an application for an institutional mem-bership might meet with more success, but Bourbaki would have none of it.[18]

The Bourbaki group gathered three times per year, once for an extended two-week period, at a youth camp, a monastery, resort, or hotel, where they made major policy decisions, drew up the table of contents for the current volume of the treatise, and delegated research. When their semiregular pub-lication of *Elements of Mathematics* became a commercial success, the roy-alties paid for travel expenses, wine, and extracurricular activities that enlivened the proceedings.[19] According to *La Tribu*, their internal newslet-ter, the Bourbaki group played chess, table soccer, volleyball, or Frisbee. They embarked with gusto on mountain hikes, bicycle excursions, or swim-ming expeditions. They caroused in bumper cars, went butterfly hunting or mushroom picking. They sunbathed, dozed off with text in hand, stuffed themselves with local delicacies and drank until royally drunk—Armagnac, champagne, rum toddies, or wine (wine being the much-needed fuel of Bour-baki's cogitation). Once under the influence, inebriated members sometimes performed a "virile French cancan or a lascivious belly dance . . . The deliberately laid-back attitude . . . gave the impression of insouciant genius."[20]

As far as the tone of the meetings was concerned, Bourbaki's biographer, Liliane Beaulieu, described them as opting "for the humorous and the ribald, on occasion ascending to the heroic contrasted with the loutish"[21]—humor was said to be their second-favorite mind game after mathematics. The mathematical discussions were not nearly as civilized as Henry F. Baker's tea parties at Cambridge. Anybody at any time could interrupt, comment, ask questions, or criticize. Dieudonné observed: "Certain foreigners, invited as spectators to Bourbaki meetings, always come out with the impression that it is a gathering of madmen. They could not imagine how these people, shouting—sometimes three or four at the same time—could ever come up with something intelligent."[22]

As the group's scribe, over the years Dieudonné came to be considered the speaker for the group as well. Between his stentorian voice and propensity for definitive statements and unchallengeable opinions, Dieudonné was known to crank up the decibel level of any conversation. It was Dieudonné who would later declare: "Down with Euclid! Death to Triangles!" He was a giant, a tall, big, and ebullient man, oft times loud and rude. He was flamboyant, with a brutal manner of expression.[23] Pierre Cartier recalled an outing to a concert hall with Dieudonné. "It was fascinating," he said. "He would look at the score in his hand and exclaim with disapproval—'OH!'—if a note was missing from the orchestra."[24] Coxeter, with comparable zeal as a musician, preferred musical scores to books for bedtime reading.[25]

The domineering Dieudonné penned first drafts of Bourbaki chapters, which were referred to as "Dieudonné's monster." From there, each chapter of *Elements of Mathematics* commonly took six, seven, even ten drafts before consensus was reached (unanimity was required, with each member having veto power).[26] And lest the enterprise be misunderstood, Dieudonné clarified: "Bourbaki's treatise was planned as a bag of tools, a *tool kit* for the working mathematician, and this is the key word which I think everybody should keep in mind when talking about Bourbaki or discussing its plan or contents."[27] Cartier agreed: "You can think of the first books of Bourbaki as an encyclopedia. If you consider it as a textbook, it's a disaster."[28]

The popularity of Bourbaki initially brought about something of a revolution in university-level mathematics. Marjorie Senechal was a graduate student in the 1950s at the University of Chicago, a hotbed of Bourbaki in America, under the auspices of Marshall Stone. Stone, strongly influenced by the ideas of Bourbaki, had made the mathematics department at the University of Chicago arguably the best in the country. He recruited the brains of the Bourbaki group, André Weil, and Samuel Eilenberg, who worked

Jean Dieudonné

closely with another Chicago professor, Saunders MacLane. "I suffered under the Bourbaki regime," said Senechal, one of MacLane's students. "Bourbaki was the method taught. I think it cost mathematics a lot of talent—a lot
of people who think visually and work visually left the profession, because
they felt they didn't have a home there anymore." Coxeter kept the spark
alive for people who wanted to continue to do concrete geometry, even if it
was unfashionable. "Coxeter was the antithesis to Bourbaki.* He was a life-
line," said Senechal, "a way of salvation from Bourbaki. Because through
him I knew there was more to mathematics, I knew there was a whole
branch of mathematics I could relate to."[29]

But the Bourbaki revolution shook more than just the universities. Bourbaki
trickled down into high schools and public schools, as mathematicians taught
by the Bourbaki method became teachers themselves.[30] Bourbaki principles

---

*Senechal elaborated to say: "If you are thinking about a mathematical idea in the Bourbaki style, you will be
working upwards from definitions and axioms and trying to continue through that logical line. If you are
working in Coxeter's style, you are also working upward, but you start with some concrete object, asking
questions about it, asking how to put that in a more general way and what that leads to. Coxeter's is a visual
and hands on approach, as opposed to a strictly logical approach."

infiltrated the "New Math" grade school curriculum reform. The New Math spread throughout the Western Hemisphere, from South America through the American heartland, into Canada, and across the Atlantic to England, Wales, West Germany, Denmark, the Netherlands, and France. New Math overhauled the traditional curriculum, ridding it of trivial problem solving and rote number juggling. Instead, schoolchildren as early as grade one learned the equations of algebra and set theory (the mathematical theory of sets, or collections of abstract objects, and the rules that govern their relationships and manipulations). Dusty and dilapidated Euclidean geometry also was forsaken—like removing Shakespeare from the syllabus and replacing it with grammar, as though one were a minor subset of the other.[31] "This tendency is not only regrettable," said Coxeter, "but unreasonable."[32]

Historically, Euclidean geometry had been under siege ever since its limited scope had been exposed. "Euclid's approach to geometry has been attacked on two grounds—that it is illogical, and that it is boring," Coxeter said in a 1967 report on the state of geometry in primary and secondary school education. "Neither criticism is new," he said, adding: "The objection that Euclid is boring is much more serious than the objection that his logic is imperfect."[33]

If Euclidean geometry was boring, Coxeter argued, this was due to the canned and ossified way it was taught. Like arithmetic, the subject had been reduced to rote learning, with teachers opening a textbook and doing the stultifying "chalk-and-talk" at the front of the classroom. Children mindlessly memorized properties of triangles and their theorems—Side-Angle-Side, Angle-Side-Angle—and regurgitated them on demand to please their teachers. They were robbed of experiencing the beauty and tricks intrinsic to heuristic learning—that is, learning through trial-and-error and making discoveries for oneself.

△ ▢ ◇ ⊕ ⊗

The French Bourbakis were one influence behind the New Math reforms; the Russians were another. When the Soviets successfully launched the Sputnik satellite into orbit, the Western world got a shock—rudely awakened to the fact that it was falling behind in science, technology, and mathematics. A colorful interpretation of events was articulated in a British report chronicling the New Math:

> It all started on that memorable day in 1957 when the Russians sent their first Sputnik orbiting the earth. Up till then the countries of the West had rather patronizingly regarded the USSR as a backward giant of a nation, hopelessly engaged in trying to educate its largely peasant people to achieve the technological

advantages of its more favoured European neighbours. The noisy "Bleep-Bleep" of the Sputnik's radio, however, quickly dispelled any notions Westerners might have that the Russians still counted on their hands or that the abacus was the sole piece of educational equipment in Soviet schools. Clearly this formerly retarded people had outstripped Britain and America in finding scientists and mathematicians of a very high caliber indeed. How had this astonishing advance in Russian scientific education come about? No one could supply the answer but it had to be admitted that Soviet schools were obviously producing more and better mathematicians and scientists than were coming from the British system of education.[34]

Following the bleeping Sputnik, the United States Congress released millions of dollars in funding for science education under the "National Defense Education Act." A flurry of international activity led to the formation of the United Nations Educational Scientific and Cultural Organization (UNESCO) and the Organization for European Economic Cooperation (OEEC). Reform of the mathematics curriculum was undertaken with urgency and idealism.[35]

The first forum of debate on the New Math was the 1958 International Congress of Mathematicians, held in Edinburgh. Then came the infamous conference at which Dieudonné whooped his war cry. Held at the Cercle Culturel de Royaumont, Asnières-sur-Oise, France, from November 23 to December 4, 1959, the conference addressed the need for reform in French mathematical education. Here the bombastic Dieudonné rose to his feet and hurled his provocatively planned statement:

"À bas Euclide! Mort aux triangles!"[36]

"Down with Euclid! Death to triangles!"

Dieudonné's statement was taken by many as a slap to geometry. Coxeter discussed it with like-minded individuals and was known to now and then unleash a scathingly critical or derogatory comment, though he did not dwell on it, nor did he let the Bourbaki venture as a whole ruffle his feathers. Dieudonné's comment seemed a succinct summary of the Bourbaki agenda—no diagrams—but the interpretation of this event by Bourbaki sympathizers diverges from a geometer's vantage point. Michel Broué, director of the Institut Henri Poincaré, who studied under Bourbaki founder Claude Chevalley in the 1960s, asserted the importance of distinguishing Dieudonné from Bourbaki. Dieudonné, by 1959, was older than fifty and therefore no longer an official member of the Bourbaki group. His "Death to Triangles!" statement is thus disqualified from representing the larger Bourbaki mandate (Broué acknowledged all the same that linking the two has become part of the Bourbaki mythology). "Others in Bourbaki were horrified," Broué said.[37] Especially since Dieudonné stuck with this opinion to his dying day. This was a source of embarrassment for some Bourbakists, such as Cartier. "I was tormented," he said. "The ideology of Bourbaki didn't match with me, it was going too far. Bourbaki was a mathematical priest—pure, pious, rigid. It was a caricature of purity. Purity creates hypocrisy, because if the rule is too strict then life forces you to break it."[38]

Chevalley, for one, espoused the no-pictures dictum, but this belied his closeted use of diagrams. He tried his best to operate by reasoning alone; he earnestly wanted to avoid intuition in mathematics. But he didn't always succeed. Cartier remembered when Chevalley was his professor, and teaching at the front of the classroom he filled the blackboard with symbols and equations. When a student raised their hand with a question, Chevalley dramatically stepped back from the board, crossed his arms, and squinted, contemplating his work through furrowed brow. He was stumped. Then he walked toward the blackboard, standing rather closer than necessary. He huddled in, with hunched shoulders, his arms creating blinders, hiding what he was up to from the class behind. "He drew a picture," said Cartier, "figured out the answer—AHA! he said—and then quickly wiped out the diagram, stood back and continued."[39]

After the "Death to Triangles!" incident at Royaumont, annual conferences on New Math were held in Denmark, Zagreb, Athens, Bologna, elsewhere in Europe, and there was a series of "inter-America" conferences in South America. The first convened in Bogotá, Colombia, in 1961. The ringleader of the New Math reform movement was Marshall Stone, who had been president of the Royaumont conference and led the way internationally.[40] As the mastermind of these international conferences, Stone delivered stirring opening addresses, calling for the modernization of mathematics:

> There are two major factors which require us to examine with fresh eyes the mathematics we propose to teach to young people in the secondary schools and in the first years at the university. One is the extraordinary growth of pure mathematics in modern times. The other is the increasing dependence of scientific thought upon mathematical methods, coinciding in time with a more and more urgent social demand for the services of scientists of every description.

> The forces exerted by these two factors on our educational system are quite clearly on the point of shattering the traditional framework of mathematical instruction and thus preparing the way for an overdue modernization and improvement of our teaching of mathematics. Like the crustacean which has to split and discard its old shell in order to grow, we must at last burst the confines of a curriculum which is plainly no longer suited to our current needs or our current conditions of life.[41]

At Bogotá, Howard Fehr, head of the Department of the Teaching of Mathematics at Columbia University, delivered a lecture titled "Reform of the Teaching of Geometry." "Euclidean geometry," said Fehr, "nowadays . . . is sterile, outside the main course of mathematical advancement, and it can be filed in the archives, without any fear, for the benefit of future historians."[42] Response was mixed. Professor Guillermo Torres, from Mexico, challenged this position and argued that the presentation of mathematics in an exclusively formal aspect "makes it appear to be an inhuman activity and with no sense at all." John Coleman, Coxeter's early student, by then the head of the math department at Queen's University, also expressed doubts during the debate following Fehr's presentation. Based on his experience, he said, students interested in mathematics were first enticed by geometry's intrinsically tactile and visual nature—geometry was the user-friendly interface of mathematics.[43]

New Math sprouted in varied forms internationally. In the United States, the main initiative was the School Mathematics Study Group (SMSG), which produced a new series of textbooks—students renamed it "Some Math Some Garbage."[44] The *American Mathematical Monthly* ran a "Letter of 75 Mathematicians" objecting to the emphasis on abstraction. The leading antagonist was NYU professor Morris Kline, who later sounded the death knell of the New Math in America with his book *Why Johnny Can't Add: The Failure of the New Maths*.[45] And the horrors of it all entered the popular culture via the genius of mathematician cum musical raconteur Tom Lehrer— he documented the debacle on his album *1965: That Was The Year That Was* with the song "New Math." The lyrics poked fun at the fact that the math was so newfangled that parents couldn't make sense of it in helping their children with homework.[46]

In France, of all places—the cradle of Bourbaki—the newspaper *L'Express* ran the headline LE CAUCHEMAR DES MATHS MODERNES (The nightmare of modern maths); "Pornography, drugs, the disintegration of the French language, upheavals in mathematical education all relate to the same process; attacking the central parts of a liberal society," the subtitle continued. And a report to the French Academy of Sciences decried: "The settheoretic option in the definition of geometry is a dangerous utopia . . . this reform misappreciates the intellectual aptitude and needs of the adolescents who attend our . . . high schools. The reform in progress seriously endangers the economic, technical, and scientific future of the Nation."[47]

In England, a telling cross section of the changes is found in the career of Sir Michael Atiyah. He was a student at Cambridge in the 1950s, when aspects of classical geometry were still hanging on as part of the university curriculum. But by the 1960s, this last bastion had languished, linear algebra having been decreed fundamental and geometry old fashioned and inessential.[48] Sir Michael's 1981 presidential address to the Mathematical Association, titled "What Is Geometry?" bemoaned this unfortunate turn in geometry's history. "Of all the changes that have taken place in the mathematical curriculum, both in schools and universities, nothing is more striking than the decline in the central role of geometry," he said. "Euclidean geometry has been dethroned and in some places almost banished from the scene."[49]

"The battle between geometry and algebra is like the battle between the sexes," said Sir Michael, contemplating the issue recently. "It's perpetual. It's an ongoing battle. And it really is a battle in the sense that these are two sides of the same story, and you've got to have both sides present." Both algebra and geometry are essential, both must be taught properly at all levels, and the resulting interaction in the highest tiers of research move the frontier forward. "It's the kind of problem that never disappears," he said. "It will never be dead and it will never get solved. The dichotomy between algebra, the way you do things with formal manipulations, and geometry, the way you think conceptually, are two main strands in mathematics. The question is what is the right balance."[50]

△ ◻ ◈ ⊛ ⊛

One outpost that kept the balance weighted toward geometry over algebra was Eastern Europe—Latvia, Hungary, and Russia. The reason is the object of speculation. Perhaps their prophylactic was the Iron Curtain—cut off from the rest of the world, and poor, they continued on with the old-world ways. Russia had a long and fine mathematical tradition all its own. Certainly, the fact that the Russians printed Coxeter's works demonstrated that they liked their classical geometry.[51]

At the 1966 International Congress of Mathematicians in Moscow, Coxeter learned of his considerable popularity in Russia. Prior to the congress he had no idea whether his books had been published there—ostensibly, Russia had agreed to international copyright laws but Russian editions, seldom the products of contractual agreement, were pirated more often than not. If mathematicians wanted royalties, they had to produce proof of publication, which was difficult if one was not in Russia. The International Congress occasioned an olive branch in the form of a book exhibit that allowed mathematicians to peruse a warehouse of all Russian mathematical publications. If they found their books, they were entitled to collect royalties on the spot. Coxeter walked around the warehouse with John Conway, who recalled that Coxeter made a lengthy list of his books and then walked away with his pockets full of rubles.[52]

Coxeter's classical geometry also thrived in Italy. Geometer Emma Castelnuovo was a Coxeter fan from afar, and vice versa. "I have all of Coxeter's books,"[53] said Castelnuovo, now in her nineties. She devoted her life not to higher math like her father, Guido Castelnuovo, but to teaching geometry in grade schools. She worked with children aged eleven to fourteen, in Italy and Africa, doing geometry "by hand," and organizing exhibits of the children's work. She attended all the congresses and commissions on mathematics education, including the "Death to Triangles!" conference in Royaumont, and worked with Piaget on the Commission for the International Study and Improvement of Education in Mathematics. In 1949, she published her first book, *La Geometria Intuitiva*, and wrote many textbooks for students. Coxeter had high praise for Castelnuovo's work and cited her as an example worth following in his report on geometry education. "In Italy today, Emma Castelnuovo has popularized and developed a [new approach to Euclidean geometry]," he said. "Her book, *La Geometria Intuitiva*, describes the teaching of geometry with apparatus resembling Meccano.* The book, beautifully illustrated, shows how geometrical shapes are used in the architecture of Italy."[54]

Another beacon was the Netherlands, where German expatriate Hans Freudenthal was credited with saving Holland from the New Math. In his 1971 article "Geometry Between the Devil and the Deep Sea," Freudenthal cast it all in lyric terms:

> Geometry is endangered by dogmatic ideas on mathematical rigor. They express themselves in two different ways: absorbing geometry in a system of mathematics as linear algebra, or strangulating it by rigid axiomatics. So it is not one devil menacing geometry as I suggested in the title of my paper. There are two. The escape that is left is the deep sea. It is a safe escape if you have learned swimming. In fact, that is the way geometry should be taught, just like swimming.[55]

Coxeter had the same sensibility: "The ability to study, grasp, and master topics in mathematics resembles in some ways the ability to swim or to ride a bicycle," he said in the geometry report, "each of which is, in a static sense, impossible of achievement. There is a trick to it, and strong motivation is needed to learn the trick. Perhaps one difference is that children seldom encounter oppressive authoritarian discipline in connection with the technique of riding a bicycle."[56] Geometry, Freudenthal said, would die of suffocation as a "prefabricated subject." It could be saved if presented as a field of wonderment and activity—folding, cutting, gluing, drawing,

---

*Meccano is the trade name for colorful metal construction toys assembled with nuts and bolts, invented in 1901 by Frank Hornby, of Liverpool, England.

Canadian Mathematical Congress, Fredericton, 1959. Top (left to right): Irving Kaplansky, Alex Rosenberg, Coxeter. Bottom: Werner Fenchel, Philip Wallace, Max Wyman, C. Ambrose Rogers, Hans Freudenthal.

painting, measuring, and fitting. "Coxeter's *Introduction to Geometry*," he said, "is a marvelous demonstration of this attitude. The author knows, in any case, exactly where the horizon is lying."[57]

△ ⬠ ⬙ ⬡ ⬢

Introduction to Geometry circulated internationally, with translations into six languages—German, Japanese, Russian, Polish, Spanish, and Hungarian. The first was the German translation, in 1963, which had a title Coxeter was very fond of: *Unvergangliche Geometrie*—Everlasting Geometry, or Geometry which Survives Everything.[58] With the publication of *Introduction to Geometry* in Japan, in 1965, architect, engineer, and geometer Koji Miyazaki became one of Coxeter's biggest fans. Also a professor emeritus at Kyoto University and Teikyo-Heisei University, Miyazaki recalled: "At that time, the name 'Coxeter' suddenly spread out in Japan as the biggest mathematician's name in the world. I am clearly remembering that time. And from that time I was thinking that Prof. Coxeter is the god of the world of geometry."[59]

A counterinsurgency against the geometry-barren New Math curriculum—as against Bourbaki—continued to take shape in all things Coxeter. From the beginning of his career through the heyday of Bourbaki, Coxeter simply averted his eyes from the antivisual antigeometry trend, and went on a crusade to bring his passion for the intuitive methods to any and all willing spectators. He lectured on "the beauteous properties of triangles," on "The Arrangement of Trees in an Orchard," on the Fibonacci numbers (with nine slides and a pineapple as a prop). On a snowy January evening, he took the night train from Toronto to Philadelphia, putting the final touches on his presentation as he traveled. The following day, he noted in his diary: "About 40 broke into spontaneous applause after my 10 min. lecture on 'Close Packing and Froth.'"* The next month he gave a version of the same lecture to seventy schoolteachers in Toronto. Two months later he spoke to a group of forty prize-winning schoolchildren on "Close Packing of Spheres," this time drawing upon an eighteenth-century book with a title he thought his young pupils might find amusing—it was called *Vegetable Statics*, by Dr. Stephen Hales, wherein Hales investigated how many peas, if as many peas as possible were compressed into a large cubic pod, would abut a central pea.[60]

In 1967, Coxeter published two more books that would become classics: *Projective Geometry* and *Geometry Revisited* (the latter with S. L. Greitzer). He churned out papers asking, "Whence Does a Circle Look Like an Ellipse?" and lectures wondering, "Why Do Most People Call a Helix a Spiral?" In another talk he issued "A Plea for Affine Geometry in the School Curriculum," and in yet another he offered simply "Reflections on Reflections," which he delivered in Pittsburgh in 1967.[61]

After his Pittsburgh talk, he traveled to Minneapolis where he was coming to the end of a long-running pet project, working for four years with a group of mathematicians on educational geometry films, *Dihedral Kaleidoscopes* and *Symmetries of the Cube* (two in a series of five films). The project, aiming to improve geometry teaching in high schools and colleges with the introduction of exciting experimental films and an accompanying series of textbooks, was part of the College Geometry Project at the University of Minnesota, well financed with a million-dollar-budget (funded entirely by the National Science Foundation; classical geometry still had its champions). Coxeter laboriously wrote and rewrote the scripts. And in *Dihedral Kaleidoscopes*, he took the role as starring geometer.[62]

The film began with Coxeter scampering across a busy street, dodging traffic, wanting to get to the other side to look at his reflection in a mirrored store window (the narrator explained: "H. S. M. Coxeter, of the University of Toronto, is a geometer. To Professor Coxeter, reflections are of particular

---

*Two months before his lecture on "Close Packing and Froth," Coxeter noted the seed of inspiration in his diary: "I saw that the no. of bubbles touching one bubble in froth should be $\frac{23+\sqrt{313}}{3} = 13.56$."

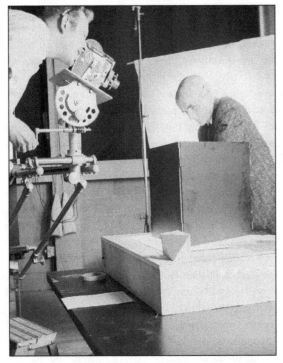

Coxeter starring in the documentary *Dihedral Kaleidoscopes*.

interest because of their implications for geometry and algebra . . ."). With a lively flute soundtrack, the film followed Coxeter as he manipulated mirrors in a darkened studio. He peered into large kaleidoscopes—constructed like tents or pens and illuminated from within,* dropping in colored paper triangles, watching as they fluttered into place, and grinning when they landed and generated pleasing psychedelic patterns on the plane. The films won many awards—in Canada and the United States, at the American Film Festival and the Golden Eagle at the CINE Film Festival, and internationally, in Belgium, Czechoslovakia, France, Italy, Argentina—broadening Coxeter's fan base even further.[63]

---

*One of the documentary kaleidoscopes was taller than Coxeter himself and equally wide, gaping jaws of mirrors (the mirrors for the kaleidoscopes were produced, after a long search, by Litton Industries at a cost of $5,500). In an outtake, Coxeter inserted his miniature daschund, Nico, into this monster Kaleidoscope to see what would transpire—Nico was puzzled if not petrified, and stood frozen in place until Coxeter rescued him (Nico died later the same year and Coxeter honored him with a dedication in his book *Twelve Geometric Essays*: "In memoriam NICO 1951–1967").

Close-up of Coxeter positioning props in a kaleidoscope.

For the most part, Coxeter's crusade was all rear-guard action. He simply continued to make his contribution in the most hands-on way he could, propagating his passion. He did, however, keep an eye on the land mines of curriculum committees with their mandates for reform.[64] He voiced his opinion and opposition, at times with uncharacteristic volume. Tim Rooney, a colleague of Coxeter's in the math department at the University of Toronto, remembered the only time he ever saw Coxeter angry: when he perceived his geometry was under attack on home turf. Coxeter was graceful and sweet, said Rooney; there existed no easier man to get along with. But when Rooney bumped into him in the hall one day in the 1960s, Coxeter was fuming. He cornered Rooney, pulled him aside, and gave him an earful about a report from a committee studying the department's roster of mathematics courses. Coxeter interpreted the report as disrespectful and denigrating to geometry; it concluded there was an awful lot of geometry on the department's course list and some of it had to go. He interrogated Rooney about it: "What's your committee doing recommending less geometry be taught?"[65]

Coxeter peeking into a kaleidoscope taller than himself.

"He really was angry," Rooney recalled. "I told him, first of all, I wasn't on that committee, at which point he cooled down a little. And then I told him I didn't agree with what the committee said about geometry, which cooled him down further." Coxeter recruited another member of the department and they tackled the chair of the committee and had a furious argument about the report.[66]

This prompted Coxeter to more directly assume the mantle of the curriculum controversy, pulling himself and the dignity of his geometry together by the frayed laces of his well-worn spectator shoes. He sat on the K-13 Geometry Committee, producing the report *Geometry, Kindergarten to Grade Thirteen* in 1967. It baldly stated: "Some recent innovations under the name of 'modern mathematics' are unsatisfactory and ought to be discontinued . . . We have in mind an excessive tendency to abstractness and rigour, a copying of procedures more appropriate to graduate school." The net effect, the report said, was that the "geometric literacy" of society was even lower than its "numeric literacy":

> The ability to visualize geometrically is a basic part of the scientist's mental equipment . . . Thus scientific literacy is founded in part upon geometric abstraction . . . Geometry is perhaps the most elementary of the sciences that enable man, by purely intellectual processes, to make predictions (based on observation) about the physical world. The power of geometry, in the sense of accuracy and utility of these deductions, is impressive, and has been a powerful motivation for the study of logic in geometry. Unfortunately, however, in the teaching of geometry the role of logic is very likely to overshadow the creative and intuitive aspect of the subject. In the past this tendency has been reinforced by the conventional attitude that visual or intuitive "qualitative" pattern work in geometry was a fit subject only for the kindergarten or lower grades.

> We wish to emphasize as strongly as possible that we do not accept this view. Visual and intuitive work are indispensable at every level of mathematics and science, both as an aid to clarification of particular problems, and as a source of inspiration, of new "ideas."[67]

Classical geometry, for Coxeter, was one of the arts—the Seven Liberal Arts, as set out by medieval universities, were the Trivium, "the three roads" of grammar, rhetoric, and logic; and the Quadrivium, "the four roads" of arithmetic, geometry, music, and astronomy or cosmology. And so it followed that the justification for studying the liberal arts applied equally to the study of classical geometry—they may seem obsolete, indulgent, and impractical courses of study, but the arts are fertile soil, fostering a freedom and breadth of thinking from which more "modern" achievements grow. A good number of the report's 120 pages contained specific suggestions for reinstating geometry and tips for teaching it in an inspiring way to primary, intermediate, and senior grade levels—complete with practical instructions for nail and plywood constructions, skeletal models made from straws and pipe cleaners, the use of shadows and mirrors, and how to draw a cube from a circular array of dots.[68]

In 1968, in a nice topological twist of history, the proper nouns "Coxeter diagrams" and "Coxeter groups" finally made their debut in—of all places—the Bourbaki volume on Lie algebras, considered by some as the most successful volume in the whole series.[69] Marjorie Senechal delights in recalling how she once looked through all the Bourbaki volumes to see for herself the depressing dearth of diagrams. Apart from the slippery-argument-caution-ahead S-curve, she found only one. It was in Coxeter's volume and it was the Coxeter diagram.[70]

Coxeter came to be included in Bourbaki after his work intrigued a Belgian mathematician by the name of Jacques Tits, now at the Collège de France.[71] Closely affiliated with Bourbaki, Tits drew the group's attention to Coxeter's work, writing the first paper ever on Coxeter groups—"Groupes et Géometries de Coxeter." The paper went unpublished until the Bourbaki volume, which Tits ghostwrote. Two-thirds of the volume is taken up with expositions on Coxeter, baptizing not only the term "Coxeter group," but also "Coxeter graph" (also known as the Coxeter diagram), "Coxeter matrix," and "Coxeter number."[72]

Coxeter was pleased with the Bourbaki nomenclature. It meant his name was writ large into the history of mathematics. With the publication of the Bourbaki volume on groups, nearly ten years had passed since Dieudonné proclaimed "Death to Triangles!" When Dieudonné visited the University of Toronto in 1969, Coxeter and others took him out for a sumptuous dinner at the Park Plaza hotel, its rooftop restaurant offering a glittering view of the city. Dieudonné was there to give two lectures, one on Lie algebras, the other on Bourbaki.[73] "It . . . seems to me," he commented, "that when examining which tools should be included in Bourbaki, a decisive element was whether or not they had been used by great mathematicians, and what degree of importance these mathematicians had attached to these tools."[74] Coxeter had certainly found success by these criteria.[75] And in another address, in 1968 at the Roumanian Institute of Mathematics, in Bucharest, Dieudonné stated, "[O]ne must never speak of anything dead in mathematics because the day after one says it, someone takes this theory, introduces a new idea into it, and it lives again."*[76] Coxeter could hardly have said it better himself.

△ ⬠ ◈ ⊗ ⊛

One decade later again, in 1980, the bright yellow cover of a publication by the Mathematical Association of America showed a hooded skeleton, the ghost of geometry, his bony finger dangling over a ratty scroll with a diagram

---

*As for Bourbaki's future, after the group's great success its productivity stalled in the 1970s during a clash with the publisher over royalties and translation rights, resulting in a protracted legal dispute, which was settled in 1980. Bourbaki then had a short resurrection, issuing revised editions of old books, and adding a few volumes to the series. "But then silence," said Pierre Cartier. "In a sense Bourbaki is like a dinosaur, the head too far away from the tail," he observed, of the subsequent generations that inevitably strayed further and further from the group's founding ideals and mandate. Just as Bourbaki members were forced to retire at fifty, Cartier joked that Bourbaki—himself or itself—should have retired at the half-century mark. Regardless, for all intents and purposes, his judgment was that "Bourbaki is dead." There is, however, an annual "Bourbaki Seminar" in Paris. And there are rumblings that further publications and revised editions might be in the works.

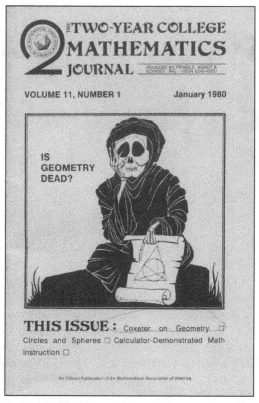

Cover of the January 1980 *Two-Year College Mathematics Journal*.

of the nine-point circle—one of the first circle theorems studied in any course of elementary geometry. The title on the cover asked: "Is Geometry Dead?"[77]

Inside the MAA volume, the first page depicted the cartoon of Coxeter as the king of geometry, wearing a crown studded with gems in the shape of the Platonic solids, followed by an article covering the 1979 Coxeter Symposium in Toronto. Eighty-five geometers traveled from all corners of the world to celebrate (a bit belatedly) Coxeter's seventieth birthday and retirement. Lás-zló Fejes-Tóth, from Hungary, a giant of a geometer in his own right, opened the conference "with a fitting and loving tribute to Professor Coxeter." And he made a presentation on "Some researches inspired by H. S. M. Coxeter," highlighting the phenomenal impact "a remark [or] a suggestion" from such an oracle had on the development of geometry over the last half-century, in-citing many a practitioner to a life's work worth of inquiry (another cartoon quipped: "My geometric broker is H. S. M. Coxeter, and Coxeter says . . ."—

Coxeter exhuming Geometry.

parodying a popular commercial for the stock brokerage firm E. F. Hutton, with the tag line: "When E. F. Hutton talks, people listen.")[78]

The account of the festivities also provided a Q&A session with the legend himself, accompanied by a caricature of Coxeter as a gravedigger, mounds of earth beside him, with a shovel and crowbar strewn about, as he cracks open the lid of a coffin with the gravestone: "GEOMETRY: 600BC–1900AD R.I.P." His interviewer asked: "If I and my colleague . . . start rhapsodizing about geometry, the reaction we frequently get is, 'Oh, well, that's a dead subject;* everything is known.' What is your reaction to that reaction?"

"Oh, I think geometry is developing as fast as any other kind of mathematics," Coxeter said. "It's just that people are not looking at it."[79]

Indeed, the cover of the same journal the following year read: "Geometry Lives!" And inside was an article by none other than Jean Dieudonné, now singing geometry's praises. There was also an article announcing a new generation of ingenious "un-Bourbakian" geometers, including future Fields Medalists William Thurston and Shing-Tung Yau, both of whom had a hand in the recent solution, by Russia's Grigori Perelman, of the Poincaré conjecture, which had eluded mathematicians for over a century.

---

*In 1981, Coxeter's friend and polymath Freeman Dyson, a professor of physics at Princeton's Institute for Advanced Study, sent him a copy of a talk he had given recently, titled "Unfashionable Pursuits." (See appendix 6 for an excerpt of Dyson's talk.) Coxeter's unfashionable path was acknowledged again some years later by University of Alberta mathematician Robert Moody, in a letter recommending Coxeter for an honorary doctorate: "Modern science is often driven by fads and fashions, and mathematics is no exception. Coxeter's style, I would say, is singularly unfashionable. He is guided, I think, almost completely by a profound sense of what is beautiful."

PART II

COXETER APPLIED

CHAPTER 9

BUCKY FULLER, AND BRIDGING
THE "GEOMETRY GAP"

Through natural selection, our mind has adapted to the
conditions of the external world . . . it has adopted the geometry
most advantageous to our species; or, in other words, the
most convenient.

—HENRI POINCARÉ, *LA SCIENCE ET L'HYPOTHÈSE*

Although Coxeter retired officially in 1977, at age seventy, he did not surrender to conventional wisdom regarding the rusting mental cogs of mathematicians beyond the half-century mark. He continued on his prolific way,
adding more than eighty items to his career bibliography, some being new
editions and reprints of earlier works, but these nonetheless required his
meticulous attention, correcting the errata that his hawk eye diligently
hunted down.[1] He had fought off pressure to retire for years—"depressing
letter from George Duff [head of the mathematics department] about my im-
pending retirement,"[2] he noted in his diary. He yielded incrementally, submitting to fractional reductions in pay. In 1975, he received notice that he
was soon to be reduced to one-third his salary,[3] the same year Buckminster
Fuller dedicated his book on the "geometry of thought" to Coxeter—
praising him as *the* geometer of the twentieth century.[4]

Coxeter and Fuller's geometric progeny met before the men did. For its
international debut, Fuller's iconic geodesic dome served as the American
Pavilion at Expo '67, in Montreal. The dome's name derived from its geometric construction: the spherical structure takes shape from a scaffolding
of struts arranged on great circles, or "geodesics"—any circle on a sphere
that divides the sphere into two hemispheres. All the Platonic solids can be
made into geodesic domes by the process of "triangulation," dividing each
face of the solid into triangles and puffing it outward until it approximates a

Buckminster Fuller's American Pavilion Dome, Expo '67, Montreal.

sphere. Fuller's geodesic dome was constructed from an icosahedron, by dissecting each of its faces into smaller triangles.[5]

At Expo, Coxeter stood gazing at the geodesic American Pavilion for quite a while, trying to calculate the frequency of its vertices. Hexagons surrounded most of the vertices where the struts met, but every so often there was a pentagon instead (six triangles were arranged around the new subdividing vertices, forming a hexagon, while five triangles were positioned around the original icosahedral vertices, forming a pentagon). Coxeter was trying to determine the number of steps from one pentagon to the closest neighboring pentagon. The variation in hexagons and pentagons lended Fuller's dome its structural integrity, and accounted for its slightly puckered or dimpled effect. Unfortunately, Coxeter did not meet Fuller at Expo, so he left baffled.[6]

When he returned to Toronto, he spent an hour looking at the Expo files in the public library without finding an adequate photograph of the dome from which to discern the frequency. Eventually he located one at the university with a professor of microbiology—it was a very big photograph with an overlaid transparency framing one pentagon with the inscription "Here I am!" But Coxeter could still not locate a second pentagon. "We were still baffled," he said in a letter to Fuller. The dome, Fuller later informed him,

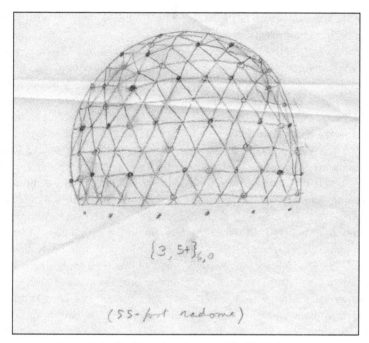

{3, 5+}_{6, 0}

(55-foot radome)

Coxeter's sketch of a geodesic dome, with highlighted vertices.

had a frequency of sixteen—starting at any pentagon vertex, one had to jump sixteen vertices in any direction before coming to another.[7]

Fuller had long been designing structurally efficient and economically affordable dwellings, hoping such abodes would alleviate the housing crisis in America. According to Fuller, "These new homes are structured after the natural system of humans and trees with a central stem or backbone, from which all else is independently hung, utilizing gravity instead of opposing it. This results in a construction similar to an airplane, light, taut, and profoundly strong."[8] Fuller wanted to build maximum shelter with minimum output—the spherical structure of the dome allowed for the largest volume of interior space with the least surface area, thus saving on building materials and expense. He wanted to "do more with less," like the honeybees.[9]

With his broad vision to "wake up humanity," Fuller was on a mission to understand the basic operating principles of the universe. For Glenn Smith, who had been a Fuller devotee[10] even before becoming Coxeter's fan, one of the elements that most attracted him to Fuller's work was his interpretation of geometry. "I think for Fuller geometry was a means to an end," said

Smith. "He was in search of the basic patterns used by nature in what he felt was a perfect comprehensive design."[11]

Coxeter and Fuller met less than a year after Expo, on March 1, 1968. Coxeter had accepted an invitation to speak at the philosophy and mathematics departments at Southern Illinois University, in Fuller's hometown of Carbondale. Coxeter's first talk, for the philosophers and other non-mathematicians, was "Geometry of Time and Space." The second was a lecture for the mathematicians, titled "Equiaffinities." Fuller attended the latter.[12] Afterward, as Coxeter recorded in his diary, Fuller "treated R and me to dinner, showed us his office and 'dome' house, gave me 4 of his books."[13] At this first meeting Fuller also asked if he could dedicate his upcoming book to Coxeter—*Synergetics, Explorations in the Geometry of Thinking*, due to be published the following year. The next month, Fuller sent a note of thanks for Coxeter's agreement, along with a copy of yet another book, *The Dymaxion World of Buckminster Fuller* by Robert W. Marks. "To Dr. Coxeter. Whose own world of mathematics has been ramified with a mastery shared by but one or two other humans of all history," Fuller wrote inside the front cover (practicing for his effusive dedication). "There are others who have made special contributions of extraordinary importance, but none with the comprehensivity [*sic*] of Dr. Coxeter. In highest admiration and joy over the priviledge [*sic*] of knowing him."[14]

From there, Coxeter and Fuller developed a simpatico rapport, on the surface anyway, sending letters and papers back and forth. On August 29, 1968, Coxeter delivered an invited lecture at the 11th Nobel Symposium, held on the island of Lidingö, near Stockholm. The focus of the symposium was "Symmetry and Functions of Biological Systems at the Macromolecular Level." Coxeter began his talk, "Helices and Concho-spirals," by asking: "What can a geometer contribute to a biochemical conference? Is there any contact between the imaginary world of geometry and the real world of living creatures? Perhaps a clue has been given by Dr. Monod in his philosophical remarks about fixity and evolution. Sharing his fondness for the five Platonic solids, I am tempted to give an account of Felix Klein's enumeration of point groups. But that is readily available in the literature (Coxeter, 1961, chap. 15). Instead, I propose to give a simple pure-geometric treatment of the following six basic theorems concerning motion and growth."[15] Coxeter sent a copy of his talk to Fuller, although under a different title—"Thank you so much . . . for your 'Man and His Environment' lecture at the 11th Nobel Symposium," Fuller replied, "as well as your truly enchanting piece on 'Mathematics and Music.'"[16]

Coxeter's alternative title, "Man and His Environment," turned out to be a tad ironic, in a calamitous sort of way. When he and Rien returned home

Buckminster Fuller

in September, having spent the entire summer abroad, they found among the backlog of mail a letter informing them that their cottage had burned to the ground in a lightning storm. Their friends who sent the bad news had watched the fiery spectacle from across the lake.[17] While Rien was heartsick with the loss of their second home, she often recalled how Donald more regretted the destruction of one of its contents: a glass and brass lamp in the form of a stellated dodecahedron.[18]

Later that September, Coxeter met with Fuller again when he came to Toronto to give a lecture at the university's stately Convocation Hall. The seventy-three-year-old Fuller, sporting his white crew cut and glasses strapped around his head with an elastic band, lectured to an audience of 1,500, mostly students, warning them: "We don't have much time." Fuller was primarily in town to open the first annual International Building Show. At that gathering he pontificated:

> Integration of the future will not be east-west as we have known it, but over the polar ice cap. The shortest great-circle routes between this part of North America and 90 percent of the rest of humanity don't go over either the Atlantic or Pacific . . .

Misunderstandings grow out of remoteness and different ways of approaching problems, but enmity and fear won't last forever. Either there will be no men on the earth or we will learn to communicate . . .

At Convocation Hall, however, the *Toronto Telegram* reported that "Bucky" was more "rapping," rattling off his ideas: "the aggregate of all humanity's consciously apprehended and communicated nonsimultaneous and only partially overlapping experiences"—riffing on the "synergy" or "energetic geometry" that "governs the physical Universe"—his bravado about humans all living on "Spaceship Earth"—and there being no such thing as "up" or "down."* The newspaper reporter, poking fun, tried to replicate Bucky's dizzying string of sound bites and then added as a kicker: "You know, he's been this way since '27 when he first started out. And for a lot of those years—people thought he was nuts. But they know he's a genius now, so it's all right." The students at Convocation Hall gave Fuller a standing ovation. Coxeter couldn't bear it till the end. He noted in his journal: "Out, disgusted, after ¾ hour."[19]

Smith, knowing both men, suspects that Coxeter respected the substance of Fuller's work, but didn't have patience for his unorthodox modus operandi. Coxeter may have found Fuller's public persona hard to take,[20] but he greatly admired his geodesic domes. Nonetheless, when asked, Coxeter did not mince words in delivering his verdict on Fuller: "Buckminster Fuller was a brilliant architect and engineer who knew very little mathematics but was very proud of himself," adding that Fuller "had overblown his stars as a mathematician."[21]

Coxeter was not the first to be frustrated by Fuller. Fuller was neither an architect nor engineer nor mathematician by training, and he became a controversial figure among experts in all those fields. He made awkward geometric mistakes, such as how many spokes are needed on a wheel to hold it rigid (Fuller said twelve instead of seven).[22] Coxeter had all the time in the world for amateurs and praised them highly. But in Coxeter's eyes, one of Fuller's downfalls was his use of preexisting material without acknowledgment.[23]

Where Coxeter sought to share credit almost to a fault, Fuller took out patents—on his triangulated icosahedron, for example—and he defended his patents vigilantly. In an unexpected triangle of interconnections, the artist M. C. Escher (whose *Circle Limit III* prints Coxeter had inspired; see chapter 11) had a run-in with Fuller via his son George. An engineer, George moved to Canada to become partner in a company that manufactured

---

*"People in China are not upside down." "Your head is out, your feet are in." "I'm going outstairs and instairs." "We're all on a space vehicle." "Like it or not, you're an astronaut."

portable domes much like Fuller's. The small company won a breakthrough contract to build these fiberglass "radomes" for the U.S. Air Force, but their victory party did not last long. They received legal notice from Fuller that he was due royalties on those and all domes the company sold. "We were livid at Buckminster Fuller for having to pay HIM royalties on the design," recalled George Escher. George was appalled that Fuller patented shapes that Plato and Archimedes had invented. "I still get hot under the collar thinking of how one would have the gall to patent a geometry so entirely settled in the public domain."[24]

M. C. Escher discussed this predicament, and Coxeter's view of Fuller in a 1964 letter to their mutual friend, chemist and crystallographer Arthur Loeb, who had a penchant for polyhedra and symmetry, and taught at the Carpenter Center for the Visual Arts, at Harvard. Escher wrote:

> About B.F. [sic], a few days ago I received some interesting information from my Canadian son George (who is now naturalized) . . . I found it troubling to hear that my friend Coxeter expressed himself in an extraordinarily negative manner with regard to B.F. in a letter to Long Sault Woodcraft Ltd. (They were in contact with Coxeter, who gave them theoretical advice.) He called B.F. a quack and wrote in such a derogatory manner about him that George cut off a part of that letter and destroyed it before permitting the letter to be filed. "In all probability, there was a considerable dose of professional jealousy; not so unusual," my son wrote to me . . . my own lay conclusion is that there are indeed charlatan-like facets to this man. A clever boy, this Bucky, for sure. While Coxeter is without any doubt an extremely skilled mathematical theoretician.[25]

Coxeter, however, formed this scathing opinion of Fuller before he met the man, and before he became intimately acquainted with his work. Over time, the geometer brought himself to overlook the polymath's indiscretions. When a reporter from *LIFE* magazine called in 1970, Coxeter gave Fuller a somewhat backhanded—but then accidentally glowing—compliment. In the article, a lengthy profile of Fuller, the reporter recounted: "I made an effort at consulting various authorities and scholars, nearly all of whom said Fuller was irrelevant to his field. To mathematicians, he was an architect; to architects, an engineer. It was like calling around Seville to check out Galileo . . . But then I called a Canadian mathematician H. S. M. Coxeter, 'the world's leading geometer,' in Fuller's estimation, and at Coxeter's urging I sent him a copy of a speech in which some of [Fuller's] lesser-known concepts were explained. Coxeter sent back a letter saying that one equation would be 'a remarkable discovery, justifying Bucky's evident pride,' if only it weren't too good to be true. The next day, Coxeter called: 'On further reflection, I see that it *is* true.' "[26] Coxeter told Fuller how impressed he was with his formula—on the cubic close-packing of balls. And he later took pleasure in proving it, noting in his diary one day in September 1970: "I saw how to prove Bucky Fuller's formula," and publishing it in a paper, "Polyhedral Numbers."[27] Of course, more than anything, Coxeter fell in love with Fuller's geodesic domes. Faced with the loss of his cottage, Coxeter had realized in a flash that a geodesic cottage would be the perfect replacement.* "It is a great pleasure to see how many people all over the world have used your structures for various purposes," Coxeter commented in a letter he addressed to "Bucky"(politely asking permission for the familiarity by asking: ". . . if I may join the distinguished company of those who call you so"). "I should think a dome (with no upstairs floor), much smaller than your house in Carbondale, would be ideal for a summer cottage. Has this ever been done?" After Fuller sent an article, "Making the Domes Available," Coxeter asked: "This article made me wonder whether it would be feasible to erect such a 4-frequency dome on a rock in the wilderness of Georgian Bay . . . Rien and I have the feeling that such a dome might be both convenient and attractive in that rural setting."[28]

As things turned out, the Coxeters decided against a geodesic-dome cottage. But Fuller's concepts continued to occupy Coxeter's mind. He noticed a connection between what he learned at the Nobel Symposium and his subsequent study of Fuller's domes. So, of course, he explored the interface of the two subjects with a paper—"Virus Macromolecules and Geodesic Domes"—and sent a draft along to Fuller.[29] In Stockholm Coxeter heard from the biochemists that they had recently refined a technique for staining viruses in such a way as to be able to observe their external shape with the electron microscope. "It must have been a thrilling experience," wrote Coxeter in his paper, "to find that, whereas some, such as measles, are long chains with a helical structure, others, being 'finite,' have well recognizable icosahedral symmetry and look like tiny geodesic domes." Coxeter went on to draw direct correlations: Fuller's 1955 dome built as bachelor officers' quarters for the U.S. Air Force in Korea seemed to be the shape of the REO (respiratory enteric orphan) virus; his thirty-one-foot geodesic sphere at the top of Mount Washington, in New Hampshire, was matched by the herpes virus and varicella (chicken pox); his U.S. Pavilion in Kabul, Afghanistan, found its twin in the adenovirus type 12; and the "radome" standing guard at the Arctic Distant Early Warning Line (a system of radar stations set up to

*Coxeter recalled reading somewhere, "the Salvador Dali built himself an icosahedral house (having fifteen faces above ground and the remaining five in the basement)."

detect Soviet bombers and missiles during the Cold War), corresponded to the infectious canine hepatitis virus.[30]

Fuller, in this intellectual gift exchange, gave Coxeter an icosahedral world map. "I had fun assembling the icosahedral world," Coxeter replied, "which now adorns my study mantelpiece."[31] And later Fuller sent him a hanging sculpture called *Tensegrity*. The word *tensegrity*, as Fuller explained in *Synergetics*, was an invention, a contraction of "tensional integrity"—it described the structural relationship between the sculpture's two components, sticks and string.[32] Coxeter was very fond of his *Tensegrity* and gave it prominent display; nestled in its own alcove by the front door, the sculpture cast a pleasingly symmetrical shadow of suspended sticks (the strings disappeared), lit from above by Coxeter's antique stained-glass Archimedean-solid light fixture from Alicia Boole Stott.

In 1975, Fuller's much-anticipated tome, *Synergetics, Explorations in the Geometry of Thinking*, was finally published with its flattering dedication to Coxeter. Off the record, Coxeter thought the book "a lot of nonsense." He suggested Fuller would have done better to consult a mathematician in the writing, rather than name-dropping in the dedication.[33] "Coxeter saw himself as a mathematician, and since Fuller attacked the traditional mathematician," noted Smith, "along with many other specializations, it would be natural for Donald to react to that attack." Smith wagered that Fuller did not think of Coxeter as a typical mathematician, out to prove theorems, but rather a mathematician who respected the intrinsic beauty in the patterns of the universe. And Fuller, in turn, sincerely respected and appreciated Coxeter's work.[34] The dedication certainly read as high praise:

> To me no experience of childhood so reinforced self-
> confidence in one's own exploratory faculties
> as did geometry. Its inspiring effectiveness in
> winnowing out and evaluating a plurality
> of previously unknowns from a few given
> knowns, and its elegance of proof
> lead to the further discovery and comprehension of a
> grand strategy for all
> problem solving.
> By virtue of his extraordinary life's work in mathematics,
> Dr. Coxeter is *the* geometer of our bestirring
> twentieth century, the spontaneously acclaimed
> terrestrial curator of the historical
> inventory of the science of
> pattern analysis.

I dedicate this work with particular esteem for him
and in thanks to all the geometers of all time
whose importance to humanity
he epitomizes.[35]

△ ☐ ◈ ⊛ ⊗

The notion of the existential role of the geometer is something that Walter Whiteley, director of applied mathematics at York University in Toronto, has also given a good deal of thought. As an applied geometer, Whiteley searches for patterns to solve problems—how the geometry of proteins affects their behavior in the body, how the shapely hood of a Mercedes is modeled in such computer programs as CAD, how a robot is instructed to reach out and grab three-dimensional objects when fed camera pictures, and what shapes and structures of buildings and bridges stand or collapse.[36]

From his work in these areas, Whiteley has come to believe in the power of the visual not only in doing mathematics, but also in applying it. Without a well-versed knowledge of the visual and the geometric approach to mathematics, society suffers what Whiteley calls a "geometry gap." To remedy the situation, he has become an ardent advocate for the visual method, and presents his ideas in a cogent presentation called "Learning to See Like a Mathematician"—it explains how mathematicians and scientists who use mathematics need to learn, or relearn, the visual and geometric languages.[37]

"The visual is central to all levels of mathematics," said Whiteley, delivering his opinion one day to small amphitheater of schoolteachers at a mathematics education conference at York University. "It changes the questions you ask, it changes the methods you use, it changes the answers, and it changes the way mathematicians communicate and teach. What you see is central to how you reason and problem solve."[38]

One of his vignettes recounted a problem during the Second World War, when the British were losing too many aircraft. Mathematician and statistical theorist Abraham Wald worked on the problem of how to save more planes; he was trying to determine where extra armor plates would be most beneficial. His first instinct was to add armor to the most damaged areas of returning planes, but after analyzing the visual data—the pattern of bullet holes on returning aircraft—Wald reached the opposite conclusion. He conducted his analysis by drawing an outline of a plane and cumulatively marking all the places where returning planes had been shot, which left almost the entire image covered, except two crucial locations. Wald then correctly surmised that the planes lost in battle had been hit in the unmarked areas—the cockpit and the tail engine—indicating it was those areas that needed more armor.[39]

Whiteley also described a scenario wherein data was poorly represented visually and this disinformation caused faulty analysis: the disastrous decision to launch the space shuttle *Challenger* in 1986. The O-rings designed to seal the joints between the rocket boosters were damaged by the cold temperatures of the launch day, but the all-important piece of information—that the O-rings' damage increased as temperature decreased—was hidden in cluttered and convoluted charts. As Edward Tufte, arbiter of the visual representation of information and professor emeritus at Yale, stated in his book *Visual Explanations*: "Had the correct scatterplot or data table been constructed, no one would have dared to risk the *Challenger* in such cold weather."[40]

In doing math, Whiteley continued, our brain is not primarily number crunching. It is seeking patterns. And as we explore, our visual cortex actually duplicates these images and patterns in the brain. This is not just a metaphor— like "the mind's eye" or "a picture in the mind." This process literally involves "thinking in pictures."* Visual images and patterns are actually built up in the brain, rather than being converted into neurological code.[41] Whiteley told the story of a bizarre old psychological experiment involving the retinotopic mapping of a monkey's brain. While the monkey stared at a visual image, a constellation of lights on a black screen arranged like a wheel with spokes, radioactive fluid was pumped into the monkey's bloodstream to follow the blood's path in the brain. The monkey was then sacrificed and its brain dissected. The location of the radioactive fluid in the monkey's brain physically reproduced, like a photocopier, the image of the constellation.[42]

For Whiteley, it all comes down to underlining how visual perception builds into reasoning in the brain. Even algebra is carried out using visual patterns within the equations and symbols—the appearance can be transformed without changing the content. "Algebra is cosmetics, not surgery," Whiteley said, displaying an algebraic equation translated into visual components of squares and circles. "Failure to do and teach mathematics visually is excluding numerous people and making mathematics harder," Whiteley concluded. And he conjectured that the dearth of the visual, the decline in classical geometry over the last hundred years, has had a regressive effect, resulting in "the geometry gap." This is much like "the ingenuity gap," a concept raised in the book of the same name—by Thomas Homer-Dixon, director of the University of Toronto's Trudeau Centre for

---

*Another weapon in Whiteley's arsenal is his deconstruction of the word *theorem*. By standard definition, a theorem is a proposition proved by a chain of reasoning. Whiteley takes it further. This word comes from the same root as *theater*, he says—from the Greek word *theorema* meaning "spectacle" or "speculation." A theorem, then, is something that has played before ones eyes, been considered with adequate speculation, to the point of epiphany when one can exclaim "I SEE!"

Peace and Conflict Studies—chronicling examples of people and societies facing a crisis of ingenuity or know-how, which leaves them unable to solve problems of their own creation. Whiteley's thesis holds that in the realm of science, the sedentary, mathematical areas of our brains, and the consequent lack of ingenuity—the inability to solve problems and make discoveries—results from an ignorance of visual and geometrical tools.[43]

"We will probably end up having to rediscover some things because we won't have people like Coxeter to make the connections," he said. Mathematicians and scientists, ignorant of historic geometric insights, will have to redo investigations from scratch, repeating the same pitfalls as their predecessors, until they reinvent the required results. In mathematics this can be part of the joy of discovery (or rediscovery). But in science, en route to urgent research, it could translate into unwelcome roadblocks and delays.[44]

Whiteley cast a concrete example of the geometry gap during a gala dinner concluding the 2002 Budapest conference, held aboard the tour boat *Europa*. After dinner, mathematicians gathered in the darkness of the upper deck as the boat passed beneath the Danube's intermittent tunnel of bridges. Seizing on perfect timing, Whiteley constructed a metaphor. He recalled a book he had read recently—*Design Paradigms* by Henry Petroski, professor of civil engineering and history at Duke University—on the theory of why and when bridges collapse, chronicling the problem of engineers failing to learn their own history. "Petroski says that within the span of forty or fifty years, the engineers who learned something from the last bridge disaster have left the field," said Whiteley, "and the next generation comes along and hasn't learned the same lessons."[45]

The geometry gap has exacerbated this situation. In the past, geometers and engineers were in constant communication—engineers knew geometry, and geometers knew engineering. When projective geometry gained popularity in the nineteenth century, for example, engineers were quick to see that the structural question of statics was "projectively invariant." Statics is the study of how forces converge, and resolve or fail to resolve—on the support structures of bridge, for example. Will a bridge retain its structural integrity when all the forces of gravity and weight are projected and converge on stress points? Projective geometry (to loosely draw the pure-to-applied analogy), is the study not of the shape and size of figures, as with Euclidean geometry, but the properties of these figures that are retained, or are invariant, under a projection—that is, when the image of the figure is projected by straight lines that converge on a plane or canvas, what qualities of the object are preserved? Engineering textbooks were filled with projective geometry until the mid–twentieth century, when geometry was in the depths of its decline. So from that point in time onward, engineering students were no longer exposed to geometry. Even the vocabulary of their predecessors was foreign, let alone any clue of which questions to ask to close the gap in knowledge.[46] And by Petroski's calculations, the world is overdue—knock on wood it shall continue to be—for a bridge to collapse.[47]

In his book *Projective Geometry*, Coxeter described this branch of geometry as a worthwhile way of "stretching our imagination."[48] And while for Coxeter stretching one's imagination was application enough, the most fundamental application of projective geometry stems from its earliest origins—the fine arts. In 1425, Italian architect Filippo Brunelleschi put forth his ideas about the geometrical theory of perspective (later consolidated into a treatise by Leon Battista Alberti, and developed further by Albrecht Dürer, and Leonardo da Vinci). From projective geometry's origin in the art world, one can easily appreciate how its properties held the attention of mathematicians like Coxeter.[49] Much as how hyper-dimensional objects are studied through projections down to lower dimensions, projective geometry explores how three-dimensional objects appear when projected onto a two-dimensional canvas, or plane. Investigations in projective geometry also consider the before-and-after relationship, the properties of the original object versus its projected image. When an artist draws a picture

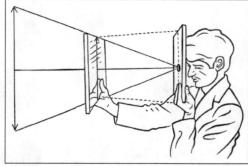

An illustration of the optical instrument used by Filippo Brunelleschi to render an accurate perspective view.

of a tiled floor on a vertical canvas, for instance, the square tiles cease to be square in the projection and become trapezoids, as their sides and angles are distorted by foreshortening (but the essential image is unchanged in the sense that the lines remain straight). Similarly, when a lamp with a circular lampshade casts a shadow, the circular rim of the shade becomes an elliptical shadow on the floor and a hyperbolic shadow on a nearby wall. "Thus projective geometry waives the customary distinction between a circle, an ellipse, a parabola, and a hyperbola," said Coxeter; "these curves are simply conics, all alike." The property of parallel lines is also altered under projection; parallelism is not preserved as parallel lines seem to meet, like railway tracks, at the horizon. The horizon, in projective geometry terminology, is called the "line at infinity" and parallel lines meet at a "point at infinity." *

Coxeter's book *Projective Geometry*, published in 1964, was heralded as a worthy contribution to the field's illustrious ancestry.[50] Reviewing the book, Gian-Carlo Rota, a philosopher of mathematics and a leader in combinatorial geometry at MIT, gave a glowing evaluation: "There is much to be said about a book that is perfect. If we were asked to pick a worthy successor

Everyday examples of projective geometry: parallel train tracks appear to meet at the horizon; a circular lampshade casts a hyperbolic shadow.

---

*Artists use the terms the "ideal line" or the "vanishing line"; and the "ideal point" or the "vanishing point."

to Euclid's *Elements*, we might choose this one. Of course, the synthetic method in geometry is now 'out of fashion.' This simply means that it will be back in fashion in another five years . . . so we might as well gear up to a few hours of high mathematical entertainment by reading his book on some cold winter evening."[51]

Over the years, Coxeter's classical geometry, as if on cue, was unearthed, a fragment here and there, like an archaeological artifact, rediscovered by mathematicians and scientists when needed. Coxeter acted as a repository of forgotten solutions, a memory bridge closing the gaps. As pockets of the old classical geometry recurred in mathematics and became relevant in the applied sciences, Coxeter was an encyclopedic sage. "In his mind he carried a lot of the connections for us," said Whiteley, "from an earlier period when this geometry was very active to a period now where it becomes active again—but in between many pieces of it were lost. He kept a culture alive." Over the course of his career, people sent Coxeter letters containing a diagram, a paper, a proof, or theorem they had recently discovered—or rediscovered—and asked, 'Have you seen this before?' Coxeter was able to look at this image, cross-reference it through his brain, and say, 'Yes, this is something that appears here and here and here.'

"It struck me how essential it was that he had in his own mind, in his own experience, the capacity to make these connections," Whiteley observed. "This is not something we can do now with Google. You can't put a diagram into Google and say find me other diagrams that are 'like this.' But this is what people of his caliber could do, in their minds. You could take Coxeter a picture and ask, 'Have you ever seen anything like this before?' And he would provide you with a geometrical metaphor or an exact reference. No computer is capable of answering those kinds of queries. How on earth are we going to replicate that," wondered Whiteley, "when we don't have Coxeter? There will be a lot of geometry that disappears into storage."

Of course, Coxeter was only human and one geometrical image he found in his filing cabinet he did not recognize. It was a particularly stunning and clever representation of symmetries by a graph. He called his friend and graph theorist William Tutte, at the University of Waterloo, and asked him if he had every seen it before. Tutte said, "Yes Donald, I have. You discovered it." Coxeter then wrote a paper about his rediscovery of his own discovery, which he titled "My Graph."[52]

△ ▢ ◈ ⊗ ⊗

In the early 1990s, Douglas Hofstadter wrote a few "Dear Professor Coxeter" letters. Hofstadter wanted to tap into Coxeter's mental archives and

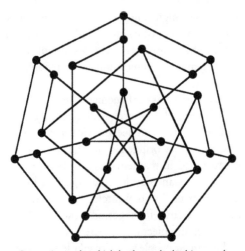

Coxeter's graph, which he forgot he had invented.

ask him if he had ever seen what Hofstadter hoped was an original discovery in projective geometry—his Garland theorem. Within the space of several months, Hofstadter sent Coxeter a trilogy of letters, each more than ten pages long, typed single-spaced and double-sided, albeit with many diagrams interspersed.[53]

In the introductory letter, Hofstadter began: "First of all, let me apologize for intruding on you with this long letter. You must have plenty to take care of besides thinking about ideas foisted upon you by a stranger. But perhaps I am not entirely a stranger. Although the letterhead gives my 'official identity' "—it's a long one: professor of computer science and cognitive science; adjunct professor of psychology, philosophy, and the history of philosophy of science; and director of the Center for Research on Concepts and Cognition, at Indiana University, in Bloomington—"you may perhaps know me—or know of me—as the author of the book *Gödel, Escher, Bach: An Eternal Golden Braid.*" Coxeter replied: ". . . clearly, you are no stranger."[54]

Hofstadter divulged he had been on a "geometry binge,"* and he wanted to tell Coxeter "of the deep debt I owe you." Hofstadter grew up profoundly in love with mathematics. He loved its "abracadabraic" and "mirabile dictu"

---

*In letter number three Hofstadter began by apologizing for once again "inflicting" on Coxeter another enormous letter: "What is wrong with me? Geometromania, I am afraid."

qualities. He made a number of modest but genuinely original discoveries in number theory as an undergraduate, and then started his graduate studies in mathematics at Berkeley, in 1966. But after a year or so, the relentless push toward abstraction and nearly total lack of imagery "brought me to my knees. I got out of math in Berkeley," as he recalled. "I was fed up. It was a horrible experience for me to be suddenly repelled by what I had always assumed would be the love of my life."[55]

Eventually he carved his niche in an area of physics, solid-state theory. Doing his PhD research, he discovered the energy spectrum he called Gplot, now known as the "Hofstadter butterfly"—the first fractal, or multifractal, ever found in physics (this preceded Mandelbrot's fractal fame). When Hofstadter graduated, however, he was ready for another switch, and he picked his long-time obsession with how the mind works. In 1979, he published his Pulitzer Prize–winning *Gödel, Escher, Bach*, and before long he had earned his status as a free agent at Indiana University; he had license to explore the workings of the mind in any way he liked. He focused most of all on the mechanisms of analogy. But one day he found himself caught up by a small problem in plane geometry and the cognitive process it generated intrigued him.[56]

In a thought experiment, of sorts, Hofstadter chose the equilateral triangle as his guinea pig. As his fascination grew, he was caught off guard by the fact that the triangle had more than one center. "I had heard words such as 'ortho-center,' or 'centroid,' or 'circumcenter,' and I sort of knew these things existed, but to tell the truth, when I found out that any triangle has *many* centers, I was really thrown," he said. "It seemed like a miracle. In fact, the triangle has an infinite number of centers, but they are not all equally interesting."[57]

At one point he made a droll analogy between this startling profusion of triangle centers and the numerous parts of the human body that might qualify as an analogy. He recounted: "Suppose I asked a bunch of people to say what they think is the 'most important'—that is, the 'most central'—part of their body. One person might shout out, 'My brain!' Somebody else might say, 'No, my stomach!' and other people might say, 'My heart' or 'My sexual organ' or even 'My belly button.' It could go on and on like that, and . . . each person could surely defend their point of view. Someone might even say, 'My kneecap!' although since there are two of them, that would break the body's natural symmetry, so maybe one would have to amend it to include both kneecaps." Looking back, Hofstadter finds this analogy flippant, but it helped him wrap his head around the fact that a triangle has not just one center but many.[58]

Hooked by his triangle investigations, Hofstadter continued on and discovered new centers of his own. Among the discoveries, there were a few rediscoveries, and at times he was crestfallen to learn that he had come along a century or more too late. He relayed his learning curve to Coxeter:

> In a bookstore, I came across the book by you and Samuel Greitzer, *Geometry Revisited*. It was electrifying. There is no better word for it. I just gobbled it up, at the same time inventing new geometric ideas on my own . . . For example, smitten with duality, I invented the idea of reciprocation in a circle before I came to the chapter in your book where it was discussed, and I naïvely thought that maybe I was the first person on earth to have come up with that idea! . . . Oh, well. In any case, all of this was astonishingly beautiful, even intoxicating, and of course your books were the vehicles that conveyed all this beauty to me.

Hofstadter had another epiphany, one he was sure was his own— "absolutely new, something stunning, deep, and beautiful." He called it the Garland theorem, for the beauty of the interconnected results; and at the core of it was a new center for the triangle, found within a cluster of triangles and circles. "All this gave me quite a bit of hope that I was the first to see this gem," wrote Hofstadter, this time in an article, "Discovery and Dissection of a Geometric Gem," a copy of which he sent along to Coxeter.[59]

In contemplating whether or not his Garland theorem was in fact a new discovery, Hofstadter noted: "Compared to a titan like H. S. M. Coxeter, I have but a minuscule storehouse of knowledge." So he asked Coxeter whether he had ever seen anything like the Garland theorem before. In response, Coxeter recommended his two books on projective geometry, which Hofstadter duly ingested. And Coxeter made a suggestion: "Before seeking publication, you should compose a conventional proof" (and offered a few hints). In the end, Hofstadter found that his Garland theorem was not the original gem he had hoped. It was at once a deflating and educational moment. "It has been a fabulous experience," he told Coxeter, "and it is you whom I credit, in large part, for launching me on this voyage and being my guide. I will never be quite the same, after having drunk so deeply from the infinite well of geometry. My life is in some central way forever changed, thanks to the mysteries and beauties of triangles and circles."[60]

Hofstadter was grateful that triangle geometry had placed him in a revealing laboratory of self-observation, a chance to watch himself in the discovery process. In his article, he discussed his belief that the visual approach and the use of analogy it generates is the primordial tool of mathematical discovery. Coxeter concurred, mentioning that he too had made profitable use of analogy in discovery, referring to his childhood paper "Dimensional Analogy" and all the fruit it bore.

Douglas Hofstadter wrote in one of his letters to Coxeter: "I am enclosing, for your amusement, this design I did, called an 'ambigram,' which is intended to be readable both right-side-up and rotated 180 degrees, as 'Coxeter.' I hope you can make it out."

In his article, Hofstadter also issued a critique, full of frustration, railing against the "enormously abstract directions that math has gone in, over the last, say, 50 or 60 years." He recounted a trip to the mathematics library at Indiana University, trying to find books or journals that might provide some hint as to whether his discovery was new. The *Journal of Geometry* was particularly dismaying. Hofstadter compiled stats on the "picture density" of the journal. From three successive journals in 1991 and 1992 he recorded fifty-two articles, thirteen of which contained pictures. "The page-level statistics are even more revealing," he observed. "In these issues there were 602 total pages, but only 39 of them had any pictures! In other words, on the average, 75 percent of the articles (39/52) and 93 percent of the pages (563/602) in the *Journal of Geometry* are pictureless. By contrast, Coxeter and Greitzer's book *Geometry Revisited*, which has 153 pages of text, has roughly 160 separate diagrams—an average of over one per page! . . . I cannot really judge the articles in the *Journal of Geometry*," he concluded. "My intuition tells me that many of them must be shallow despite their air of depth, but surely some of them *are* genuinely deep and important. Sometimes I feel positively daunted by the remoteness and incomprehensibility of the whole journal, and I feel a kind of childish admiration for anyone who can think at such abstract levels. But I oscillate between respect and disgust. It is a very strange and uncomfortable feeling."[61]

For Hofstadter, the visual is the absolute crux of mathematics. "I always have to have a picture," he said recently. "And I don't like to make it sound as if, when I say visual, that a blind person couldn't have it. I think a blind person could have it just as much. I feel the visual means thinking about space—however one represents space in one's head.[62] A bat may be blind but certainly has a wonderful understanding of space; I think a bat would be capable of geometry if it had the intelligence. So it's not the eyes so much. The word 'visual' for me means something that goes on in the brain that has direct contact with space and distance in space."[63]

The lack of the visual in modern mathematics Hofstadter can't quite explain, though not for lack of trying. It's not quite a conspiracy, or deliberate dishonesty, he said. He offers this analogy: "I feel that mathematicians have developed a paranoid fear of non-rigor or of intuition. They almost don't want to admit that they are human. It's a little bit like what's happened since September 11th in the United States where airport security has gone up and up and up, out of perhaps some genuinely correct fear but also out of paranoia of terrorists . . . to the extent that you can't even carry a nail clipper onto a plane. . . . It's as if mathematicians have this mania for turning everything into prickly formal symbols and using as many symbols as possible. Even words like 'if,' 'then,' and 'is' are routinely replaced by symbols, so that what *could* be understood if you wrote it in words looks very technical and forbidding, and as if it had much deeper meaning than it really does. It pushes people away, even people who love math; not only the English majors who maybe would never have been attracted to math, but also people who are good at math. My feeling is that there is a great deal of obfuscation, of obscurantism in mathematics. I don't mean to say that rigor doesn't have its place. But I feel as if hidden behind what mathematicians say, there are often pictures. And that mathematicians are frightened of showing these pictures. Why they are frightened I don't know."[64]

Hofstadter feels on the whole that the pendulum is swinging back in favor of the visual. As he said in one of his letters to Coxeter, "I have a sense that we are on the verge of something of a turnaround. By distributing my article and getting back various responses, I have learned that a number of important mathematicians are 'closet Euclideans.'"[65] Walter Whiteley, then, is in decent company singing the praises of geometry's visual tool. He is joined by a chorus of mathematicians who quietly or vociferously support a more prominent place for the visual perspective in math and science. "The point of view of people like Coxeter, and many mathematicians like myself," said Sir Michael Atiyah, "is that geometry trains the imagination—it is sport for imagination and inspiration—and that thinking about geometrical things is very important not only for mathematicians but scientists and engineers in their attitudes toward the three-dimensional world and how we see it." After geometry disappeared from the curriculum, university engineering departments complained that students couldn't understand three-dimensional geometry—they didn't know where to start in building machines because they didn't know how to draw things in three dimensions. As a backlash to these deficiencies, combined with the campaigning of Coxeter and others, a movement is afoot to reintegrate geometry into the syllabus in a modernized way, focusing on the intuitive as well as the formal side of mathematics. However, to say the pendulum has swung back is an exaggeration. "Pendulums tend to wobble around a bit, and this is not an easy task to get right," said Sir Michael. "It keeps changing with every generation and it is still very much a live issue."[66]

In 2001, the Royal Society of London, the apogee of all mathematical and scientific wisdom, published a report on the state of geometry education. It, too, argued for reinstating the study of geometry to its deserving stature. One recommendation urged the development not only of an awareness for the historical and cultural heritage of geometry in society, but also the development of skills for applying geometry in real-world contemporary contexts. That as many students as possible fully develop their mathematical potential through geometry, the report stated, "is a matter of national importance."[67]

CHAPTER 10

C₆₀, IMMUNOGLOBULIN, ZEOLITES, AND COXETER@COXETER.MATH.TORONTO.EDU

> You see, which parts of mathematics are applicable—not applied
> but applicable—it is very hard to tell in advance. . . . And why
> mathematics can be applied to other things, to physics—
> that's a mystery.
>
> —HENRI CARTAN

One day in the early 1960s, Gord Lang, a communications man, visited Coxeter in his ivory-tower office with an applied query about sphere packing—without classical geometry and the age-old question of how best to pack spheres, like a grocer's display of oranges or a cannoneer's stack of cannonballs, the information transmitted through cyperspace would be garbled beyond recognition. In one of Coxeter's papers there is a very simple explanation of sphere packing, starting with the related "kissing number" problem: "Every intelligent child knows that a penny on a table can be surrounded by exactly 6 others, all 'kissing it' and that when the pattern is continued the pennies arrange themselves in straight rows. Tangents at all the points of contact form a tessellation of regular hexagons, one surrounding each disc. In other words, the 2-dimensional ball-packing problem is completely solved. Packing billiard balls in 3-space is less obvious."[1]

Packing spheres in higher dimensions kept geometric minds reeling for centuries. The problem dates to 1611, when the German astronomer Kepler conjectured that the densest way to pack spheres was by following the method of the grocer stacking oranges. Kepler was unable to prove his hypothesis, as were many mathematicians through history. Eventually, Thomas Hales at the University of Pittsburgh proved Kepler correct in 1998. After six years of work Hales produced a proof generated by three gigabytes' worth of computer programs, also known as "proof by exhaustion" or the "brute force method."[2]

The grocer's arrangement allows twelve balls to congregate around a central ball, which leads us back to the kissing number problem, in dimensions higher than two. In 1694, Sir Isaac Newton bounced this problem around with Oxford mathematician David Gregory. They debated whether a rigid material sphere could be brought in contact with thirteen other such spheres of the same size. Gregory thought yes, Newton no.[3] In 1727, Stephen Hales (no relation to Thomas) tried to figure it out by compressing several fresh parcels of peas in the same pot, keeping the pot closed and adding water (approximating a force of as much as 1,600 pounds). He observed that the peas dilated and formed "into pretty regular Dodecahedrons" but his results were ultimately inconclusive.[4] Various mathematicians eventually solved the three-dimensional kissing number problem in the nineteenth century—Newton was correct.[5]

When Gord Lang first visited Coxeter at the University of Toronto, he was in the early stages of developing a modem. Lang had previously overseen the design of a centralized purchasing system for Trans Canada Airlines, which was installed at the airline's Toronto office. And he had been in charge of a communication system for Datar, a Canadian Navy project. Lang noted proudly that Claude Shannon, the progenitor of communications technology and author of "Mathematical Theory of Communication," reviewed the design concepts for Datar, which was considered the world's first digital data network. Shannon was completely satisfied and noted that it was the first practical application of his information theory.[6] Shannon's theory, published in 1948, addressed the problem of information capacity—the capacity of a channel to transmit information in a manner that ensured messages were received as error free as possible. He posited that the design of such a communication system was analogous to the sphere-packing problem of the geometer—sphere packing was a strategy for efficiently storing and encoding data to eliminate errors.[7]

During Coxeter and Lang's many meetings, it was Coxeter's job to point Lang in the direction of any sphere-packing research that might be useful in developing a modem. But first, it was Lang's part to explain to Coxeter this application of geometry, to give the pure mathematician a sense of what he was looking for.[8]

The challenge of sending information digitally is to do so with the least possible amount of ethereal noise contaminating and confusing the data traveling down a modem line, or the voice through a telephone line. All the information that travels down the wires and through the skies is encoded as digits, as zeros and ones (or more complicated symbols). The coding must be done in a way that minimizes waste—waste of money and power—and in a way that minimizes distortion. "So 'cheese' is not distorted to sound like 'choose' or 'geez,' " said Neil Sloane, at the AT&T Shannon Lab. "Once you start trying to minimize distortion, you very quickly come up against the questions of packing balls into a box in high dimensional space."[9]

Think of the information signals as points, Sloane suggested; for information points to be transmitted accurately and clearly, they need to be far apart, not interfering with one another. So imagine that the points are positioned at the exact center of a billiard ball. If the information points are then packed together and sent down the wires, the outer circumference of the billiard ball's sphere would serve as a protective shield, insulating the central information from disruption or distortion; such an arrangement would absolutely guarantee that points cannot be too close together and that information would not be garbled.[10]

"Let's solve the problem of how to pack lots and lots of balls, all the same size, in 1000 dimensions, a 1000-dimensional box," said Sloane. "Why 1000 dimensions? The number of dimensions simply corresponds to the number of numbers—ones and zeros—in each coding, like a bar code. Say your mother is talking into the microphone and you want to convert her voice to zeros and ones. What you do is take a little snippet, take one second, and you cut up her voice signal into little pieces. You chop it up so you have little sausages coming out of her mouth. And each sausage is one second long. Now, in order to send that voice code, you need to sample it. You look at its values 1000 times a second—think of that as the coordinates of a point [of information] in 1000 dimensions."* And so, by figuring out the best arrangement of billiard balls in a thousand-dimensional box—the number of dimensions corresponding to the number of coordinates, such as (x, y, z . . .) up to one thousand—an answer presents itself as to how information should be best encoded and sent down the wires. In our present-day digital reality, however, the sample is done continuously, producing the problem of packing spheres in infinite dimensions—one sample after another after another after another, ad infinitum, chugging along endlessly, thus producing the coordinates of a point in infinite-dimensional space.[11]

In the early days of communications research, telephony scientists like Lang focused on much lower dimensions, such as eight. When Lang came calling on Coxeter, he was interested in a packing called the E8 lattice—lattice

---

*Taking a thousand samples a second would not be enough to get a good rendering of your mother's voice, Sloane said. The question, then, is how fast do you have to sample the voice, at what frequency? One of Shannon's theorems stated that the sample must be taken at twice the bandwidth. "If the frequencies in your mother's voice are typical—she is a bit shrill at times—but typically you don't have to do over 3000 Hertz to get pretty good fidelity of a voice," said Sloane. "So double [3000] and that's the sampling rate: 6000 times a second."

because the center points of packed spheres align in such a way that when connected they produce a crisscrossing lattice of points. It was then Coxeter's job to bring to Lang's attention any relevant sphere-packing research, useful fodder for his modem design. Coxeter provided Lang with both historic research and scoops on ongoing research, the cutting-edge developments on the ancient sphere-packing problem. And Lang convinced Coxeter to do some pure research that might help Lang's applied pursuit. Coxeter was reluctant, but after some coaxing he finally decided he might be able, and willing, to contribute.[12]

To this end, in 1963, Coxeter made a contribution to the ball-packing problem that had been undertaken by Kepler, Gregory, and Newton. Coxeter showed that the "12 around 1" arrangement*—known to produce a hexagonal arrangement of balls—could be shifted a bit, or translated, by rolling the balls ever so slightly, and in doing so the balls reconfigured into an icosahedral arrangement. Buckminster Fuller later named this process the "jitterbug transformation." Although Coxeter's amount of wiggle room wasn't enough to accommodate a thirteenth ball, it did permit the twelve balls to be arbitrarily rearranged (as Conway and Sloane later proved). Coxeter wrote the paper "An Upper Bound for the Number of Equal Nonoverlapping Spheres That

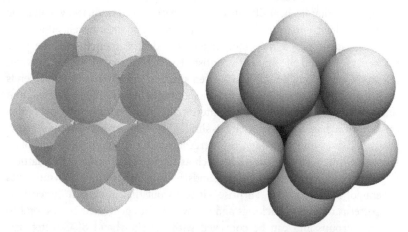

Coxeter showed that Kepler's hexagonal close-packing of spheres, left, could be transformed into the icosahedral arrangement by simply rolling the balls around.

---

*As Coxeter liked to point out, "The 12 centres of these balls form the vertices of a quasi-regular polyhedron, the cuboctahedron, which was described by Plato." Buckminster Fuller invented a new term for the cuboctahedron, calling it the "vector equilibrium" (Coxeter did not approve).

Can Touch Another Sphere of the Same Size," and gave Lang an early copy before publication.[13]

When Lang refined Coxeter's results for his purposes, he used the "brute force method," cobbling together time on many computers, in Toronto and overseas. After months and months, a box was shipped with the results in punch card format—the Paleolithic precursor of computer disks—and in the early 1970s Lang submitted a proposal for an E8 lattice modem.[14] He then went to see Coxeter with his proposal to show him results of their collaboration.[15] "Lang was very pleased with himself, and he was determined to show Coxeter how he had put the abstract theories to good use," said Bob Tennent, who went along for the ride that day (he was then an undergrad studying engineering science at the University of Toronto, working a summer job as a computer programmer for Lang). "Lang made an appointment with the great man and dragged me along to show him the results on the computer print outs," said Tennent, now a professor in the School of Computing at Queen's University. "I think the only reason I went with Lang to see Coxeter is that the former wanted a witness for his (expected) triumphal demonstration and appreciation by the latter. Coxeter was polite but noticeably cool to Lang's exposition. Eventually Lang stopped and asked Coxeter whether he felt gratified that a practical application had been found for his hitherto purely abstract geometrical theorizing. Coxeter calmly replied that, on the contrary, he was appalled that his beautiful theories had been sullied in this way. Lang, the engineer and entrepreneur, departed disappointed and uncomprehending."[16]

The incident was a nice sampling of the different mind-sets, the pure mathematician and the applied scientist. The bare bones of sphere packing, the study of packing points into a given space as far apart as possible, lends itself naturally to application. "But that's not the reason I'm interested, and that's not the reason Coxeter was interested," said John Conway, who also on occasion has found himself at similar to-apply-or-not-to-apply crossroads. Coxeter and Conway are interested in the sphere-packing problem because the nuances are exquisite, such as the symmetries. In the E8 lattice, for example, Conway said the E stands for the "exceptional" symmetries generated. And—surprise, surprise—it turns out that in certain dimensions the patterns of sphere packings and the symmetries generated correspond to Coxeter groups and can be conveyed with the shorthand of Coxeter diagrams.[17]

The fact that Coxeter responded as he did to Lang's work was in part a demonstration of his selective disinterest in applications. But he didn't turn up his nose at every application.[18] Coxeter may have found Lang's modem so off-putting in the moment, when its potential to link and feed supercomputers was laid out before his eyes, because Coxeter despised, almost more than anything, the onslaught of the computer age. "I deplored the attention that people gave to computing," he said once. "I was afraid they might neglect other subjects." He was relieved when computer science at the University of Toronto was separated from the math department. "I didn't want to have it in mathematics," he asserted. "Lots of young people are tempted away from pure mathematics when they find that computing is such an up-and-coming field."[19] Coxeter never used a computer, let alone a modem. Although, not wanting to be out of touch with the world of fans that wanted to be in touch with him, he had his son-in-law, Susan's husband Alfred, send e-mails on his behalf.[20] And, ironically enough, the computer server in the University of Toronto math department has been named in Coxeter's honor—coxeter@coxeter.math.toronto.edu.[21]

△ ▱ ◈ ☺ ◉

The alliance of the computer and geometry ultimately only served to bolster classical geometry's cause. Over the years, mathematics departments have increasingly aligned themselves with the computer science—the University of Innsbruck, in Austria, has a Department of Engineering Mathematics, Geometry, and Computer Science as part of its faculty of civil engineering. And NASA's Langley Research Center would not be complete without its Geometry Laboratory, providing "expert assistance in the construction and analysis of computer based geometry," focusing on the engineering program Computer Aided Design (CAD). Its main services, according to the "GEO-LAB" mission statement, is its "availability of high capacity workstations on which to develop and visualize the geometry" and "a team of geometry specialists with the expertise to apply the tools and techniques to specific applications." One of those applications involves studying the arc of flight paths, and another simulating the velocity of the airflow between the wheels of a plane's landing gear, which contributes to the noise during takeoff and landing. In the latter study, numerical data is mapped by the "data viz group" into three dimensions, simulating a landing gear in operation. Experimental results can then be displayed in stereo on a large screen, giving the researchers the benefit of "immersive visualization"—they can rotate the model to see directly between the wheels, for example, generating a better understanding of the airflow field and modifications to the gear geometry that will reduce the amount of noise produced.[22]

Even the average home computer contains a graphics card that creates images by projective geometry—images projected from one domain or dimension to another, allowing higher dimensions to look like three-dimensional images on the two-dimensional canvas, or computer screen. This technology is used in creating the convincing animation that flies you around in a three-dimensional video game, and movies by Pixar, such as its Academy Award winners *The Incredibles* and *Geri's Game*, a short film about an old codger who plays a game of chess against himself.[23]

Tony DeRose,[24] a senior scientist and member of the "Tools Group" at Pixar, often gives a lecture he calls "How Geometry Is Changing Hollywood" (sometimes he calls it "Math and the Movies"). One example DeRose gives is the use of projective geometry in transforming the three-dimensional image of Bob Parr onto the two-dimensional screen.[25] Pixar animation scientists also use Euclidean geometry to assemble independent scenes—each little set piece and character is modeled in its own separate Euclidean space, and then, by a geometric transformation, the separate components are assembled into a common environment, or a scene in one of the films.[26]

The message in DeRose's talk, however, is not only that geometry is an alive and vibrant field, by illustrating that a lot of old geometry, from hundreds or thousands of years ago, is nowadays used in animation. Rather, he wants to entrance kids with the fact that animation is a field pushing the boundaries of new mathematics, and new geometry. "Subdivision surfaces" is a new geometric tool that allows animators to efficiently represent on a computer complex surfaces that "deform"—when a character is speaking—say, the elderly Geri—the surface of his face must be represented in a lifelike way as it deforms during speech. "Classical differential geometry tells you how to look at curvatures with all sorts of theorems that allow you to analyze the surface," said DeRose, "but it doesn't really give you any constructive tools—I want a surface that looks like this, how do I represent it with a computer in a constructive fashion that I can animate quickly, that I can render quickly? That's the problem that subdivision surfaces address."[27]

Originally proposed in 1978 by Ed Catmull, the president and cofounder of Pixar, subdivision surfaces operate by starting with a coarse set of polygons stitched together into a seamless surface. Realistic human skin or clothing is then created by repeatedly splitting and smoothing the coarse polygons. This splitting and smoothing process, together with other tools such as wavelets, provides Pixar animation scientists with compact ways of vividly representing highly detailed surfaces—to create characters that have a comic-book aura, all the while looking as convincing as people; characters who, as the Pixar literature boasts, have "organic translucent skin that makes them subtly glow from within."[28]

Perhaps the most stunning by-product of the computer age, as far as the Coxeterian perspective is concerned, is that his visual approach to geometry has received a tremendous boost from the exactitude and certainties of computers; the mouse and the computer screen have considerably improved

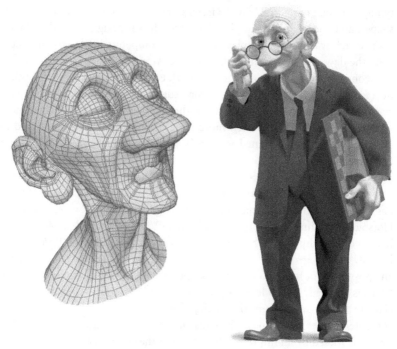

A computer graphics tool successively subdivides a "control mesh,"
left, to create the face of the Pixar character Geri. © Pixar.

upon the raw materials of pencil and paper. As a result, computers and computer programs have instigated a classical geometry renaissance in the classroom.[29]

One program in particular, The Geometer's Sketchpad, has happily overthrown the geometry kit, taking it into the realm of virtual reality. It hit the market in 1991, after five years of development with a grant from the National Science Foundation—a small group of mathematicians and educators had again been rattling the cages at NSF, saying, "What about geometry?" and insisting that more needed to be done to rejuvenate the subject in classrooms.[30] The computer program was designed quite literally to provide a visual sketchpad to draw and construct geometric shapes that could be stretched and moved, keeping their special properties intact (such as lines being perpendicular), allowing students to discover and verify geometric facts in an interactive way. "It was a great leap forward," said Doris Schattschneider, the senior geometer on the project, and an emeritus professor at Moravian College in Bethlehem, Pennsylvania.[31] The program was mouse driven (no typing, typing, typing in BASIC code, entering coordinates to produce a crude shape), and had high-tech graphic demands—it was designed for the Macintosh, which was just then replacing Apple II computers in classrooms (these were the days ruled by Commodore 64 and Atari).[32]

Geometer's Sketchpad, now in its fourth edition, has met with enthusiastic response. According to educational surveys and studies of secondary classrooms, Sketchpad is the software that mathematics teachers find most valuable for students; it is used more than any other computer tool; and in 1993 a version was designed for the Windows platform, at IBM's request and expense.[33]

Coxeter, of course, never worked with Geometer's Sketchpad. But many of his fans, fans of classical geometry, are avid users. As a research tool it has penetrated the academic scene, with William Thurston and Douglas Hofstadter lauding its utility.[34] And Walter Whiteley finds that the moving images of Sketchpad take his visual perspective to an entirely different level. Indeed, having geometric diagrams produced by the computer eliminates any possibility of flaws, to which visual perception might otherwise succumb. "This is suddenly a new level of precision," said Whiteley. "The mathematician no longer spends any time going down the wrong track because the computer makes it obvious when reasoning in the diagram is wrong. We might be fooled by an imprecise diagram, but the computer will show the error." Almost as important, in Whiteley's opinion, is the process of "learning what to expect" and "learning what to look for"—by observing a sufficient range of examples, dragging a sketch around, or creating a series of sketches to expand one's experiences. "At some point, you are able to 'image' a new example, or rerun an old example, in your brain," said Whiteley, "—without the computer."[35]

"When you work with Sketchpad," Whiteley effused, "the image in your brain is actually altered to have a different focus, a different precision. And then when you reason without the computer you make different choices. It really does alter the kind of reasoning you do with images. You have a different sense of what holds and what doesn't. And you may have a different conviction about your results. You do a sketch, you drag the sketch around . . . and you may be utterly convinced that the thing is true without ever having seen a proof. And the absence of a proof doesn't shake your conviction. Because you know visually, and viscerally, you are on the right track."[36]

△ ◻ ◈ ◉ ⊗

In his liaisons between math and science, Whiteley has often consulted with biochemists and biophysicists, participating in geometrical mathematical modeling of how protein molecules function in the human body. Given the

shape of a protein, Whiteley investigates how it will interact with the body
or with a specific drug. He tries to determine whether a protein's regions
will be rigid or flexible, because this is the property that dictates how a protein interacts. Working in the York Math Lab,[37] Whiteley and his students
devise computer algorithms that shorten the biochemist's search, tinkering
with the geometric models, adjusting their struts and nodes, trying to discover how many rigid and flexible vertices each sample protein structure
might have.[38]

There are several reasons why the shape of proteins matters in the body.
"The body expects a protein to fold, to take shape," Whiteley said, opening
and closing the flexible model cage of an icosahedron that can be flattened
and expanded as desired. If the body sees a protein that is not taking shape,
the protein is assumed to be defective and the body then rips it apart and
reuses it. That's what causes cystic fibrosis, an inherited disease of the mucus glands. A protein made by the CFTR gene—cystic fibrosis transmem-
brane conductance regulator—fails to take shape and fails to function,
eventually resulting in a buildup of mucus in the lungs, pancreas, and other
organs.[39]

The opposite occurs when a protein is too rigid and lasts too long in the
body—proteins, on average, last a matter of seconds, or as long as one day.
"There's a normal cycle where things are formed and then taken apart again
and reused," said Whiteley. "If something is too stable, that also poses a
problem. Something that is too stable becomes a problem in the same way as
gallstones become a problem." Proteins that are too stable cause mad cow
and Alzheimer's disease. In mad cow, misfolded proteins known as prions
refuse to break down, become sticky, and pile up in the form of plaques
that wreak havoc on normal brain function. In Alzheimer's, the amyloid
precursor protein jams mitochondria—the so-called cellular power plants—
causing the suffocation and death of neurons. Our bodies are biochemically
fine-tuned: proteins cannot be sloppy and flexible or they will fail to function, but neither can they be too rigid, or they won't be broken down and recycled.[40]

Our immune system works according to this geometric jigsaw puzzle. The
dynamics of molecular interaction are either a "lock and key," where two
rigid molecular shapes have to be exactly right to click and work together, or
an "adaptive fit," where at least one protein is flexible, functioning like a
catcher's mitt and changing shape, enclosing on the incoming protein.[41]

Immunoglobulin, the primary molecule in our immune system, is highly
complicated and flexible. Its structure is described by biochemists in terms
of a "body" with "arms," "elbows," "hands," and "fingers" that grab or
recognize antigens. These molecular extremities are extremely dextrous and

limber, their double-jointedness allowing them to latch on to what they need, wherever it is. When attacked with the icosahedral common cold virus, for example, the fingers change shape rapidly, producing both lock-and-key and adaptive-fit receptors, trying to make a match with the viral antibodies. "When you've built an immune response," said Whiteley, "it means you've found the production machine for the things [antibodies] that will respond and detect and bind to the viruses. It has to be so specific because you don't want the immune system to attack the rest of your body—that's what arthritis is, an autoimmune disease, it starts eating your own body up. It's a delicate balance: too active and you're in trouble, too passive and you're in trouble. It's a dance."[42]

Knowledge about the shape of molecules and proteins influences the design of drugs used to treat disease—successful "drug docking" depends on getting the fit precisely right. In Whiteley's opinion, some of the most interesting geometry in the last quarter century has been in this field of analyzing and designing molecules to fit prescribed purposes. Many drugs have mirror opposites, or molecular "chirality," which often have a very different effect, sometimes used to treat an entirely different disease. One form of ritalin inhibits attention deficit disorder, the other form is an antidepressant; one form of ketamine is an anesthetic, the other a hallucinogen. Everyday examples of this "stereochemistry" (the study of the three-dimensional arrangement of atoms in molecules) occur in flavors common to the taste buds: limonene's left-handed molecule is found in lemons, whereas the right-handed molecule is found in oranges; and the mirror opposite of aspartame is a bitter substance.[43]

Over years of collaborating with scientists and engineers to find geometric solutions to applied problems, Whiteley accumulated observations that led him to formulate his theories about the crucial importance of "seeing like a geometer," and the phenomenon he calls the "geometry gap." All too often, students coming into these fields do not have the visual geometrical background they need. "The nature of geometry as Coxeter did it is really important in the way geometry rises up in these areas and is unavoidable," said Whiteley. "People have to be able to recognize, 'Yes, we are dealing with such and such a shape from geometry here. Yes, these are the historical roots that are relevant to us.'"[44]

Whiteley recalled that at a conference on zeolites, the Coxeter entities were bandied about with great regularity. Zeolites are naturally occurring crystals, porous like the nooks and crannies of a kitchen sponge, and, to cite one application, are used to refine gasoline. When the petroleum comes out of the ground, some of its molecules are too big, too thick. Refinement filters out the small molecules that will burn faster in your car, and cracks the bigger

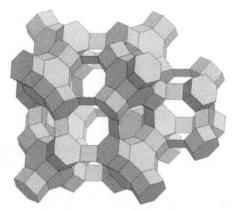

A mineral zeolite, Tschörtnerite, showing its polyhedral building units.
The shape-selective properties of zeolites make them useful in filtering pollutants
from the atmosphere and water.

molecules into small pieces. Zeolites, referred to as molecular sieves, do both
simultaneously. The current problem involves refining heavy oil, which is
made up of even bigger molecules and requires a zeolite with bigger holes. But
such a zeolite does not exist naturally, and material scientists are trying to devise an artificial zeolite that will do the job. "It's a multimillion-dollar industry
and chemists are busy with major funding," said Whiteley. Yet without a
background in geometry, they find themselves scrambling to find solutions,
mining Coxeter's classical geometrical oeuvre for patterns of infinite honeycombs, hyperbolic tilings, crystallographic clues for new zeolite prototypes.[45]

"There are so many layers and hierarchies of geometry," said Whiteley.
"And knowing which kind of geometry you should be looking at may be the
central choice you make on the way to a solution. In the absence of that
knowledge—knowing which geometry to choose—you are lost in a morass
of detail and may never see the solution. Several times in my own work I
have had to trick applied mathematicians into sitting down and looking at a
problem in terms of a simpler geometry, where all of a sudden they would be
able to see what the solution to their problem is."[46]

"Coxeter," he said, "was a bridge that has lived through the period of
the low points of geometry and brought us through to the current day
where there is a rising interest in it again, with these essential problems we
are grappling with. The geometry gap that we're still living with, and in
some ways will struggle with to a greater extent [now that] Coxeter is no
longer around to be the bridge, is the gap between all the pieces of geometry
which practitioners have done all throughout history."[47]

Geometry will continue to be more and more relevant to scientific problems, and the question is, as scientists are working on these problems, will they be able to recognize the geometric content? If scientists do not have the ability to experience these epiphanies themselves, will there be any geometers walking the earth in fifty years to consult, to point them in the right direction? Or will scientists waste precious time reinventing, spinning their wheels in past pitfalls, before they arrive at solutions to urgent problems?

Whiteley's perspective does seem to be penetrating, by osmosis perhaps, the collective consciousness of the mathematics and scientific communities. This is evident at the funding level, where the approval of proposals is increasingly hitched to one criterion: collaborative ventures between mathematicians and scientists. Both Canadian and U.S. grant agencies have allotted considerable chunks of money for collaborations. "They are saying, 'Come to us with a team that includes a mathematician and a biologist—only by having new mathematics come in do we think the problems are going to be solved. Therefore we are going to pay you to collaborate.'" The sad side of this equation, from Whiteley's standpoint, is that scientists in the natural course of things are not learning enough mathematics and geometry to be able to provide the relevant insights themselves.[48]

△ ◻ ◈ ◉ ◈

The geometric knowledge gap also came to bear in astrophysics and chemistry with the long hunt for the shape of a carbon molecule composed of sixty carbon atoms—known as C_{60}—which ultimately earned its discoverers the 1996 Nobel Prize. Previously, chemists had been aware of two forms of carbon—graphite (used in pencil lead), in which the atoms are stacked in hexagonally ordered sheets; and diamond, with atoms lying in a three-dimensional array linked by tetrahedrally oriented bonds. Chemists had speculated there was another form of carbon, having measured its vibrations indicating there should be sixty atoms, but the subsequent laboratory search for the geometric structure was a tricky and protracted endeavor.

Sir Harry Kroto, then a professor of chemistry at Sussex University (now at Florida State University), and his codiscoverers, Robert Curl and Richard Smalley, from Rice University in Texas, worked long and hard to ascertain C_{60}'s shape—the arrangement that would allow sixty carbon atoms to wind themselves together and agglomerate in a hollow cage. The search would have been easier had they known Coxeter's classical geometry, and, in particular, his book *Regular Polytopes*. "We were trying to figure out the rules of engagement," said Kroto, who admitted: "I knew nothing about the geometry."[49]

The earliest murmurings about large carbon molecules in the black clouds of the Milky Way galaxy came in the 1960s. Advances in molecular radioastronomy had indicated these clouds were like archives, storing secrets of the universe—molecules that played a crucial role in the birth of stars and planets were floating around in outer space, making their presence known by emitting radio waves at very specific and identifiable frequencies.[50] Of particular interest to Kroto were long-chain molecules with alternating single and triple carbon bonds. Kroto was one of the early adventurers, continuously upping the ante, gambling that larger and larger extragalactic compounds of carbon might exist, and training his scientific ear to the frequencies these molecules emitted from the depths of constellations, such as Taurus.[51] These ventures initially were considered long shots—in the 1970s, molecules with more than three or four "heavy atoms" (such as carbon, nitrogen, or oxygen) were believed to be rare and undetectable. But Kroto and others played the long odds, and sure enough he found compounds with chains of five and then seven carbon molecules in the interstellar realm. The next quest he set for himself: "Solving the puzzle of how they got there in the first place."[52]

By the early 1980s Sir Harry wondered whether these carbon chains had been blown out of giant red carbon stars, stars known to pump vast quantities of carbon chains into space. He seized on an opportunity in 1985 to simulate the atmospheric chemistry of these stars, working with Curl and Smalley in Texas, fitting the project between higher-priority applied research on semiconductors.[53]

Setting out, Sir Harry was sure the simulation experiment would detect carbon atoms of twenty-four or thirty-two molecules. As trial runs progressed, the team confirmed that conjecture but, in addition, they found something altogether and amazingly different. The measurements indicated that one carbon cluster was particularly strong, peaking consistently at 720 atomic mass units, which corresponded in weight to a species of carbon compounds with sixty atoms. As Sir Harry recounted: "What might this special 'wadge' of carbon be?"[54]

Soon they reached a consensus that C$_{60}$'s structure might be that of a spheroid—a figure not quite perfectly spherical, like a soccer ball. For Sir Harry, this brought back memories of Buckminster Fuller's geodesic dome, which he, like Coxeter, had so closely examined and admired at Expo '67. Since then he had collected a file full of photographs of the dome and cobbled together a homemade model with his children. He now wished he could get his hands on the model, but it was across the ocean at home. Over the next day or two, Smalley and Curl fashioned makeshift models, matching the properties of Fuller's dome—sixty vertices, with twenty hexagonal and

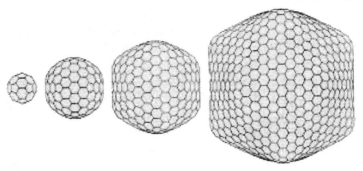

Schematic diagrams of Fullerenes, from left: C_{60}, C_{240}, C_{540}, and C_{960}.

twelve pentagonal faces. That is how they found the structure of C_{60}, or as it is more commonly known, the Buckminsterfullerene.[55]

Sir Harry and the others agreed on the name Buckminsterfullerene because Fuller's work had been the inspiring point of reference.[56] "I knew Buckminster Fuller's work, but I didn't know the Coxeter connection," said Sir Harry. "Certainly, Coxeter's book *Regular Polytopes* would have been helpful." It would have helped find the shape in the first place, reducing the grappling in the dark, and later it would have been useful in producing and confirming the structure of C_{60}. Sir Harry didn't find Coxeter's book until subsequent research, when he became bewitched by even larger carbon compounds. He purchased a molecular model set, and using *Regular Polytopes* as his instruction manual, constructed C_{240}, C_{540}, and C_{6000}—the family of giant Fullerenes.[57]

With the catchy name Buckminsterfullerene, and the Nobel Prize, C_{60} grabbed the limelight outside scientific circles. In December 1991, confusion over the composition of the Buckminsterfullerene and its future played out in Britain's House of Lords.

LORD ERROL OF HALE ASKED HER MAJESTY'S GOVERNMENT: "What steps [are they] taking to encourage the use of Buckminsterfullerene in science and industry?"

THE PARLIAMENTARY UNDER-SECRETARY OF STATE, DEPARTMENT OF TRADE AND INDUSTRY, LORD REAY: "My Lords, the Government have been following with interest the emergence of Buckminsterfullerene and support research currently being undertaken at Sussex University through the Science and Engineering Research Centre. However, it must be left to the judgment of firms whether they wish to pursue research into commercial applications of Buckminsterfullerene and other Fullerenes."

BARONESS SEEAR INTERRUPTED: "My Lords, forgive my ignorance, but can the noble Lord say whether this thing is animal, vegetable or mineral?"

LORD REAY: "My Lords, I am glad the noble Baroness asked that question. I can say that a Buckminsterfullerene is a molecule composed of 60 carbon atoms known to chemists as C_{60}. Those atoms form a closed cage made up of 12 pentagons and 20 hexagons that fit together like the surface of a football."

LORD RENTON: "My Lords, is it the shape of a rugger football or a soccer football?"

LORD REAY: "My Lords, I believe it is the shape of a soccer football. Professor Kroto, whose group played a significant part in the development of Buckminsterfullerenes, described it as bearing the same relationship to a football as a football does to the earth. In other words, it is an extremely small molecule."

LORD CAMPBELL OF ALLOWAY: "My Lords, what does it do?"

LORD REAY: "My Lords, it is thought that it may have several possible uses; for batteries, as a lubricant or as a semi-conductor. All that is speculation. It may turn out to have no uses at all."

EARL RUSSELL: "My Lords, can one say that it does nothing in particular and does it very well?"

LORD REAY: "My Lords, that may well be the case."[58]

Now that technology has caught up to C_{60}, some of its more promising applications are emerging. One employs C_{60} as a superconductor—superconductors are used to make powerful electromagnets, such as MRI machines, as well as digital circuits and microwave filters for cell phone base stations. A nanomedicine company, named C Sixty Inc., was formed to investigate biopharmaceutical applications of Fullerenes. Their current research explores C_{60} as a vehicle for drug delivery. In aqueous solutions, Fullerenes are fairly stable, which suggests they might be useful for carrying precise amounts of medication through the C_{60}'s cagelike structure and depositing their contents at exactly the right site. C Sixty Inc. is focusing its research on the delivery of anesthesia and contrast imaging dyes. And the company is working with the drug company Merck to test Fullerenes as antioxidants, for C_{60} seems to be adept at soaking up cell-damaging free radicals, the by-product of oxygen reacting with other chemicals in the body. The small scale of Fullerenes facilitates their passage through the blood-brain barrier—a defense structure that blocks possibly poisonous molecules in the blood from brain tissue—and thus creates potential for using their antioxidant properties to treat degenerative neurological conditions, such as Alzheimer's and Lou Gehrig's disease. Other medicinal applications include binding C_{60} to antibiotics to target resistant bacteria, or cancer cells such as melanoma, and even AIDS.[59]

Sir Harry keeps himself apprised of C_{60}'s applications, to an extent. He is a fundamental scientist—as Coxeter was a pure mathematician. "People like me," he said, "in a sense spend a lifetime avoiding applications."[60]

Coxeter, however, was a bit more old-school, not hunting applications but appreciating and exploring them when they came along. According to Whiteley, prior to the twentieth century, there was no such chasm dividing pure and applied, with geometers working on a range of scientific problems. Archimedes was an engineer, devising his Archimedean screw (still used for irrigation in developing countries), and Euclid worked on applied problems pertaining to optics. James Clerk Maxwell was a geometer and a physicist, highly regarded as the nineteenth-century scientist who had the greatest impact on twentieth-century science. With the modern emphasis on specialization and fragmentation of the disciplines, the tradition of the hybrid mathematician-scientist, or scientist-mathematician, is slipping away—parallel to the loss of geometry is the loss of connections, bridges from geometry and mathematics to fundamental and applied science.[61]

CHAPTER 11

"COXETERING" WITH M. C. ESCHER
(AND PRAISING OTHER ARTISTS)

For some minutes Alice stood without speaking, looking out in
all directions over the country . . . "I declare it's marked out just
like a large chessboard . . . all over the world—if this is
the world at all."

—LEWIS CARROLL, *THROUGH THE LOOKING-GLASS*

Whereupon the Plumber said in tones of disgust, "I suggest that
we proceed at once to infinity."

—J. L. SYNGE, *KANDELMAN'S KRIM*

The penultimate conference Coxeter attended was "Aspects of Symmetry,"
at the Banff Centre for the Arts in 2001. As he prepared to speak, he flicked
on the overhead projector and slid on his first transparency. Due to a minor
malfunction, at that moment his entire being was bathed in a gigantic projection of colored fish on a Poincaré disc, shrinking smaller and smaller,
seemingly never ceasing as they rounded the vanishing line of the sphere's
horizon. "The topic of my paper," Coxeter began, "is one that has intrigued
me and preoccupied me for nearly five decades. It's about what I call the 'intuitive geometry' of my friend M. C. Escher."[1]

Coxeter found in Escher a soulmate based on their mutual affinity for infinity, and they collaborated after a fashion (their methods being slightly
at odds). The geometer's intersection with Escher demonstrated Coxeter's
humanism in the very broadest sense, for he disregarded the endemic arts-science divide, which was much more prevalent a half century ago than today. "Coxeter wasn't just doing mathematics," observed one of his students,
Ed Barbeau, now an emeritus professor of mathematics at the University of

Toronto. "He saw himself as playing a role in the advancement and preservation of human knowledge. And so all during his career he had contact with people outside the narrow field of mathematics—architects, musicians, artists. He really felt that mathematics was part of the humanities as well as science.[2] And this came through in his courses. He was doing geometry in a way that made it live."[3]

△ ▢ ◈ ⊛ ⊗

Coxeter first set eyes upon Escher's work in September 1954, at the International Congress of Mathematicians (ICM), in Amsterdam. While Coxeter attended lectures, and delivered his own on "Regular Honeycombs in Hyperbolic Space," Rien joined other spouses on a tour of the city. One stop was Amsterdam's recently renovated Stedelijk Museum, where the ICM had sponsored an exhibit featuring Dutch artist M. C. Escher.[4] Alongside the Van Goghs hung Escher's hallmark drawings of reptiles, birds, fish—periodic tilings of the plane in the manner of an interlocking jigsaw puzzle, with each puzzle piece a congruent creature (there were also a few of Escher's carved wooden balls, and his warped, if not impossible, perspectives such as "House of Stairs" or "Relativity").[5] Rien chatted in Dutch with Escher at the exhibit, and she mentioned the similarities between his art and her husband's math; she may not have been the least bit mathematically minded, but she was well tuned to the intellectual desires of her husband. Coxeter noted in his diary a few days after his lecture: "R showed me the Escher drawings and sculptures."[6]

In a glossy exhibit catalogue, Dutch mathematician N. G. de Bruijn prefaced the show by saying: "Mathematicians will not only be fascinated by the geometric motifs. Even more important, perhaps, is the same playfulness which one finds everywhere in mathematics and which accounts for the charm that a great many mathematicians find in their profession."[7] Escher also delivered a lecture at the ICM conference, though Coxeter was unable to attend: "I wanted to come to your lecture," he later apologized by letter, "but unfortunately it clashed with other engagements."[8]

By this stage in his career, Escher was able to pack lecture halls throughout Europe, even though he considered himself a poor public speaker and frequently recycled his speeches. Considering the close timing to a talk he delivered to a physics society one year prior, Escher may well have reused much of the same material yet again at the ICM, beginning with a smattering of information on his materials and how he came to think of them as extra appendages—his wood blocks, copper plates, and lithographic stones, as well as his press, ink, and many types of paper for printing. "Meanwhile, all this technique is merely a means, not an end in itself," Escher said. "The end he [the artist] strives for is something else than a perfectly executed print.

His aim is to depict dreams, ideas or problems in such a way that other people can observe and consider them."[9]

Patrons at Escher exhibits characteristically displayed nothing like "the usual solemnity and silent incomprehension" of visitors at the average modern art exhibit. They laughed out loud in appreciation, or in awe, of Escher's whimsy.[10] Forever uncertain of where his art fit, Escher was not so sure of his public. "The artist's ideal is to produce a crystal-clear reflection of his own self," he once said. "[T]here is little chance that we will succeed in getting through to a large audience, and on the whole we are quite satisfied if we are understood and appreciated by a small number of sensitive, receptive people." Escher espoused a theory about two types of people: "feeling people," that is artists, interested in interpersonal relations, and "thinking people," the scientists, focused on the language of matter, space, the universe and its objective existence. "Fortunately, there is no one who actually has only feeling or thinking properties," he said. "They intermingle like the colors of the rainbow and cannot be sharply divided."[11] Escher nonetheless was troubled that there was a divide at all, the two camps holding in common a suspicion, irritation, and devaluation of the other's work. He wished a better understanding and rapport could exist between the arts and sciences,[12] a theme C. P. Snow would articulate in his famous lament on "The Two Cultures" in 1959.[13]

Before the close of the congress in Amsterdam, Coxeter purchased two Escher prints.[14] Not long after, Coxeter's and Escher's meeting of minds flowered—their own small rebuttal to the isolation of their respective fields. Coxeter developed an intense appreciation of Escher's intuitive sense for geometry, and Escher found Coxeter's mathematical illustrations to be just the catalyst he needed to spark his creativity. "By keenly confronting the enigmas that surround us, and by analyzing the observations that I have made," Escher once remarked, "I ended up in the domain of mathematics. Although I am absolutely innocent of training or knowledge in the exact sciences, I seem to have more in common with mathematicians than with my fellow artists."[15]

△ ▢ ◈ ◉ ◈

Following the Amsterdam meeting, Coxeter wrote to Escher, asking permission to use two of his drawings—his winsome tessellations of beetles and horsemen—to illustrate a paper he was preparing, his 1957 presidential address to the Royal Society of Canada. His paper was titled, "Crystal Symmetry and its Generalizations," as a part of the gathering's larger theme, "A Symposium on Symmetry." Escher granted permission, though he could never have expected how this decision would affect his work in the years to come.

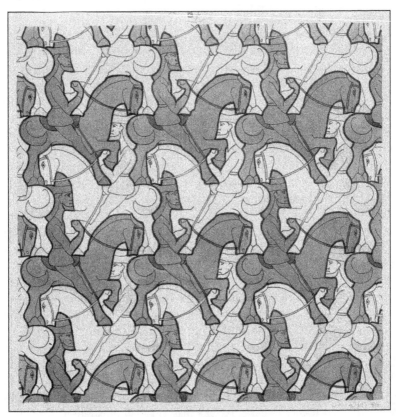

M. C. Escher's regular division drawing of horsemen, 1946.

With his "regular division of the plane" drawings, as Escher called his jig-saw symmetry creations that had caught Coxeter's attention in Amsterdam, the artist had worked out his own mathematical ground rules, his own ad hoc methods for springboarding seamlessly from two dimensions into three, or metamorphosing from the rigid lockstep order of terra firma into a bird's freedom of flight, and then swooping down again into the shimmering fluid-ity of fish in a lake below.[16] At first, Escher admitted, he had no idea how to systematically build his interlocking congruent figures. The process gradu-ally came to him. He studied the literature on the subject, thought through all the possibilities, and formed his own layman's theory. "It remains an ex-tremely absorbing activity," he said, "a real mania to which I have become addicted, and from which I sometimes find it hard to tear myself away."[17]

In geometry, the process of tiling the plane, or tessellation, involves covering a flat surface—such as a floor, or more technically speaking, the

Euclidean plane—with a collection of shapes that fit together without any spaces or overlaps. Among the infinite array of regular polygons, it turns out that only three fit the task of tiling the plane by congruent copies of themselves: the equilateral triangle, the square, and the hexagon* (analogously, in three dimensions, there are the five regular solids but only one of these, the cube, can tile three-dimensional space in a similar manner, with no overlaps or spaces).[18]

Escher had begun tiling as a child, carefully choosing the shape, size, and quantity of cheese pieces to perfectly fill his slice of bread,[19] but his formative influence came later when he visited, and revisited many times, the Alhambra in Spain.[20]† The time-honored art of filling a two-dimensional plane with a repeating pattern of polygonal shapes reached its peak in thirteenth-century Spain with this sprawling citadel-palace built by the Moorish monarchs of Granada—a complex begun by Mahomet Ibn al-Ahmar, founder of the Nasrid dynasty, and continued by his successors in the fourteenth century. The Alhambra is a temple of geometric tilings, etched into every surface, blanketed like a full-enclosure vinyl-wrap advertising campaign. The Moors utilized repetitive patterns of geometric complexity, in tribute to the infinite power of their God. Their unwavering preference for abstract patterns, Coxeter noted, was due to their strict observance of the Second Commandment—"You shall not make for yourself a graven image, or any likeness of anything that is in heaven above, or that is in the earth beneath, or that is in the water under the earth; thou shall not bow down to them or serve them . . ." But Escher, Coxeter said, "being free from the Moors' scruples, makes an ingenious application of these groups by using animal shapes for their fundamental regions."[21]

The practice of tiling the plane, undertaken by Escher and Coxeter for its aesthetic patterning appeal, provides another example of pure geometry finding inadvertent application.[22] Tilings are often found on floors, wallpaper, and brick walls. But, as Coxeter instructed in *Introduction to Geometry*, within his chapter devoted to two-dimensional crystallography, the tessellation of the plane finds relevance in the natural world in the science of crystallography. Coxeter went so far as to say, "Mathematical crystallography provides one of the most important applications of elementary geometry to physics."[23]

The field of crystallography—a multidisciplinary science, occupied by physicists, biologists, chemists, as well as mathematicians—investigates the

---

*An infinite number of nonregular polygons can tile the plane, and similarly, any number of abstract shapes.

†One would expect Coxeter to have visited the Alhambra as well. But he did not; he boycotted Spain, bull-fighting being the bridge too far for his beliefs in animal rights and pacifism.

internal structure of crystals, the geometric patterns of their molecular makeup. The atoms of crystals form a lattice arrangement with uncompromising regularity.[24] For example, the double-helical structure of DNA was deduced from crystallographic data, and the tetrahedrally oriented bonds of a diamond produce the gem's prized hardness.[25]

Crystals, in fact, are classified by their seventeen planar symmetry groups (planar meaning 2-D; in 3-D there are 230 crystallographic space groups), the collection of all motions—translations, rotations, reflections, glide-reflections, screw motions, and rotary reflections—that, when they act on the crystal structure, leave the structure invariant. For that reason, Coxeter groups and Coxeter diagrams make themselves useful in crystallography.[26]

In discussing crystallography in *Introduction to Geometry*, Coxeter used as illustrations the same two Escher drawings—the beetles and the horsemen. When Martin Gardner reviewed Coxeter's book in *Scientific American*—a book Escher referred to as Coxeter's "abracadabra high abstractions . . . Of course, I don't understand one syllable, except for some funny and profound observations at the beginning of each chapter"—Gardner allocated a quarter of the review to Coxeter's two pages featuring Escher. And a colorized version of Escher's regular division bird print was featured on the cover of the magazine.[27] (See appendix 7 for more on crystallography, and Sir Roger Penrose's tiling, once allegedly pirated for toilet paper quilting.)

△ ▢ ◈ ⊗ ✦

By the time Escher met Coxeter, the artist's regular division tessellations—most often variations on birds or fish, but sometimes lizards and butterflies—had become very popular and profitable. He was dazed by the success. It meant the labors of his signature pieces would consume him for the foreseeable future, leaving scant time for new work. After years and years of his mania for devising regular divisions, however, Escher's curiosity had begun to wander.[28] He wanted to break free from the Euclidean plane and portray a more convincing infinity—infinity was obscured by illusion, yet he felt its pull as irresistibly fathomable and approachable.[29] "The flat shape irritates me," Escher said of his regular division drawings. "I feel as if I were shouting to my figures, 'You are too fictitious for me; you just lie there static and frozen together; *do* something, come out of there and show me what you are capable of!' "[30]

When Escher's inventiveness stalled, he tackled the obstacle much as a mathematician does an intractable problem—with obstinacy.[31] Mathematicians pose questions that nag and pester, they keep chipping away at a problem, until the truth, a solution, presents itself (or the enterprise crumbles and proves impossible). Escher plugged away with tireless inquisitiveness, and in that sense he had a mathematician's soul.[32] He logged weeks and months of brainwork before he ever reached the reward of manually cutting the woodblocks, and the rhythmic relief of printing. Escher's son, George, recalled how his father's work ethic dominated the atmosphere of their home:

> A new concept could take months, sometimes years of incubation before it led to a print . . . [h]is moods changed between irritated abstraction and relaxed discussion of some small problem, between restless pacing behind his closed door and sudden announcement that he had found some satisfying solution. During this period of gestation father demanded complete quiet and privacy. The studio door was closed to all visitors, including his family, and locked at night . . . One day the expectant atmosphere in the home would ease. The studio door opened, and we were invited to look at the new design, still on paper . . . In the weeks that followed . . . from the studio emanated light-hearted whistling and the ritual sounds of woodcutting and printing.[33]

Escher at work in his studio.

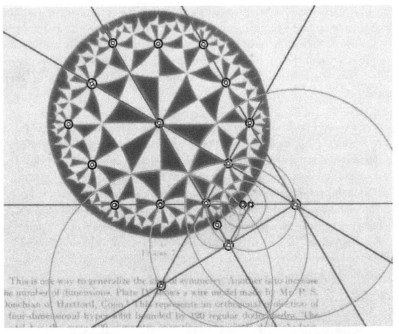

Escher's pencil tracings (enhanced) overtop Coxeter's diagram showing a tiling
of the hyperbolic plane.

After Escher's long hunt to capture a more convincing infinity,[34] one day
in 1958 he swung open his studio door and claimed victory in his battle to escape the flat Euclidean plane. A letter from Coxeter had dropped on his drawing table like a creative bomb—it "gave me quite a shock,"[35] he later told the
geometer. After using Escher's prints in his "Symposium on Symmetry" presidential address, Coxeter had sent Escher a copy of his paper as thanks. When
Escher opened the package, he was proud enough to see the reproduction of
his regular divisions, the beetles and the horsemen. However, when he set eyes
upon the other illustrations—some mathematical figures Coxeter used to
evoke non-Euclidean symmetry, in the hyperbolic plane and on the sphere—
Escher's long-awaited epiphany came to him with a jolt.[36]

Escher immediately set to work trying to glean from Coxeter's paper, "a
method for reducing the size of a pattern from the centre of a circle to the periphery, where the figures get progressively closer and smaller."[37] The only way
Escher could figure out the method was through the intuitive hands-on approach that he used for all his work. Using trial and error, he tinkered until he
hit upon something that worked. He used a compass and traced over the figure

in Coxeter's paper—a symmetrical pattern of black and white triangles filling Poincaré's disk—attempting to decipher the pattern of circle centers whose arcs outlined the triangles. He constructed a geometric scaffolding on Coxeter's original figure, and then graduated to his own reproduction, producing a large drawing of intersecting circles.[38] In a letter to his son, George, Escher raved about the discovery:

> [Coxeter's] hocus-pocus text is no use to me at all, but the picture can probably help me to produce a division of the plane which promises to become an entirely new variation of my series of tessellations. A circular regular tiling, confined on all sides by infinitely small shapes, is really wonderful . . . At the same time, it seems as though I am distancing myself from whatever might be successful with the public. But what can I do when a problem pulls at me so much that I can't leave it alone? It is not as easy as it looks. Try it. Put one (or four) squares of whatever size in the middle of a circle (for instance, separated by two straight lines through the centre) and make them smaller leading outward, something like chess boards. It won't work with only fourfold axes; you have to alternate them with sixfold ones in a most peculiar way, which is normally not possible on a flat surface. The borders are only partly straight lines (only three crossing centre lines) and the rest are all circles. Without Coxeter's model I never would have thought of it.[39]

Soon thereafter, the artist's first woodcut "inspired by the Coxeter system" was finished. Escher called it *Circle Limit I*—"to me it is the most beautiful one that I have made of the 'smaller and smaller' type."[40]* He could not stop gazing at the circular "all-encompassing limit of infinitely small shapes, all so logical and ordered . . . I am anxious to hear the reaction of Mr. 'Cokeseater' himself, to whom I sent a copy."[41]

Escher's creation impressed Coxeter, who wrote back and offered advice on how the pattern could be continued in the same manner (he indicated this with a red dot on Escher's enclosed grid of circles, which he sent back). Coxeter also answered, with great mathematical panache, a question Escher asked as to whether other systems, besides this one, could reach a circle limit. "I say yes, infinitely many!"[42] Coxeter replied. And he elaborated with more of the "hocus-pocus text"[43] that Escher found so useless. "He is so frightfully clever in his answers," Escher scoffed, "and throws symbols around whose meaning I can hardly understand; but fortunately he has added a few drawings . . ."[44] Escher, through sheer artistic grit and tenacity, willed himself toward the result

---

*It is interesting to note that the symmetries of Escher's *Circle Limit* patterns display the same symmetries of the Platonic solids, simply inflating them from spherical to hyperbolic geometry.

he envisioned. "Maybe Coxeter could help me with a single word," he wrote, "but I would prefer to find it myself . . . also because I'm so often at cross-purposes with those theoretical mathematicians, on a variety of points. In addition, it seems to be very difficult for Coxeter to write intelligibly to a layman."[45]

Still, Coxeter was the consummate teacher. He delighted in drawing even amateurs' attention to the most sophisticated subtleties of what they were doing. No matter whom he engaged, and regardless of whether they were struck dumb or disinterested, Coxeter was always determined to share what he was seeing—the mathematically significant insight.[46] "He was always explaining to other people the mathematics of what they did, even when they had no clue that stuff was there," said Doris Schattschneider, an Escher scholar. "He just saw that as his mission in life. He couldn't help himself."[47]

Coxeter no doubt thought he had been helpful in answering Escher's question. In a pattern that was to be repeated by his encounters with other artists, his mathematical "help" often generated flummoxed reaction.[48] But his technical assistance at once forged the artists' self-reliance, giving them a glimmering of what they might be onto, and leaving them to figure things out for themselves. This jibed with the heuristic approach to geometry that Coxeter advocated in the classroom: his belief that the best path to learning was hands-on experience, leading to the thrilling intellectual buzz of a self-made discovery.[49]

△ ◻ ◇ ⊗ ⊛

If ever asked why he did what he did—indeed, why he kept doing it into old age when many of his colleagues had retired—Coxeter delivered a curt retort: "No one asks artists why they do what they do. I'm like any artist. It's just that the obsession that fills my mind is shapes and patterns."[50] Coxeter was artistically inclined in discrete ways. His mother being a painter and his father a sculptor meant that he was more than familiar with the calling. As much as Escher and other artists worked mathematically, Coxeter as a mathematician worked artistically, in his intuitive, visual, tactile methods.

This said, Coxeter was not an aficionado of fine art. He enjoyed a visit to an art gallery, but the artwork on his walls included portraits and landscapes by his mother, a few Escher prints, and varied renderings of polyhedra and polytopes. The latter collection came to include geometrical works by artists and other geometrical amateurs who orbited Coxeter from hither and yon, including the startlingly profound and frequent offerings from George Odom, a resident of the Hudson River Psychiatric Center, in Poughkeepsie, New York.

"I've been solipsistically sealed in my own world here for thirty years, which is what I wanted," said Odom. "The only reason I've stayed here as long as I have is because there is a minimum of small talk here—I hate small talk. Here I could pursue my own interests, do my painting, my mathematical models, my sculpture, with a minimum of socialization. This is the loneliest place on Earth I could find."[51] Odom suffers from flattening bouts of depression, though he tells the story of how he ended up where he is with the rat-a-tat-tat pacing and nonchalant tone of someone very content with his fate. Framing his story in terms of the trajectory that led him to Coxeter, Odom begins at age eleven, when in the early 1950s he visited the Museum of Modern Art and was taken by the polyhedral sculptures of Buckminster Fuller. At seventeen, he dabbled in the gay scene in New York, became disenchanted, and went in search of "something more substantial, something with lasting value." The search led him to Coxeter, who, Odom said, for three decades dating from the 1970s, was one of only four contacts he had with reality and the human race.[52] The other three men Odom handpicked as correspondents, with whom he chronicled the development of his thoughts, were his brother; his psychiatrist Charles W. Socarides, a clinical professor at the Albert Einstein College of Medicine/Montefiore Medical Center in New York (with Socarides, Odom discussed the connection between the seemingly unrelated notions of Freud's Oedipal complex and mathematical idealism); and Father Magnus Wenninger, a mathematician and Benedictine monk at St. Augustine's Monastery and School in Nassau, the Bahamas (as it so happened, Wenninger was another of Coxeter's friends from afar).[53]

Odom withdrew from the world and rented a furnished room in Yorktown Heights where he started reading Western philosophy, Bertrand Russell, and Edna Kramer's book *The Main Stream of Mathematics*. He called Kramer in New York, and she sent him a reading list on polyhedra, indicating the person he wanted to get in touch with was a professor at the University of Toronto. After a suicide attempt, and a stay at New York Hospital, Westchester Division, where the nurses and physicians marveled over Odom—"He talks like a college professor"—Odom finally put down roots at the Hudson River Center, and from there he began his correspondence with Coxeter.[54]

"Coxeter was an incredibly cultivated man," said Odom. "A totally civilized human being." They corresponded on several subjects—on biblical interpretations, philosophy, psychology, and metaphysics, even Odom's flirtations with death, regarding which Coxeter cited an apropos quotation from Lewis Carroll's *Through the Looking Glass*.[55] But mostly their discussions focused on mathematics. With nearly every letter Odom sent a model, and both men segued from their mathematical thoughts—with masterful gravitas—to the dismal state of the world. Said Odom in one letter, "The enclosed two drawings pretty much reduce polyhedral symmetry to bare structure (I'm afraid most people would not think it beautiful—they would prefer the mathematics of the atom bomb)." And on another occasion, Odom said: "I thought you might enjoy having this primary structure I discovered some time ago—I'm convinced that geometry is 'not of this world'—i.e. it is transcendent and aristocratic—even though 'the world' has used it shamefully without any sense of the sublime or gratitude. If mathematics is 'the queen of the sciences' what is the king?" To which Coxeter responded: "Maybe the King of the Sciences is Ecology. I hope you agree that it is shameful for USA to be the only one, among hundreds of countries, not to pledge to save endangered species." They also bonded artistically. A postcard arrived from Odom featuring *The Clown* by Henri Matisse, with a short note: "Many thanks for the magazine and article. You are a wonderful person and I love you madly.* I didn't know your parents were artists. That explains a lot."[56]

Odom often signed his letters "your admiring student." But Coxeter, he said, treated him as an equal. "Thank you for the copy of your historic study—you flatter me as I'm not a mathematician as you well know. I'm an artist but you and I are both interested in order and that leads us both to Beauty . . . Art and Science are quite different in that art involves a lot of phenomena that Science would ignore as 'trivial'—but both are interested in the same end—of Beauty, Truth, Love—that is if the art is really art—and the science is really science . . ." Odom considered Coxeter his mathematical "Other." He was the sounding board against which Odom developed his ideas, and Coxeter was his biggest promoter.[57] Odom discovered a construction for the golden ratio and Coxeter, after recognizing its unexpected and beautifully simple method, formulated it as a problem and sent it in for publication, in Odom's name, in the *American Mathematical Monthly*.[58]

Odom also discovered a compound of ten cubes, which he sent along to Coxeter. Coxeter, once again, was delighted—another remarkably simple and beautiful discovery. It demonstrated that the rotations of the cube into itself exactly correspond to the permutations of four colors. He wrote Odom and promised to use the model in his keynote address to the Mathematics and Art section of the International Congress of Mathematical Education, in Quebec City in August 1992—"giving credit to you, of course."[59]

△ ⬡ ⬦ ⬗ ⬢

Odom's third major discovery was a construction of four hollow interlocking triangles. Again, he sent Coxeter a model.[60] A remarkable likeness of this structure crossed Coxeter's desk a short while later, a gift from another geometrically inclined and artistic soul, English sculptor John Robinson. Upon the recommendation of a friend, Robinson had sent Coxeter a book of his

---

*On another occasion, Odom closed by telling Coxeter: "There are very few people I admire—You are lucky enough to be one of them."

Coxeter with Odom's discovery of four hollow interlocked triangles.

most recent sculptures—*Symbolic Sculpture, The Universe Series.*[61] Coxeter appreciated what he saw: exquisite executions in bronze, wood, and wool tapestry, of many geometrical concepts; the golden rule, the Archimedean spirals, golden spirals, cones, knots, pyramids, triangles, ovoids, Möbius bands, circles, and tangents.[62]

Robinson had begun his career with "representational" pieces—children playing, as well as busts of President Ronald Reagan and Queen Elizabeth II.[63] He described his shift to mathematically inspired works by citing the words of Auguste Rodin: "I have come to know that Geometry is at the very heart of feeling, and that each expression of feeling is made by a movement governed by Geometry. Geometry is everywhere in Nature. This is the Concert of Nature."[64] And he looked to Carl Jung's writings in *Man and His Symbols:* "The Artist is, as it were, not so free in his creative work as he may think he is. If his work is performed in a more or less unconscious way, it is controlled by laws of nature that, on the deepest level, correspond to laws of his psyche, and vice versa."[65]

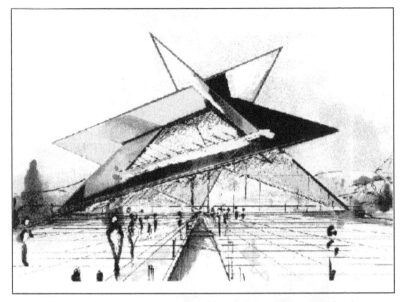

Robinson insisted the inherent rigidity of his *Intuition* sculpture would make a
self-supporting roof for a building he hopes someday will be built; Coxeter believed
such a building would collapse of its own weight.

In particular, Robinson's piece called *Intuition* caught Coxeter's eye. Like
Odom's model, it was an orderly tangle of interlocking hollow triangles, but
Robinson's sculpture had only three. And while Odom's was a brightly
painted cardboard model that fit nicely into the hands, Robinson's was at least
quadruple the size rendered in shiny stainless steel, or a gargantuan 6-foot by
9-foot sculpture in wood.[66] "Your title 'Intuition' was well-chosen," Coxeter
informed Robinson, "because, although I am quite sure that George never saw
any work of yours, there is something uncanny in their several points of re-
semblance: he and you had a similar 'intuition.' "[67] For Robinson, the sculpture represented "a knotted core of stability within the centre of knowledge,
from which comes sparks of originality and invention, often for no apparent
reason. We call these sparks INTUITION. The sparks shoot in all directions,
but come from the core of experience."[68] For Coxeter, these two constructions
represented a piece of mathematical serendipity that deserved a paper.

Coxeter conducted a comparison of the two structures,[69] wrote up his results, and sent them along to Robinson (his article was published in the
Mathematical Intelligencer, with Odom's homemade model on the cover).[70]
Robinson's response to the mathematical hieroglyphs was much like Escher's. "I must confess," he wrote in response, "that I don't understand the

mathematics of your essay, but I do get immense satisfaction in looking at the equations and knowing that they relate directly to something that has 'popped' into my brain . . . The act of 'popppin' is why I called the sculpture INTUITION."[71]

△ ▢ ◈ ⊛ ⊛

Escher, similarly undaunted by Coxeter's "hocus pocus" suggestions, continued with his quest to capture infinity, fiddling with his *Circle Limit I* to correct the shortcomings in his first hyperbolic approach. He called the process of working on his *Circle Limits* "Coxetering"—as in, "Today I finished my first printing of 14 impressions of my new 'Coxetering,' "[72] or "I think that 'Coxeterings' are the best solution to the plane-filling patterns. I should use this exclusively for the time that is left to me, only it is so much more difficult than my earlier puzzles."[73]

Escher was elated with his Coxeterings, though he feared others would not see them as he did. "I have tried to explain the 'smaller and smaller' print to several visitors," he said, "but it is clear that most of them are uninterested in the beauty of this infinite world in an enclosed plane. Most people have no idea what it means. It saddens me because I am busy with the next print, which will be much, much better."[74] Escher's "meaning" was "capturing the infinite." He could wax philosophical about infinity, governed as it was by an indiscernible set of laws:

> There is something in such laws that takes the breath away. They are not discoveries or inventions of the human mind, but exist independently of us. In a moment of clarity, one can at most discover that they are there and take them into account. Long before there were people on the earth, crystals were already growing in the earth's crust. One day or another, a human being first came across such a sparkling morsel of regularity lying on the ground or hit one with his stone tool and it broke off and fell at his feet, and he picked it up and regarded it in his open hand, and he was amazed . . . We never succeed in achieving completely that perfection which haunts the spirit: a perfection we can only see with the inner eye.[75]

Coxeter might have nodded and said "Ah, yes. That's quite nice!" to this articulation of infinity. But by contrast, Coxeter's articulation of infinity was much more literal, a large numerical entity—as demonstrated by his "Up late, we washed ∞ dishes"[76] notation in his diary after a dinner party, or his observation that "Rien made ∞ phone calls."[77] He dismissed out of hand the notion that infinity somehow translated into the hereafter; the only afterlife he believed in was on the molecular level of corporeal decomposition.[78] And thus, Coxeter may not have identified so easily with Escher's more poetic musings:

> Human beings can't imagine that the stream of time could ever come to a halt . . . That is why we clutch at a chimera, an afterlife, a purgatory, a heaven, a hell, a rebirth, or a nirvana, all of which would then be eternal in time and endless in space. . . . Deep, deep infinity! Rest, dreaming removed from the nervous tensions of daily life; sailing over a calm sea, on the bow of a ship, toward a horizon that always recedes; staring at the waves that go by and listening to their monotonous, soft murmuring; dreaming away toward unconsciousness . . . [79]

The correspondence between Escher and Coxeter contains no discussion about Escher's philosophical visions. Instead, the geometer's letters with the artist, and his series of Escher-inspired papers, are regimented by Coxeter's interest in the mathematical substructure of the drawings, quantifying their geometric foundations and marveling at the unorthodox method that brought his friend to such precise results.[80]

Escher finally achieved his ideal and sent Coxeter a print of *Circle Limit III* in May 1960—inscribed "With gratitude, M. C. Escher."[81] "It is intended to be an elaboration of the first black-and-white print which you received before . . . it certainly succeeded better this time," Escher said to Coxeter, continuing in his somewhat awkward English: "The whole area is filled up with series of theoretically an endless number of fish, swimming head-to-tail in the same color. The white curved lines through their bodies accentuate the continuity of every series."[82] Coxeter responded: "The picture is very successful, both interesting and beautiful."[83] But still not picking up on Escher's illiteracy with the mathematical translations, Coxeter sent back three pages of analysis—about the picture's symmetry group, generated by which rotations, the angles of the vertices, and providing references to two of his books, *Regular Polytopes* and *Generators and Relations*.[84] "It's a pity," Escher said to his son, George, "that I can't understand a word of it."[85]

Later that year, Escher had plans to visit his son in Montreal, and Coxeter arranged invitations for him to give two talks in Toronto—one at the Ontario College of Art, and another at the Art Gallery of Ontario. Coxeter made sure that mathematicians attended the latter, and he arranged a reception afterward at his house, where Escher was staying. The day after the talk, Escher sat in on Coxeter's lecture on non-Euclidean geometry at the university, for all the good it did him.[86]

In between visits (the Coxeters stopped in at the Eschers' in Baarn on a number of occasions, once receiving a tour of Escher's studio[87]), Coxeter began his series of Escher papers and lectures, which easily amounted to more

Escher's *Circle Limit III*, 1959.

than a dozen: his interpretation of the artist's work matured, insight by insight.[88] In Coxeter's first analysis of *Circle Limit III*, for example, he stated his opinion that the woodcut would have been still more beautiful without the white arcs artificially dividing each fish into two unequal parts. These arcs, Coxeter said, "have no mathematical significance."[89] Three years later, when he allowed this paper to be reprinted in the book *The World of M. C. Escher*, he had changed his mind and deleted his assertion about the mathematical irrelevance of those white arcs.[90] Subsequently, he wrote two papers celebrating the arc's mathematical virtues. "Of all Escher's pictures with a mathematical background," Coxeter began in a 1979 paper in the arts and sciences journal *Leonardo*,

> the most sophisticated is his 1959 woodcut, *Circle Limit III*, which used four colours in addition to black and white. Queues of fishes of each colour are swimming along white arcs that cut the peripheral circle at a certain angle . . . [We] shall see why all the white arcs "ought" to cut the circumference at the same angle, namely 80° (which they do, with remarkable accuracy). Thus Escher's work, based on his intuition, without any computation, is perfect,

even though his poetic description of it (. . . "perpendicularly from the boundary") was only approximate.[91]

In coming to this conclusion about Escher's exactitude, Coxeter had recruited a student to meticulously measure each of the arcs in *Circle Limit III*. He liked to recall how one of Escher's angles initially seemed to be off by a few degrees, suggesting a sloppy, amateur error. Just to be sure, Coxeter had the arcs double-checked, and it turned out not to be Escher's error, but the student's error. More specifically, what Coxeter discovered from his microscopic analysis of the print was that the arcs were not hyperbolic lines, as he and others assumed, but rather branches of equidistant curves that cut through corresponding vertices of the octagons of the underlying tessellation. "Escher did it by intuition, I did it by trigonometry," Coxeter proclaimed wondrously.[92]

In that revelation, Escher's work brought the geometer to a new mathematical understanding, seeing the hyperbolic plane in a way he hadn't before considered. This new perspective continued to ferment, and nearly two decades later, in 1996, at the age of eighty-nine, Coxeter wrote another paper on those same white arcs, this time providing a more elementary and pleasing proof. In his first paper, Coxeter used a non-Euclidean method (hyperbolic trigonometry) to prove the angle measure of the arcs to the boundary circle, even though the parameters of the model Escher illustrated were, in fact, Euclidean. "I think it bugged Coxeter that his first proof used hyperbolic geometry even though the Poincaré disc model of hyperbolic geometry is really a Euclidean model," said Schattschneider. "It's a Euclidean model for hyperbolic geometry. I think Coxeter wanted to show you could do it with only Euclidean geometry and Euclidean trigonometry—that was his second paper, showing that all you needed was ordinary Euclidean trigonometry. And I think he was really quite proud of that."[93]

Coxeter took infinite pleasure in his string of Escher papers, but he regretted that the artist died (in 1972) before his papers devoted to *Circle Limit III* were published.[94] One might assume that Coxeter's approval hardly would have mattered to Escher. He had suffered no small amount of anxiety over the years, worrying that the artistic merit of his work, the perception of his creativity, was diminished by its scientific content. Escher always considered himself an outcast in the art world, never quite belonging. He noticed a review of his work, with three prints, in the *Saturday Evening Post* in 1961 by the eminent E. H. Gombrich, an art historian at the University of London. Gombrich was critical of his work on aesthetic grounds, but Escher noted hopefully, "He is still moved by it because he goes on and on—more than three columns. Furthermore, the publishers are even paying $140 for the reproduction rights for those three pictures! This could snowball."[95]

Coxeter kept in his Escher file a review by *Globe and Mail* art critic John Bentley Mays. The headline read, HIGHER DOODLING AND OTHER GIMMICKS. The derisive review of an Escher exhibit in 1996 at Canada's National Gallery began by saying, "Ottawa hippies have less than a week to bathe in nostalgia about the really trippy kicks they got by staring at Escher's puzzle-images while stoned on acid, and before they grew up . . . It's enough to make a grown-up art critic weep."[96]

Escher was distressed with the enthusiasm shown for his work by those young hippies who made him so popular at university poster sales. They saw in Escher's work not his sense of wonder at the cosmos: they saw a disorder and chaos that he never intended.[97] But neither was Escher a mathematician. He pondered abstract mathematical concepts and liberated mathematicians' intellectual fantasies—his prints illustrated obvious concepts such as groups, symmetry, and infinity, as well as more the subtle concepts of reflection, duality, recursion, topology, and relativity.[98] "But the sad and frustrating fact remains that these days I'm starting to speak a language which is understood by very few people," Escher wrote, betwixt and between about where he stood. "It makes me feel increasingly lonely. After all, I no longer belong anywhere. The mathematicians may be friendly and interested and give me a fatherly pat on the back, but in the end I am only a bungler to them."[99]

△ ◻ ◈ ◉ ◍

Sculptor John Robinson didn't worry so much about where he fit in. For Coxeter's ninetieth birthday, a sculpture of *Intuition* was installed in the front garden of the Fields Institute, in Toronto (an identical Robinson sculpture graces the garden of the Isaac Newton Institute for Mathematical Sciences at Cambridge). As a more personal birthday gift, Robinson created another sculpture, inspired by the geometer's fondness for mutually tangent circles. With a bit of guidance in a letter from Coxeter—an "easily solved" equation (of course, not so easily for Robinson; he was relieved the letter also provided a solution)—Robinson produced a five-sphere sculpture he called *Firmament*.[100]

"I called the sculpture Firmament," said Robinson, "because it reminds me of the marvelous 19th century working models of the Solar System that fascinated me as a child in the London Science Museum. I didn't understand what I was looking at then, just as I don't understand Donald Coxeter's mathematics now. What I do understand is that the Universe is a Miracle . . . and that this kind of Mathematics is part of the Miracle."[101]

CHAPTER 12

THE COXETERIAN SHAPES OF THE COSMOS[1]

Spirit of the Universe! Whither are we drifting, and when,
where, and how is all this to end?

—J. J. SYLVESTER, 1867

Coxeter passed the milestone of his ninety-fifth birthday demonstrating the same alacrity with which he had caught trains in Cambridge and scraped through traffic in Toronto: as he described (the former) in his diary, "by the skin of my teeth."[2] Another fête was planned in his honor at the Fields Institute, and a few days before the big event—on his actual birthday, February 9, 2002—Asia Ivić Weiss, Coxeter's last PhD student, threw a small luncheon at her downtown Toronto home. She baked Coxeter a birthday cake in the shape of a hedgehog, with almonds stuck in the icing for its coat. It was a lovely, intimate occasion, until the horrible moment during dessert when Coxeter suffered a heart attack and slumped from the table, unconscious. CPR administered by Susan did nothing to resuscitate him; he appeared to have left the land of the living. But then, as Weiss's husband called 911, Coxeter revived himself, choking on an almond from the cake. Despite surviving his heart spell, he was admitted to the hospital for a visit his family truly expected would usher him into the hereafter.[3]

Coxeter lay in hospital as guests arrived in town for the Fields party. Three guests—John Conway; Marc Pelletier, a geometric model-maker from Boulder, Colorado; and geometry lover Glenn Smith from Texas—were staying at his house, empty of the legend himself. No one knew quite how to deal with the apocalyptic scenario.[4] Having traveled all this way to celebrate Coxeter, they were greeted by his daughter, Susan, and relics upon relics of his career: seventy-five years' worth of five-year-at-a-glance pocket diaries, rows of filing cabinets full of A–Z correspondence, Alicia Boole Stott's polytope scrapbook, and his entourage of models worthy of a Smithsonian collection. Smith recalled that it played like a scene from *Zorba the Greek*, when the old woman dies, has no heirs, and her belongings are up for grabs. Coxeter's heirs had no attachment to his dusty memorabilia, so the devoted disciples in attendance were left discussing, questioning, where Coxeter's prized geometric possessions would go.[5]

Coxeter handled the predicament with good-humored equanimity: "I'm not ready to disappear from this life yet."[6] He rallied, with quiet and stoic determination. On the day of the bash, he discharged himself from the hospital with a day pass and arranged to be chauffeured by ambulance to the Fields Institute, Toronto's internationally renowned mathematical think tank. He rolled in by wheelchair. As gallant as ever, he received greetings from colleagues and fans. Festivities began in the atrium, the fireplace lit for the occasion. Atop a wooden double-helix staircase, Coxeter's birthday present, a five-foot geometric wire sculpture, donated by a generous and anonymous benefactor, was unveiled hanging from the ceiling—a welded stainless-steel projection of the four-dimensional hyperdodecahedron into three dimensions, made by Pelletier.* Coxeter gazed upward with pure joy, hardly the dozy appreciation of an old man at death's door; his polytopal muse still filled him with wonder. Staring at the hyperdodecahedral mobile, Coxeter told Conway, standing beside him, about his latest polytope ideas, and his plans for the paper he was due to deliver that summer in Budapest.[7]

△ ▢ ◈ ◉ ✪

Coxeter fielded many compliments after his Budapest talk on four mutually tangent circles. "That really was a proof from 'The Book' you gave this morning," said Karoly Bezdek, the secretary of the conference's program committee, chatting with Coxeter later in the day when Coxeter turned up at Bezdek's afternoon lecture. "The other day I opened your book on *Regular Polytopes* with my son," continued Bezdek, "because he had just seen *Star Wars* and he wanted to draw multi-dimensional shapes." His motivation for this remark may have been science fiction, but Bezdek called Coxeter's work a "prototype for discovery."[8]

Having successfully delivered the opening lecture, Coxeter had the remainder of the conference to relax and enjoy other presentations. The morning after his talk, he awoke energized. He reinstated his exercise regime, abandoned only months before—push-ups (six) with his hands in his slippers

---

*In total, Pelletier has made seven such sculptures, also called the 120-cell—a Waldorf school in Texas is home to one, and Princeton's math department another.

to protect them from the hard floor, sit-ups (maybe a dozen), arm circles backward and forward, and upside-down "air bicycle," hips in his hand and legs pedaling through space. His circulation revved and ready to go, Coxeter sat on the edge of his bed in his pajamas and planned his day, reading over the conference program and selecting the lectures he wanted to attend.[9] Over the course of the week there were several: Weiss's, John Ratcliffe's, Ernest Vinberg's "Hyperbolic Reflections Groups," Igor Rivin's "Geometry of Polyhedra," and "Hyperbolic Coxeter Groups of Large Dimension" by Poland's Tadeusz Januszkiewicz and Jacek Swiatkowski.[10] At all of them, as he sat in the front row, his namesake Coxeter entities were also front and center—cropping up repeatedly in the banter and scribbled on overhead projectors, summoning their omnipresent symmetries.

On the second day of the conference one lecture in particular caught Coxeter's attention: "Visualizing Hyperbolic Geometry." He arrived just in the nick of time and flumped down into his seat. Soon enough, the audience was cocooned in a darkened lecture hall, peering at a multimedia screen, their heads tilted congruently upward. It was the common conference scene, except this time the mathematicians were wearing stereoscopic 3-D glasses,

Coxeter watching Jeff Weeks's presentation in Budapest.

with one red lens and one blue lens fitted to a boxy white cardboard frame. They might well have been watching a 1950s 3-D-craze flick such as *House of Wax*, with Vincent Price and Charles Bronson. But before them stood Jeff Weeks, a freelance geometer from Canton, New York, and the recipient of a 1999 MacArthur fellowship. Weeks is also the author of *The Shape of Space*, a book exploring the possible shapes of the universe. At his Budapest talk, he presented his custom-made computer-generated model of his latest hypothesis: the universe, he conjectured, may be shaped like a dodecahedron. Plato might have been right after all.[11]

Weeks has been wondering about the shape of the universe since he was a teenager. "It's the sandbox we're born into," he said. But his search was always theoretical—using mental images and sketches, and later numbers and equations as his tools. The attraction was pure aesthetics. "The different possible shapes for the universe are so beautiful," he said. "The appeal of geometry is describing real space, and the universe is the ultimate space."

Coxeter's work appealed to Weeks for the same reasons. "Donald Coxeter is obviously a brilliant mathematician. But what really makes him special even among other brilliant mathematicians, is his excellent taste, his sense of beauty and simplicity," said Weeks. "He works in these four-dimensional and curved spaces where it takes a leap of the imagination to get there to begin with, but then when he gets there he doesn't get lost; he doesn't fall into the trap of proving lots of obscure theorems. He proves good theorems. He looks at things in very concrete and very simple ways, sort of the equivalent to the beauty of a honeycomb pattern that bees might make with lots of symmetry, except he's doing something similar in four dimensions or in curved spaces."[12]

The intersection of Weeks and Coxeter's geometric sensibilities was obvious from the subject matter of Weeks's three-dimensional movie: a computer-generated dodecahedral universe, a magnified mass of honeycomb, each side of each cell flashing in multicolor, the Earth spinning inside and the viewer rotating and gliding through the nexus, as if traveling within a spaceship through a background of endless black.

Weeks came upon his "What is the shape of the universe?"[13] project quite out of the blue. He received an e-mail from a cosmologist asking a technical question about the vibrational modes of a spatial manifold. Weeks did not have the answer, but he offered to find it. When he discovered cosmologists were expecting hard data from outer space that would allow them to test the actual shape of the real universe, his usual wide-eyed demeanor widened some more. "This was a dream come true for a theoretical mathematician," he said, "to finally have some data on the way." Before he knew it, he was collaborating with an international coterie of cosmologists.[14]

Weeks's task, through continuous exchange with the cosmologists, was to

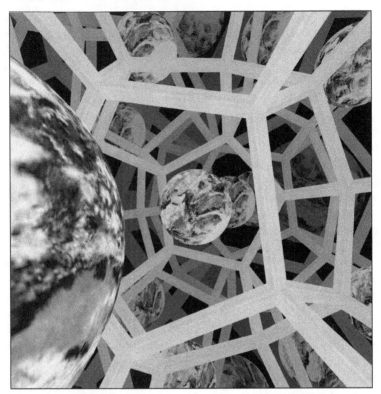

Within the dodecahedral model of the universe, computer-generated by Jeff Weeks.

provide the raw materials. First, he determined which geometrical structures were plausible shapes for the universe by playing around within the existing classes. Currently, there are three classes of shapes considered to be contenders. The standard and favored model is an infinite and flat universe, forever expanding under the pressure of an as yet inexplicable "dark energy." The other two are the hyperbolic model (saddle shaped, with negative curvature, causing parallel lines to eventually diverge), and the spherical model (with positive curvature; such a closed universe eventually stops expanding, then contracts in a "big crunch"). The dodecahedron—a sphere shaved slightly flat to form its twelve faces—fits within the spherical class.[15]

Weeks also calculated how each shape would behave in space, and these first two tasks involved a lot of sitting around, thinking with pen and paper (pens make darker, firmer lines than pencils, he finds), fiddling with the mathematics of the models.[16] Then he worked up formulas to prove his hypotheses. Finally, he devised computer programs to run his formulas. Cosmologists plugged Weeks's geometric formulas into simulations of the physics of the universe. The results would be compared to data expected from NASA's Wilkinson Microwave Anisotropy Probe (WMAP). The WMAP probe was sent up to map cosmic microwave background radiation, the echo of the origin of the universe—the assumed big bang—and provide data about its early history and scale.[17]

One particularly useful indicator of universe topology is the temperature fluctuations of radiation emanating from the big bang. In an article in *Nature* magazine, Weeks and his colleagues explained these fluctuations by comparing them with the sound waves of musical harmonics:

> A musical note is the sum of a fundamental, a second harmonic, a third harmonic, and so on. The relative strengths of the harmonics—the note's spectrum—determines the tone quality, distinguishing, say, a sustained middle C played on a flute from the same note played on a clarinet. Analogously, the temperature map on the microwave sky is the sum of spherical harmonics. The relative strength of the harmonics—the power spectrum—is a signature of the physics and geometry of the universe.[18]

When the WMAP data arrived in February 2003, it only partially confirmed the prevailing infinite-flat model of the universe. All the small and medium-size temperature waves were present as predicted, but the model failed to find any of the broad wavelengths that should exist in such a large and infinite universe. One explanation, said Weeks, is that outer space simply isn't that big and thus could never produce such large waves in the first place. "A violin is never going to play the low notes of a cello because a violin's strings aren't long enough to support such a long sound wave," he said. "It's the same with the universe. Its waves cannot be larger than space itself." Enter the finite-dodecahedron model. The behavior Weeks predicted for a dodecahedral universe matched all the WMAP data. "It was a very pleasant surprise. Our model fit even better than we expected."[19]

The future of dodecahedral space still faces major challenges. The model's calculations of spatial curvature must be compared to more precise data from the Planck Probe, scheduled to launch in 2007. The results of the probe could either fine-tune Weeks's model or refute it entirely. His model must also pass what's called the "circles-in-the-sky" test. If the dodecahedron model is correct, a computer-coded search should be able to detect six pairs of matching circles across the cosmic horizon—echoes from the big bang vibrating against the twelve faces of the dodecahedral universe. So far, no circles have been found.[20]

While they await the final verdict, Weeks and the cosmologists are at once holding out hope and exploring other options, such as the possibility of a universe that is finite in some directions and infinite in others. "You don't want to ignore the other possibilities," said Weeks. "But personally, I'm not ready to declare the circles missing."[21]

The dodecahedron model, if it holds up, has implications for quantum mechanics and theories about the big bang. It could affect an exquisitely more insightful understanding of the blinking night sky, and crack open a sliver of potential for traveling into its furthest depths. "Hypothetically," said Weeks, "if you head off into a dodecahedral universe you would travel in a straight line and come back to the starting point. But it would take a long, long time." The dodecahedron model will also open a new conundrum of a question: If the universe is finite, what is beyond? "Nothing," said Weeks. "But it is a very profound nothing. The best way to answer that question is to make the question go away."[22] And with some elaborate epistemological reasoning, he can.

At the end of the Budapest presentation, enamored by Weeks's dodecahedral universe, Coxeter stood from his seat, made his way toward him, and shook his hand.

"Very nice!" Coxeter said.

"Here's the man responsible for a lot of the images you saw," said Weeks (he was wanting to give Coxeter at least as much credit as he deserved, and, out of reverence to the legend before him, perhaps even a little more). The audience lingered while Weeks replayed his movie. Coxeter left behind his cane and held his glasses to his face with both hands, moving closer to the video screen to get an amplified 3-D view. "It's quite nice," he said. "It looks like you're inside some sort of network of lines and polygons. Quite impressive!" He'd seen that sort of thing before, the honeycomb tiling of space. But he thought it was quite something to see these familiar shapes alive in such dynamic form—a polytopal universe that he could not only delve into with research but almost literally dive into himself.[23]

△ ▢ ◇ ⊛ ⊗

A few months after Weeks's presentation, the *New York Times* ran an article in its science pages exploring the possibility of an infinite flat model for the "macroscopic dimension" of the universe. "Very curious. Very surprising," Coxeter said. "I would have thought otherwise. I'd expect it to be like the surface of a hypersphere, elliptic in the geometric sense."[24] The shape of the universe as a whole is still pretty much anybody's (or any cosmologist's) guess. But a related realm of study is the interior structural topology within the universe, or the "microscopic dimension." Coxeter's work plays out here as well. Coxeter and John Petrie's discoveries of the regular skew polyhedra in 1926 resurfaced more than a half-century later to find application in astronomy. "The astronomy application [of the regular skew polyhedra] is a nice punch line!" said J. Richard Gott III, the astrophysicist at Princeton responsible. "I think the regular skew polyhedra have been somewhat overlooked . . . not as well publicized as the regular polytopes."[25]

Gott became acquainted with the regular skew polyhedra almost forty years after Coxeter and Petrie made their discoveries. As an eighteen-year-old student at Mayme S. Waggener High School, in Louisville, Kentucky, Gott rediscovered the same figures. "The first one I found was hexagons-four-around-a-point," he said. "I noticed four hexagons could join in a saddle-shaped surface and this could be continued to make a repeating sponge-like structure." He did a project on his discovery, winning him first place in mathematics at the National Science Fair–International in 1965. As an undergraduate at Harvard he wrote up his findings and sent them to the *American Mathematical Monthly*. "I don't know who the referee was—but I've always fancied that it might have been Coxeter himself!" The referee informed Gott—quite to his surprise—that this class of figures had been discovered earlier, and that Coxeter later proved these were the only three regular such figures by his criteria. Gott, however, had discovered seven, employing less restrictive criteria.[26] The referee accepted the additional findings and Gott's first scientific paper was published on the topic in 1967, titled "Pseudopolyhedrons."[27]

Pseudopolyhedrons were reprised in Gott's later investigations on the cosmos when he became an astrophysics professor at Princeton. In 1986, he was investigating the topology of the structure within the universe. "At that time there was a debate over the topology," he said. "In one model there were isolated clusters of galaxies in a low density background (a meatball topology), while in the other model there were isolated voids with a honeycomb structure (a Swiss-cheese topology), galaxies being located on walls surrounding isolated voids." From his high-school work, he knew there was a third possibility—a sponge-like topology, since that was the topology of the regular skew polyhedra. "I realized that this had to be the correct answer for the universe, because the theory of inflation predicts that the clustering pattern of galaxies we see in the universe today should have originated from random quantum fluctuations in the early universe. Such random fluctuations have the property that positive and negative fluctuations are equivalent . . . Therefore, the topology, the shape of the high-density regions must initially be the same as that of the low-density regions. This was not the case for the meatball topology or the Swiss-cheese topology.[28]

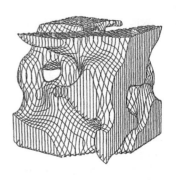

J.R. Gott's spongy universe.

Gott, however, knew that a sponge's structure was the same on the interior and exterior, since this was true of the regular skew polyhedra. A marine sponge is permeated by a series of tunnels, allowing water to trickle through. If you pour concrete into those passageways, let it set and harden, and then dissolve the marine sponge with acid, you are left with a concrete sponge. "This is the pattern of clustering galaxies we can observe," he said, "In other words, what a number of studies have now confirmed is that we see a sponge-like pattern of galaxy clustering, with great clusters of galaxies being connected by filaments, and voids being connected by tunnels to make a giant sponge." Gott published this theory in the *Astrophysical Journal*—titled "The Spongelike Topology of Large Scale Structure in the Universe"—complete with pictures of the Coxeter-Petrie regular skew polyhedra to illustrate the argument. The *New York Times* picked up the story and ran it on the front page: RETHINKING CLUMPS AND VOIDS IN THE UNIVERSE.[29]

△ ▢ ◈ ◉ ⊛

In coming to grips with the omnipotence of geometry, a statement by Brian Greene, a superstring physicist at Columbia University in New York,* stands out: "There is perhaps no better way to prepare for the scientific breakthroughs of tomorrow than to learn the language of geometry."[30]

Implicitly Greene was referring to the long-standing conundrum of modern physics with which Einstein wrestled in the latter half of his life: the search for a grand unified theory, a single theory that can explain the fundamental physical laws of the universe, on the big scale of our galaxy and all

---

*Greene is the author of the Pulitzer Prize–nominated book *The Elegant Universe*, and its sequel, *The Fabric of the Cosmos*.

the other galaxies, and the small scale of a nano-size speck of nothing. Such a theory seeks to unite Einstein's general theory of relativity, explaining the large-scale properties of the universe, with quantum theory, explaining matter and energy on an atomic and subatomic level. The trouble with these two theories of existence, currently held true, is that they are incompatible—they cannot both be correct (quantum theory, which describes the behavior of elementary particles assumes gravity is negligible, whereas the theory of general relativity, stating that gravity equals space-time geometry, holds that quantum mechanics is not needed in the description of the laws of nature—the mathematics do not mesh).[31]

String theory replaces particles with strings, open-ended or closed as loops, and in doing so resolves the incompatibility issues between quantum mechanics and general relativity (it isn't obvious how this resolution occurs, but it is interesting to note that the purpose of uniting these theories is all aesthetics—it would be simpler, more neat and tidy, if the laws of the universe could be reduced to one formulation, if the component parts were synthesized). The strings are so tightly curled into minuscule multiple dimensions—ten or eleven dimensions, or even twenty-six, which are constrained by precise symmetries—that they are invisible to the naked eye and even the eye armed with the best scientific technology.[32]

String theory has become more complicated since it was introduced a few decades ago, with a number of variations on the original theory. These theories have been brought together in a grand multi-universe plot, known as M-theory. Edward Witten, based at Princeton's Institute for Advanced Study, is the mathematical physicist behind M-theory (some call him the "pope of strings"). "M stands for magic, mystery, or matrix," Witten said. "I think they speak for themselves, except possibly for matrix; matrices are used in an approach to understanding M-theory. In addition, our present understanding of M-theory is murky."[33] That is to say superstrings and M-theory are still just theoretical children, joyous yet chaotic puzzles.

Coxeter was peripherally aware of string theory. It came to mind one day when discussing *Through the Looking-Glass* by Lewis Carroll.* This was one of Coxeter's favorite books due to its logically nonsensical nature. He especially liked the "Jabberwocky" passage and would say that word—"Jabberr-wOckAy!"—with such enjoyment. He could recite entire stanzas with the same dramatic intonation. Coxeter often dipped into his ratty copy of *Alice,* and after many readings he hadn't tired of it at all. "It's like reading

---

*Lewis Carroll was the pen name for mathematician C. L. Dodgson. He wrote mathematics books under his own name, but invented his pseudonym by translating his first two names, Charles Lutwidge, into Latin, producing Carolus Lodovicus, which he then Anglicised and reversed in order.

about a part of mathematics that you know is beautiful," he explained, "but that you don't quite understand. Like string theory. That's as much a mystery to me as it is to anyone else who can't make head nor tails of the eleventh or sixteenth dimension."[34]

Therein, unwittingly, he was onto something. The enduring problem with string theory is that string theorists themselves can't even explain it. "Don't ask me to explain what I just said," demurred Stanford string theorist Lenny Susskind, speaking off the top of his head at the Strings05 conference in Toronto. Among some string theorists, the murkiness seems to be causing more consternation than joy. String theory may hold promise to unify all the forces of nature, but just as easily the enterprise may prove to be so grandiose and esoteric that it is beyond human intelligence. Participants at Strings05 gathered one evening for a session pondering when "The Next Superstring Revolution" would occur. Eight panelists, speaking in alphabetical order, professed to their embarrassment that they had no idea when or how the next breakthrough would come.[35]

Witten rose from the audience during the Q&A session and offered a "cautionary tale about predicting the future."[36] Even if string theory does not turn out to be the be-all-and-end-all model of the physical universe, he expects the theory and all its infinitesimally small dimensions will ultimately evolve into a new branch of geometry. "[S]tring theory does appear to contain a lot of rich geometrical ideas, so far not so well understood, which I believe will have a lot of influence in mathematics over time," he said. "One reason I think so is that the little pieces that are so far discovered have already had a considerable influence."[37]

One little piece of the puzzle that has helped decipher the mysteries of string theory is mirror symmetry. Brian Greene and Ronen Plesser, associate professor of mathematics and physics at Duke University, discovered the mirror symmetry of string theory and in so doing revealed how elements previously thought to be totally unrelated were in fact intimately interconnected.[38] As a result, a horrifically complicated calculation, previously considered impossible even for the best mathematicians, became breathtakingly easy in a mirror-opposite space. They called their discovery "mirror manifolds." And, with the impasse broken, subsequent calculations proceeded with "relative ease." Greene elaborated in *The Elegant Universe* with a tangible analogy:

> It's somewhat as if someone requires you to count exactly the number of oranges that are haphazardly jumbled together in an enormous bin, some 50 feet on each side and 10 feet deep. You start to count them one by one, but soon realize that the task is just too laborious. Luckily, though, a friend comes along

who was present when the oranges were delivered. He tells you that they arrived neatly packed in smaller boxes (one of which he just happens to be holding) that when stacked were 20 boxes long, by 20 boxes deep, by 20 boxes high. You quickly calculate that they arrived in 8,000 boxes, and that all you need to do is figure out how many oranges are packed in each. This you easily do by borrowing your friend's box and filling it with oranges, allowing you to finish your huge counting task with almost no effort.[39]

Or consider Greene's analogy with Coxeterian tools. Imagine you are peering into the same pile of oranges, but then realize the pile isn't as big as you thought. You blink, tilt your head for a new perspective, readjust your focus, and you see that it is actually a smaller pile of oranges replicated again and again in an almost invisible mirrored box, like one of Coxeter's finite kaleidoscopes. In order to make the calculation, then, all you have to do is count the oranges in the immediate box, and then multiply that number by the mirrored reflections to get your total.

"Mirror symmetry is a nonclassical relation between spacetimes," said Witten, "that are quite different in Einstein's General Relativity but turn out to be equivalent in string theory. This enables string theorists to do calculations that would otherwise be out of reach, and has turned out to have surprisingly interesting applications in geometry."[40]

The mirror symmetry breakthrough was a boost for string theory and mathematics alike. But rumblings are that if a bigger breakthrough doesn't occur soon, and in the form of streams of empirical evidence, string theory will at best be deemed a branch of mathematics or philosophy, but not part of physics.* Data is desperately needed to confirm or refute the fundamental string theory hypothesis: Within these microscopic eleven dimensions resides a new species of subatomic particles, known as supersymmetric (SUSY) particles, or "sparticles." Physicists are trying to detect traces of these sparticles at such places as Fermilab, in Batavia, Illinois, and the CERN, in Geneva—the latter, home of the world's largest particle accelerator and the center of the universe for determining the content of the universe in its first trillionth of a second. And an even more powerful particle accelerator is being built at CERN—the Large Hadron Collider (LHC), which, when it is switched on in 2007, will probe even deeper into matter and smash nuclei together with even more collision energy. String theorists are crossing their fingers; they

---

*In the meantime, while the string theorists await the next breakthrough, Amanda Peet, from the University of Toronto, proposed string theory become a "faith-based initiative." Whereas ever-the-jester Susskind said, "There's nothing to do except hope the Bush administration will keep paying us."

hope the LHC will prove the existence of string theory's supersymmetric particles. The hunt for sparticles, like the hunt for C_{60}, is on.[41]

This raises a question: it sounds far-fetched, but could Coxeter's templates of symmetries for shapes in multiple dimensions possibly unlock part of the puzzle of the supersymmetric unified theory of everything? After conducting a survey of the existing literature—with Google, searching "Coxeter and M-theory"—the answer is yes. The search revealed a paper by Marc Henneaux, a specialist in black holes, at the Free University, Brussels, and director of the Service de Physique Théorique et Mathématique. It is titled, in big bold letters,

PLATONIC SOLIDS AND EINSTEIN THEORY OF GRAVITY: UNEXPECTED CONNECTIONS[42]

It turned out it wasn't so much scholarly paper as it was a PowerPoint presentation intended for a general academic audience (though Henneaux has written scholarly papers on this subject as well). His presentation read, in part:

GRAVITATION = GEOMETRY

Einstein revolution: gravity is spacetime geometry

General relativity has proved to be remarkably successful . . . but there are . . .

PROBLEMS

General relativity + Quantum Mechanics = Inconsistencies
(e.g., infinite probabilities!)

Synthesis of both should shed light on the first moments of universe
(« big bang »), on black holes, and on the problem
of why the vacuum energy is so small.

Towards a solution: string (M-)theory?

SYMMETRIES: THE KEY?

Symmetry = invariance of the laws of physics under certain changes in the
point of view

What are the underlying symmetries of M-theory?
Platonic solids: the golden gate to symmetry

Platonic solids, of course, are those "toys" Coxeter played around with so often in his work. And sure enough, a little further into the presentation, Coxeter's work was cited:

Coxeter groups may thus signal a much bigger symmetry[43]

"Coxeter's work does make an unexpected appearance in Einstein's theory of gravity," confirmed Henneaux. He and his collaborators[44] found that Coxeter groups "popped out" when studying particular solutions of Einstein's equations.* "One can show that the generic solution of Einstein's equations contains singularities, places where some fields become infinite. We studied the Einstein equation in the vicinity of such singularities—in the cosmological context you have the big bang singularity and you want to understand how the fields behave as you approach the big bang."[45]

By analogy, Henneaux and his colleagues showed that the dynamics of the gravitational equation are the same as the dynamics of a billiard ball moving on a hyperbolic billiard table. "You would have a billiard ball moving in some portion of hyperbolic space in higher dimensions," he said, "and when the ball hits the border of the billiard table you would have a reflection. We can show that all the possible reflections when the ball hits the walls would generate a Coxeter group—the whole thing conspired to give you a Coxeter group." The team came upon this finding somewhat by accident. They computed the trajectory of the billiard balls, and found that it worked out to a very nice geometry. And they computed the angles between the walls of the hyperbolic billiard table, and found that the angles were pi divided by an integer—pi over three, or pi over two—just as in Coxeter's kaleidoscopes, with angles separating mirrors being fractions of pi.[46]

This research helps the hunt for the fundamental formulation of string theory because the existence of the symmetries of a Coxeter group might be an indication that the theory itself has a huge symmetry group. This in turn indicates there are beautiful and elegant structures underlying string theory, and lends credence to the theory's potential. "We believe that exhibiting, explicitly, this huge symmetry will help in understanding string theory and the

---

*These are Einstein's equations pertaining to gravity, not the famous $E = mc^2$. The Coxeter groups that pop up in these equations—that is, the Coxeter groups that here prove to be useful mathematical tools—are those informing the mysterious infinite continuous symmetries that seem to underpin all existence (specifically, the hyperbolic Coxeter groups: A{1}++, pure gravity; and E{10}, 11-dimensional supergravity).

generalization of string theory known as M-theory," said Henneaux. "[P]re-sumably if we understand better the symmetries, we would be able to make a step toward a deeper formulation of the theory. And we believe that maybe this is a way to attack the problem."[47]

The pope of strings also gives his blessing to Coxeter groups and their potential role in unraveling the puzzle. "Maybe some unfamiliar infinite-dimensional groups like E{10} will be important in string theory—at times people have made this conjecture—and if so maybe it will be helpful to un-derstand the Coxeter groups," said Witten.[48]

"String theory is one place where there are a vast amount of links to dif-ferent parts of mathematics," said Sir Michael Atiyah. "That's what makes it one of the more exciting things going at the present time . . . And if it hadn't been for that, one might have worried, because there are no experimental re-sults for string theory, and if there were no connections to mathematics, the critics might just say, 'Well, these guys are doing crossword puzzles.' "[49]

Sir Michael, whose realm of modern geometry also overlaps with the physics of string theory, has been instrumental in bringing the two factions together. "Over the last 25 years, I've been learning some physics, talking to physicists and helping to bridge the gap. It's a difficult gap to bridge," he said, "because traditional mathematicians tend to view with a lot of suspi-cion the work of physicists. The physicists' work is full of hypothetical ideas for which there are no proofs . . . Physicists' intuition is based on the world of experiments," Sir Michael explained. "They think conceptually when they think about physics because it's about electrons and particles and forces and fields—the electric field and the gravitational force and the magnetic field. By definition these are meant to be forces that you can imagine. They have ideas in their mind when they talk about the things they are discussing, and that is very similar to the way a geometer thinks—in terms of space and pictures and so on . . . When they get to the stage of converting intuition into formal arguments, to writing papers, then physicists tend to resort to al-gebraic formulae . . . Then a lot of their conceptualizations are converted to formulae and they write down the equations to say how the particles and forces behave. So there are two sides to the story in physics, just as there are in mathematics."[50]

But when the physicists started thinking about string theory, the kind of geometry they needed wasn't at the intellectual ready. "The geometrical thinking did not come very easily to physicists when they started getting in-volved with string theory," said Sir Michael. "They can think conceptually in terms of physical terms, particles and fields and forces, but when it came time to translating these into mathematical geometrical concepts, they weren't too sure of their ground. Because the old fashioned simple straight

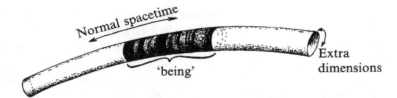

The "hosepipe" model of string theory, drawn by Sir Roger Penrose, shows how strings might be curled into small extra dimensions. A "being" who inhabits this world straddles the extra dimensions and is therefore unaware they exist.

line geometry of Euclid wasn't appropriate, and they hadn't learned the new modern curved and complex geometry."[51]

Subsequently, after physicists pulled at modern mathematics for what they needed, the physics of string theory, in turn, spurred on modern geometry, producing a complicated set of links that physicists predicted, but which mathematicians would not have otherwise considered. "When you have unexpected links, they turn out to involve lots and lots of things crisscrossing each other," said Sir Michael. "Mathematics is a very intricate pattern of beautiful designs. You discover unexpected links and you increase the pattern, you explore and you draw more analogies and you draw more pictures. It becomes one big mosaic. And it goes on and on and on."[52]

△ ▢ ◈ ◉ ◍

By the end of the Budapest conference, continuing on and on with geometry was, for Coxeter, top of mind. As he roamed the halls of Hungary's Academy of Sciences, visiting one lecture room after another, Coxeter noticed the many mirrors in the building—lining the elevator, on the landings of grand sweeping marble staircases, in hallway alcoves, and in the library. As he wandered by yet another mirror, he recited a passage from G. K. Chesterton's *Manalive*, which had become his mantra of late: "There is something pleasing to a mystic in such a land of mirrors," he said. "For a mystic holds that two worlds are better than one. In the highest sense, indeed, all thought is reflection."[53]

Coxeter wasn't at all mystical about his land of mirrors; he was driven, obsessed, with a steely passion to experience everything he could about symmetry and polytopes. And he wanted to keep at it. One evening in Budapest, he received a call in his hotel room from Gyorgy Darvas, a local geometer and director of Symmetrion. Darvas was off to another mathematics gathering a couple of hours away, but he'd heard Coxeter would be in town and he did not want to miss the chance to meet him. Coxeter slowly made his way downstairs to the lobby for an impromptu meeting with Darvas. "I've wanted to meet you for 50 years," Darvas said. He presented Coxeter with two copies of a journal he edits, *Symmetry: Culture and Science*, published by the Symmetry Society. Darvas added that he would be happy to publish anything Coxeter sent his way. And he gave Coxeter an invitation to a conference, Symmetry Festival 2003 *, in Budapest the following summer.[54]

"Oh, how lovely!" Coxeter said, adding with a guffaw: "But I won't be alive in 2003!" Darvas gently pointed out that 2003 was only one year away.

"2-0-0-3," said Coxeter slowly. "Those numbers look so odd."[55]

By the last day of the conference, the rejuvenated Coxeter wasn't feeling his mortality so much and had reconsidered Darvas's invitation for a return visit. Departing for the airport, bidding his good-byes, he told colleagues, fans, and friends gathered for his send-off that he would see them back in Budapest the next year.[56]

---

*At Symmetry Festival 2003, Hungary's Sándor Kabai and Szaniszló Bérczi were due to give a talk exploring the usefulness of polyhedra in "Space Stations Construction and Modelling."

PART III

Aftermath

CHAPTER 13

FULL CIRCLE SYMMETRY

And summer's lease hath all too short a date.

—SHAKESPEARE, SONNET 18

"We have now reached the end of our journey,"[1] wrote Coxeter, as if coming to the end of a fable in the epilogue to *Regular Polytopes*—a book that had become, some mathematicians reckoned, the most quoted geometry text of the century.

After his return from Budapest, Coxeter acquiesced to the reality that he was nearing his end. He was more gaunt and ascetic looking than usual, his mind increasingly betrayed by his body. After one particularly calamitous day, Coxeter invoked a limerick from nonsense poet Edward Lear, lamenting: "I feel like the Old Man from Thermopylae, who never did anything properly." Mathematicians around the world had long since put Coxeter on their "longevity watch."[2] And nearly all his contemporaries had predeceased him, which gave him the minor advantage of knowing, somewhat, what to expect.[3]

Coxeter and Pólya, his friend from Princeton, had kept in touch into old age. In 1978, ninety-year-old Pólya wrote, "Dear Coxeter, You know everything about geometry, elementary or otherwise, *n* or more dimensional . . . I have not seen you for a long time, but perhaps we can meet sometime, somewhere, somehow . . ." They did not, but they continued to write. Coxeter sent some of his latest papers and received Pólya's final return letter in 1985: "I am close to 100 years, and for my age not too badly off." Pólya died later that year, at ninety-seven.[4]

Coxeter had closely marked the birthdays of many of his friends and colleagues. In 1992, on John L. Synge's ninety-sixth, he wrote a mathematically laden letter of congratulations:

Dear Jack,

We send you our warm congratulations on your 96th birthday. It is splendid to have lived so long. And 96 is a very fine number, being the number of vertices of the regular complex polygon 4 {4} 3. It has been known only for a few months that these 96 points have the symmetrical coordinates . . . this is one of the results in the new paper I am writing . . .

On Synge's ninety-seventh, Coxeter commented on his success at having reached "a nice prime number!" And for his ninety-eighth, Coxeter penned his best wishes on Synge's actual birth date, only seven days before his death in March 1995. Synge, in his letters back to Coxeter, described his decline with fine detail: "The pulse-rate at rest is normally 48 per min, but mine lies between 30 and that. I have a great lack of energy." Just before he died, Synge remarked with characteristic wit: "I'd be delighted to wake up in the morning and find I'd died in the night." His last days were in a nursing home, and he told his daughter, the much-decorated mathematician Cathleen Synge Morawetz, "Be sure to tell Coxeter to move into a nursing home. It's the best thing."[5]

Despite the recommendation, Coxeter intended to avoid a nursing home at all cost. When he was in his nineties, he nursed Rien through the horrors of Alzheimer's, bathing, dressing, and feeding her with superhuman compassion and devotion. Coxeter moved Rien into a home only in her final months, after a fall had broken her hip and her will to live. Over the protracted period of his wife's illness, Coxeter allowed himself little respite. A nurse came to the house only one day a week, a day Coxeter used for trips into his office at the university campus to catch up on mail.* He received a much-treasured piece of mail from an eleven-year-old polytope prodigy, Clifford Zvengrowski, living in Calgary, Alberta. Enclosing a photograph of a skeletal polytope model he had constructed (nearly as tall as the boy himself), Clifford wrote to say: "Dear Professor Coxeter, My three most favorite people that live or lived are Archimedes, Brahms, and yourself. Most kids I know would like to get an autograph of some rock singer or baseball player, yet I am hoping for yours."[6]

---

*One piece of mail he tended to on August 10, 1997, pertained to a recent royalties statement. "Dear Ms. Falaster, Thanks for the royalty statement. I regret that only 13 copies of TGE [Twelve Geometric Essays] were sold. The trouble seems to be that few people have heard of the book, although it contains some of my most original articles, frequently cited by other authors. Could you do something about promotion? Tell the public that I have just been awarded, by the Royal Society of London, this year's 'Sylvester Medal,' which is the mathematical equivalent of a Nobel Prize. Yours sincerely, Donald Coxeter."

Coxeter also managed to keep on top of the world's political injustices. In 1997, he gladly hand-delivered a petition to the president of the University of Toronto, signed by one hundred professors, protesting plans to bestow an honorary degree on former president George H. W. Bush.[7] The event went ahead as planned, but scores of faculty "robed up" for the convocation ceremony anyway, and then walked out en masse during the proceedings. "We chose a point when the encomium to Bush was waxing odious," recalled mathematics professor Chandler Davis. "The ceremony was in the Great Hall of Hart House, and by pre-arrangement the exiting profs marched straight out of the building by the south door where over a thousand supporters had gathered to cheer us." More than half the total of sixty attending faculty exited. Coxeter, then age ninety, preferred to simply stay home and boycott.[8] He once said he never thought there could be anything worse than President Bush until a second President Bush came along.

When Rien died in 1999, Coxeter enjoyed a second wind of sorts. He threw around his weight a bit when drafting his will. Initially, he had decided to donate his Rosedale house (by then valued at $1.5 million) to the University of Toronto. He reconsidered his offer, however, when he perceived a change in the university's pedagogical approach—shifting from "learning for its own sake" to "learning for opportunity." Coxeter thought it particularly egregious that research chairs were endowed by corporations. He rescinded the offer to the university and stipulated instead that the proceeds from the sale of the house be divided equally between the Fields Institute and his alma mater Trinity College, Cambridge.[9]

With his daughter Susan as his helpmate, he once again filled his itinerary with international engagements, such as a trip to Stockholm for the Symmetry 2000 conference. There he spoke on another of his signature subjects, the "Rhombic Triacontahedron," and he planned to use a new type of model invented by his friend, geometer and geophysicist Michael Longuet-Higgins, at the Scripps Institution of Oceanography, UCSD. Called RHOMBO, the model's component parts are six-faced solid blocks that click together by a patented system of magnets. Longuet-Higgins had sent Coxeter samples as he developed the model, and Coxeter became an enthusiastic promoter, saying: "Your RHOMBO blocks are very much more than a toy!" And on another occasion he informed Longuet-Higgins: "I took both balls to the NATO meeting on Polytopes, where many participants enjoyed pulling them apart and re-assembling them." At Symmetry 2000, in Stockholm, Coxeter delivered his lecture, using the blocks, with Longuet-Higgins sitting in the audience. In the middle of his presentation, Coxeter fumbled the model and dropped it on the floor, causing it to fly into pieces. Longuet-Higgins rushed to the rescue and reassembled it. "Did Donald drop the blocks on purpose? I believe he did," said Longuet-Higgins, "so as to give me a chance to demonstrate my RHOMBO. That would have been just like Donald."[10]

Coxeter also made one last trip to his native Cambridge, where he was anointed an honorary fellow of Trinity College (the Master of Trinity, Amartya Sen, wondered aloud why it had taken the College so long to do so[11]). He managed back-to-back conferences in Vancouver and in Madison, Wisconsin, even though he was immobilized and on painkillers with a cracked pelvis. He made two summer appearances at the Canada/USA Mathcamp for math whiz kids; went to Banff for his talk on Escher; and finally visited Budapest.[12]

And then there was a journey of another sort, the crowning celebration of his career, his ninety-fifth-birthday party at the Fields. After the dedication ceremony with the unveiling of the 120-cell, John Conway delivered a touching yet humorous ode to his mentor. "My aim is to try to tell Donald Coxeter something about polytopes that he doesn't already know. I'm not at all confident that I can pull it off. But I am going to try," he said, trying to reassure Coxeter that the impact of his oeuvre was enduring.

"But before I do," said Conway, "I want to step back 25 or so years, if I may, to Donald Coxeter's 70th birthday celebration. Nearly all the people there were students of Coxeter's, or grand-students or great grand-students, and they were all getting up and saying how this man had been such a great inspiration in their lives. Well, I rather brashly thought I'd do something different. I stood up and said I was there to forgive Professor Coxeter for having tried to murder me. I then told a story which actually has a few elements of the truth about it."[13]

"A long time ago Professor Coxeter came to Cambridge to give a lecture when I was a student there in the late fifties. I didn't realize it during the lecture, but that was his attempt to murder me. He chose as his weapon something Agatha Christie never thought of: a mathematical problem—he ended his lecture asking for a solution to a problem he'd been pondering." It was a problem about geometrical groups, the rotational polyhedral groups, and Coxeter groups. Conway walked out of the lecture room, and crossed Trumpington Street, the main road in town. "Just as I was in the middle of the road," said Conway, "the solution to Professor Coxeter's problem hit me. Right when he calculated it would, I figure—he judged its level of difficulty precisely. Because as it turned out, it was not the only thing that hit me. At the same time as the solution hit me, or a few microseconds later, a garbage truck also hit me. Fortunately, it didn't do too much damage; it was an unsuccessful attempt at murder. And after being shouted at for being a damned fool by the men hanging off the back of the truck, I limped back to

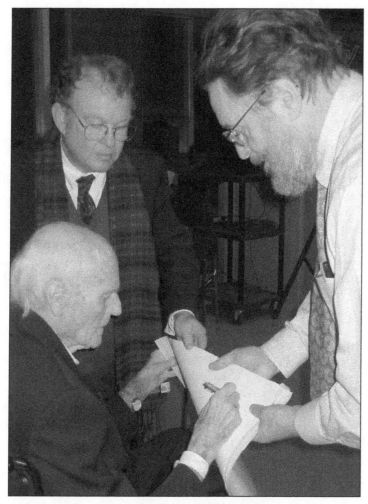

Coxeter autographing a portrait of himself, with Glenn Smith (left) and John Conway,
at Coxeter's ninety-fifth-birthday celebration, hosted by the Fields Institute,
Toronto, February 2002.

the room and told Professor Coxeter all this and gave him my solution,
which to this day I refer to as the 'The Murder Weapon.' "[14]

"I'm one of the greatest Coxeter lovers," said Conway in closing. "He has
a certain way with presentation that is elegant and carries the reader along.
With mathematics what you're doing is trying to prove something and that
can get very complicated and ugly. Coxeter always manages to do it clearly

and concisely, with beauty. Coxeter kept a little flame of geometry alive by doing such beautiful works. There is a quotation from Walter Pater's book *The Renaissance*. Pater was describing art and poetry. He refers to a hard, gem-like flame: 'To burn always with this hard, gem-like flame, to maintain this ecstasy, is success in life.' Somehow," Conway said, "that always makes me think of Donald Coxeter."[15]

△ ▢ ◈ ⊛ ⊛

Coxeter enjoyed his Indian summer while it lasted. He identified with John Galsworthy's interlude with that title in his book *The Forsyte Saga*. "People treated the old as if they wanted nothing," wrote Forsyte. But Coxeter, like Forsyte's character "Old Jolyon,"

> ached a little from sheer love of it all, feeling perhaps, deep down, that he had not very much longer to enjoy it. The thought that some day—perhaps not ten years hence, perhaps not five—all this world would be taken away from him, before he had exhausted his powers of loving it, seemed to him in the nature of an injustice, brooding over his horizon. If anything came after this life, it wouldn't be what he wanted.[16]

Spells of resignation, grumpiness, and sadness came over Coxeter as he set out for the trip home from Budapest. He grudgingly allowed Susan to whisk him around the airport in a wheelchair and through the "diplomats only" passport check. All the while, he grumbled about the injury Susan had inflicted upon the pages of his lecture: "This is what's left of my precious Budapest paper," he said, the sheets of his talk, minus a jagged torn-off bottom half of the opening page, sitting atop his briefcase on his lap, his fingers rapping away in annoyance.

"I'm sorry, Dad. I didn't realize what it was. I just needed something fast to calculate the tip so I ripped off a piece . . ."[17]

He could hardly be denied his rancor. But the cabin-fever antics between father and daughter only escalated after Budapest. He precipitated music wars with his daughter—Coxeter listened to Bruckner, which was like fingernails on a blackboard to Susan, who blasted her country tunes in retaliation. Not long before he died, the tension broke when Susan and her father shared a big laugh. Susan had made her father his usual breakfast, toasting the rice bread twice (as recommended for his digestive problems), buttering it with nondairy spread (he had become lactose intolerant), cutting it twice into triangles (he was finicky, and triangles being far superior to rectangles), only to have him take one bite and say he didn't much feel like toast. This pushed Susan past her limit.[18]

"Dad if you don't want f*#!ing toast, tell me you don't want f*#!ing toast!" The same happened later that day with lunch when she made falafels. "Dad! Tell me if you don't want f*#!ing falafels!"* Midafternoon, Susan left her father on his bed and said she was going to her room, in the maid's quarters of Coxeter's historic house, for a nap. Twenty minutes, that's all she asked. She pulled her blind, reclined, and closed her eyes. At that very moment, her father began blowing his orange "emergency" whistle, slung at the ready around his wrist. Muttering her frustration under her breath, Susan was back at his side. "Susan," he said, not even opening his eyes, "could you please remove my *f*#!ing* shoes?" It was first time he'd used the word in his life. "I had taught him to swear at 96," recalled Susan. "And he taught me how horrible it sounded."[19]

While his mind remained active, Coxeter had clearly entered the homestretch. He occupied himself daily with putting the final touches on his Budapest paper, readying it for publication. And he began to pull books from his library on "convexity," preparing his paper for Symmetry 2003. But soon his ailments and mishaps began to multiply. He thought it terribly unfair, given his pure lifestyle and his long track record of near-perfect health. What did him in was too many falls, out of bed or down the flight of stairs from his bedroom. One tumble in particular warranted numerous stitches and a Band-Aid the size of a rectangular half piece of toast. With that, his bedroom moved from upstairs to the main floor—a makeshift bedroom in the dining room. Coxeter insisted the arrangement was temporary, that he would move back upstairs when he regained his strength. He still climbed the stairs to his bedroom and down to the basement each day for exercise, his arms clutching the railing, pulling his bowed body along behind him like a mountain climber scaling a peak. A visitor commented that his task did not look easy. He said, "It's been my experience that nothing worth doing in life is."[20]

On a Saturday at the end of March, Coxeter put the final touches on his Budapest paper. He so relished making corrections that he could not quite believe no more "errata" were to be found.[21] Having polished his thoughts on four mutually tangent circles to perfection, Coxeter died two days later, on March 31, 2003, with his cat Amy curled on his stomach. He had no illusions about infinity or hyperspace materializing in the hereafter. He had reached mortality's event horizon. And he stipulated there be no funeral. Susan and Edgar poured his ashes under a tree to the west of the front door at

---

*Usually he was easy to please: another day, for lunch, he ate pea soup from the can and declared: "Ahhh! Pianissimo!"

the Rosedale house, his final nod to symmetry, balancing the spot where he left Rien's ashes to the east.[22]

One final Coxeterian act of symmetry had been set in motion prior to his death. Coxeter traveled to McMaster University, in Hamilton, Ontario, where neuroscientist Sandra Witelson has accumulated a "brain bank"—she acquired a specimen of Einstein's brain,[23] and arranged to acquire Coxeter's (she also invited John Conway's brain, and he accepted[24]). Witelson investigates anatomical manifestations of genius. "Behind every beautiful mind is a beautiful brain," she told Coxeter when he visited to hear about results so far. "Sir Donald," as Witelson called him, had undergone some pre-postmortem tests—a neuropsychological analysis and a full-brain MRI. Dr. Witelson explained that she was interested in his parietal lobe in particular—the region of

The last portrait of Donald Coxeter, photographed at home in Toronto, November 2002.

the brain crucial to conjuring intuitive concepts, visual images, multidimensional images of time and space. "Einstein's parietal lobe was twice as large as that of a normal brain,"[25] said Witelson.

Since Coxeter's death, further testing on his brain has so far confirmed the preliminary research. The results seem to reveal that Coxeter had a bilateral expansion of his parietal lobe—not quite the same type of enlargement Einstein displayed, but similar, and a large parietal lobe nonetheless. Indeed, as Witelson had pointed out to Coxeter, one could see the bump on the top of his bald head. She also explained that the right and left hemispheres of average brains do not display mirror symmetry. Coxeter's and Einstein's brains, however, were "more mirror-image" than usual.[26] Coxeter was clearly pleased. "So," he said, "my interest in symmetry has not been misplaced."[27]

APPENDIX 1

FIBONACCI AND PHYLLOTAXIS

The pineapple displays a botanical phenomenon Coxeter was very fond of: phyllotaxis, literally meaning "leaf arrangement," but pertaining generally to buds, and also present in sunflowers, daisies, and pinecones,* which grow in patterns described by the golden ratio.

The golden ratio is derived from a celebrated sequence of numbers discovered in 1202 by Leonardo of Pisa, also known as Fibonacci (his father was nicknamed Bo Nacci, or "the good natured guy," so he was "the son of good nature"). Fibonacci noticed this special sequence of numbers by watching rabbits reproduce. "He assumed that rabbits live forever, and that every month each pair begets a new pair which becomes productive at the age of two months," Coxeter recounted in his book *Introduction to Geometry*. "In the first month the experiment begins with a newborn pair of rabbits. In the second month there is still just one pair. In the third month there are 2; in the fourth, 3; in the fifth, 5; and so on." The Fibonacci numbers 1, 1, 2, 3, 5, 8, 13 are special, because each successive number is the sum of the previous two; and the golden ratio is obtained by dividing any number in the set of Fibonacci numbers by the previous number, which always gives a result in the neighborhood of 1.618.[1]

In the botanical application of the Fibonacci numbers, plant outgrowths seek an optimum amount of living space and in so doing sprout in a pattern of intercrossing "whorls." In a sunflower, where the buds become seeds, one family of 55 clockwise whorls intersects another family of 89 counterclockwise whorls—55 and 89 being successive Fibonacci numbers. With a pinecone, there are 8 "dextral" and 13 "sinistral" whorls (these terms refer to spirals like a corkscrew, and like the mirror image of a corkscrew, respectively). With a pineapple, the dextral and sinistral whorls are always Fibonacci numbers, but not always the same Fibonacci numbers. The ratios of alternate Fibonacci numbers measure the fraction of a turn between successive leaves, buds, or organs, emanating from a plant's stalk—Coxeter gave the examples of ½ for elm and linden, ⅓ for beech and hazel, ⅖ for oak and

---

*Soaking a pinecone in water for thirty minutes causes the buds to close and accentuates the phyllotaxis pattern.

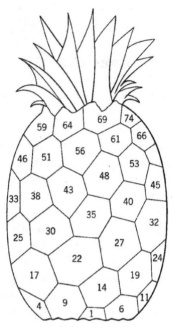

Coxeter's diagram investigating a pineapple's pyllotaxis.

apple, $3/8$ for poplar and rose, $5/13$ for willow and almond.[2] A similar scenario occurs with pineapples, pinecones, daisies, and so on.

In a lecture titled "Chirality and Phyllotaxis," Coxeter stated that the botanical application of the Fibonacci numbers was introduced by Kepler in 1611 with two paragraphs of his book *A New Year's Gift*, which Coxeter quoted: "We may ask why all trees and bushes—or at least most of them unfold a flower in a five-sided pattern, with five petals. In apple- and pear-trees this flower is followed by a fruit likewise divided into five . . . Inside there are always five compartments to hold the seeds . . ."[3]

Coxeter concluded his exposition in *Introduction to Geometry* by saying: "[I]t should be frankly admitted that in some plants the numbers do not belong to the sequence . . . Thus we must face the fact that phyllotaxis is really not a universal *law* but only a fascinatingly prevalent *tendency*."[4]

APPENDIX 2

SCHLÄFLI SYMBOLS
OF THE 3-D AND 4-D REGULAR POLYTOPES[1]

3-D Regular Polyhedron	Faces	At Each Vertex	Schläfli Symbol
Tetrahedron	4 triangles	3 triangles	{3, 3}
Cube	6 squares	3 squares	{4, 3}
Octahedron	8 triangles	4 triangles	{3, 4}
Dodecahedron	12 pentagons	3 pentagons	{5, 3}
Icosahedron	20 triangles	5 triangles	{3, 5}

4-D Regular Polytope	Faces	At Each Edge	At Each Vertex	Schläfli Symbol
5-cell	5 tetrahedra	3 tetrahedra	4 tetrahedra	{3, 3, 3}
8-cell	8 cubes	3 cubes	4 cubes	{4, 3, 3}
16-cell	16 tetrahedra	4 tetrahedra	8 tetrahedra	{3, 3, 4}
24-cell	24 octahedra	3 octahedra	6 octahedra	{3, 4, 3}
120-cell	120 dodecahedra	3 dodecahedra	4 dodecahedra	{5, 3, 3}
600-cell	600 tetrahedra	5 tetrahedra	20 tetrahedra	{3, 3, 5}

APPENDIX 3

COXETER DIAGRAMS

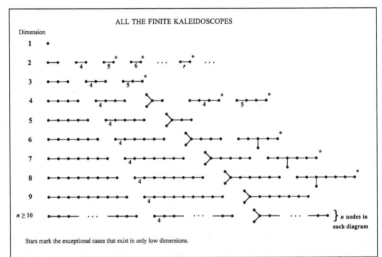

ALL THE FINITE KALEIDOSCOPES

Stars mark the exceptional cases that exist in only low dimensions.

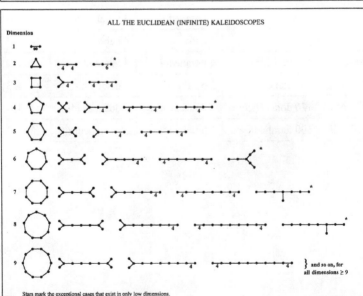

ALL THE EUCLIDEAN (INFINITE) KALEIDOSCOPES

Stars mark the exceptional cases that exist in only low dimensions.

COXETER GROUPS

Coxeter groups are the algebraic equivalent of Coxeter diagrams. The Coxeter diagram for the icosahedron, having three mirrors or nodes, would be translated into symbolic algebraic terms, the mirrors, or the mirror reflections, being represented by x, y, and z.

The algebraic language created by Coxeter groups is governed by the equivalent of grammatical rules. But in fact, the rules are better described by a mathematical analogy—the way in which reflections combine with one another in Coxeter groups is much like a multiplication table.

For example, let's take a Coxeter group of order 4, which is generated by just two mirrors.

Here, the symmetries generated by the reflections in the two mirrors (which are hinged at a 90° angle) can be denoted by the lower case letters e, a, b, c, where e commonly describes the identity symmetry, corresponding to the real image. The two mirror images would be a and b, and the peculiar fourth image behind the seam of the mirrors would correspond to c. Here the magic appears: the seam image c is equal to the image in mirror a being reflected by mirror b—image c is the result of the real image bouncing first in mirror a and then mirror b. In order to get the fourth image c, you "multiply" a times b. The "multiplication table" of symmetries fills in like this:

	e	a	b	c
e	e	a	b	c
a	a	e	c	b
b	b	c	e	a
c	c	b	a	e

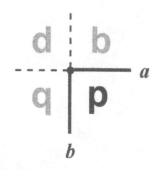

The chart lays out how the images reflect in multiple fashion in the mirrors. The identity element, or the original image, results from reflecting only in mirror a, and again in that same mirror. As we saw previously: $a^2 = 1$ (or $a^2 = c$). But when the original image is reflected through mirrors a and b, another image results, namely, image c. The entire chart can be completed in this manner. So again, if you have one reflection in mirror a, you see just your mirror image; if have one reflection in mirror b, you just see your mirror image. But when you look at that image behind the crack between the mirrors, the light is actually bouncing first off mirror a, then off mirror b, and back into your eyes, and it is that double bounce that gives you the very unexpected extra image. And somehow c is different in a geometrical way, as you can check by waving your right hand at images a, b, and then c.[1]

Thus, the Coxeter group of order 4 is a collection of the symmetry transformations e, a, b, and c, which comes equipped with a multiplication table governing how the transformations interact.

Coxeter groups of higher order progress in the same manner, with larger multiplication tables, adding to the algebraic language. Each Coxeter group has its own algebraic characterization, which, when memorized, evokes the symmetry properties of that particular group.[2] Or mathematicians can look it up in *Regular Polytopes*, where Coxeter has already done all the work.[3]

MORLEY'S MIRACLE

A quintessential sampling of Coxeter's mathematical spirit is found in his book *Introduction to Geometry*. His opening chapter was titled "Triangles," which focused first on Euclid (perhaps a logical place to start but also a nice volley in response to the "Down-with-Euclid-Death-to-Triangles!" mantra). Therein he laid out Morley's theorem, aka Morley's Miracle, a theorem Coxeter much appreciated.[1]

He also included Morley's Miracle in his book *Geometry Revisited*, coauthored with Samuel Greitzer, which featured a purple and yellow construction of the theorem on the cover (but oddly, the construction is faulty, not possessing exact trisectors).

"One of the most surprising theorems in elementary geometry"— surprising because it was so simple and went undiscovered for two thousand years—"was discovered about 1899 by Frank Morley,"[2] wrote Coxeter. A shy but deliberate man, Morley did not go public with his theorem and it was first published by another party, F. G. Taylor and W. L. Marr, in 1914.[3] Morley was born in England, graduated from Cambridge in 1884, and later moved to the United States and became a professor of mathematics at Haverford College, in Pennsylvania; after his triangle discovery, he was appointed a professor at Johns Hopkins, in 1900.[4]

The theorem Morley discovered states: "The three points of intersection of the adjacent trisectors of the angles of any triangle form an equilateral triangle."[5] According to Conway, "The property of equilaterality surprises everybody."[6]

Morley's son, Frank V. Morley,[7] remembered his father's discovery: "I was a schoolboy when my father, who was almost forty years older than I was, sketched for me, free-hand, a penciled diagram of the simplest form of the above-discussed theorem in plane geometry. I tested it once with my own drawing instruments. No matter what shape of the original triangle I started with, there in its midriff was an equilateral triangle, picked out by the trisectors. It was wizard, it was weird—and it was True!"[8]

As Coxeter told the tale, Morley mentioned the theorem to his friends, who in turn spread it around the world as mathematical gossip. It was heralded as one of the most astonishing and unexpected theorems in mathematics, and a gem whose sheer beauty allows few rivals. After twenty years, Morley published his theorem in Japan. The first two proofs of the theorem included a trigonometrical proof by M. Satyanarayana and an elementary proof by M. T. Naraniengar. The theorem continued to evoke proof after proof—150 within 50 years—and still does.[9]

John Conway invented the latest proof in 1995, which he first announced by e-mail to a geometry newsgroup. His proof is widely appreciated because it avoids trigonometry, making it unnecessary to handle all six triangles separately. Here is Conway's proof (* in the proof corresponds to + in the diagrams):

Path: world!forum.swarthmore.edu!gateway
From: conway@math.princeton.edu (John Conway)
Newsgroups: geometry.puzzles
I have the undisputedly simplest proof of Morley's Trisector Theorem.
Here it is:
Let your triangle have angles 3a,3b,3c and let x* mean x+pi/3, so that
  a+b+c=o*. Then triangles with angles
o*, o*, o*
a,b*,c* a*,b,c* a*,b*,c
a**,b,c a,b**,c a,b,c**
exist abstractly, since in every case the angle-sum is pi.
Build them on a scale defined as follows:
o*,o*,o* - this is equilateral - make it have edge 1
a,b*,c* - make the edge joining the angles b* and c* have length 1
- similarly for a*,b,c* and a*,b*,c
a,b**,c (and the other two like it) - let me draw this one:

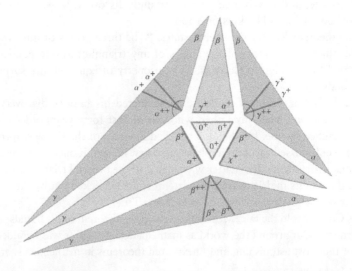

(Note: in these pictures alpha = a, beta = b, gamma = c)

Let the angles at A,X,C be a,b**,c, and draw lines from X cutting AC at angle
b* in the two senses, so forming an equilateral triangle XYZ.

[Z and Y are where the red lines meet the CA line in the bottom triangle, X is
its other vertex]

Choose the scale

so that XY and XZ are both 1.

Now just fit all these 7 triangles together! They'll form a figure like:

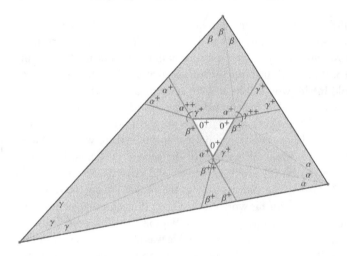

To make it a bit more clear, let me say that the angles of APX are a (at A), b*
(at P), c* (at X).

Why do they all fit together? Well, at each internal vertex, the angles add up to
2pi, as you'll easily check. And two coincident edges have either both been
declared to have length 1, or are like the common edge AXZ of sorry - AX
of triangles APX and AXC.

But APX is congruent to the subtriangle AZX of AXC, since PX=ZX=1,
PAX=ZAX=a, and APX=AZX=b*.

So the figure formed by these 7 triangles is similar to the one you get by trisecting the angles of your given triangle, and therefore in that triangle the
middle subtriangle must also be equilateral.

John Conway[10]

"Triangle geometry refuses to die," said Conway, who is currently completing
The Triangle Book. Containing hundreds and hundreds of theorems on triangles, *The Triangle Book* has been in the works for a decade or more. It is a collaborative effort with award-winning high school mathematics teacher Steve
Sigur, of the Paideia School, in Atlanta, Georgia. Conway intends that the book
will be produced in the shape of a triangle. "It is going to be the standard book
on triangles forever," said Conway. "Or at least for a very long time."[11]

APPENDIX 6

FREEMAN DYSON ON
"UNFASHIONABLE PURSUITS"

In 1981, Freeman Dyson sent his friend Coxeter—"one of my favourite people"—a recent lecture he had delivered, "Unfashionable Pursuits." One of the most famous physicists (famed for both his physics and his elegant essays), based at Princeton's Institute for Advanced Study, Dyson met Coxeter through Leopold Infeld (who had been at the institute and subsequently the University of Toronto), and they maintained a leisurely correspondence over the years[1]— leisurely in that they conversed, of course, about polyhedra, but for Dyson this was just a sideline. Dyson is a polymath if ever there was one. He is known widely for his books on science for the general public, such as *Weapons of Hope*, on the ethical problems of war and peace (Coxeter also enjoyed his book *Disturbing the Universe*, about the people Dyson encountered over his scientific career); for his work on the Orion Project, proposing space flight using nuclear propulsion; and for his "Dyson Tree," a genetically engineered plant capable of growing on a comet. He was also the winner of the Templeton Prize for Progress in Religion in the year 2000, netting him 795,000 sterling.

Dyson delivered his "Unfashionable Pursuits"[2] lecture at the institute, and sent it along to Coxeter, "With all good wishes from one unfashionable character to another." A short way into his talk, Dyson recalled the early days of his career:

> It has always been true, and it is now more than ever, that the path of wisdom for a young scientist of mediocre talent is to follow the prevailing fashion. Any young scientist who is not exceptionally gifted or exceptionally lucky is concerned first of all with finding and keeping a job. To find and keep a job you have to do competent work in an area of science which the mandarins who control the job-market find interesting. The scientific problems which the mandarins find interesting are almost by definition, the fashionable problems . . . It is no wonder that young scientists who care for their own survival keep close to the beaten paths . . .

> Our Institute here is no exception. When I first came here as a visiting member thirty-four years ago, the ruling mandarin was Robert Oppenheimer. Oppenheimer decided which areas of physics were worth pursuing. His tastes always

coincided with the most recent fashions. Being then young and ambitious, I came to him with a quick piece of work dealing with a fashionable problem, and was duly rewarded with a permanent appointment . . .

The running of young scientists after quick success and quick rewards is not in itself bad. The concentration of their efforts into narrow areas of fashionable specialization is not necessarily harmful. After all, the fashionable problems become fashionable not by the whim of some dress-designer but because a substantial majority of scientists judges them to be important.

Nonetheless, Dyson argued that scientists and mathematicians would do well to employ more creative free rein and less pragmatic ambition in directing their studies, and that the sciences on the whole would benefit from conscientiously making room for the unfashionable characters, such as Kurt Gödel, "an independent and recalcitrant spirit . . . one of the few indubitable geniuses of our century, the only one of our colleagues who walked and talked on equal terms with Einstein." And he pointed to a few invaluable examples in "ancient history."

As an example of a great mathematical physicist whose work is of crucial importance to the development of physics at the present time, I mention the name of Sophus Lie. Lie has been dead for 80 years. His great work was done in the 1870's and 1880s, but it has come to dominate the thinking of particle physicists only in the last twenty years. Lie was the first to understand and state explicitly that the principles of mechanics, which in his day were synonymous with the principles of physics, have a group-theoretic origin. He constructed almost single-handed a vast and beautiful theory of continuous groups, which he foresaw would one day serve as a foundation of physics. Now, a hundred years later, every physicist who classifies particles in terms of broken and unbroken symmetries is, whether he is aware of it or not, talking the language of Sophus Lie. But in his lifetime Lie's ideas remained unfashionable . . . A more recent example of a great discovery in mathematical physics was the idea of a gauge field, invented by Hermann Weyl in 1918. This idea has taken only 50 years to find its place as one of the basic concepts of modern particle physics.

Dyson continued, casting a glance around the world of mathematics to find some unfashionable ideas in the moment which, by his foresight, might later emerge as essential building blocks for the physics of the twenty-first century. "Roughly speaking," Dyson said, "unfashionable mathematics consists of those parts of mathematics which were declared by the mandarins of Bourbaki not to be mathematics." One example of such accidental beauty, an isolated curiosity seemingly not leading anywhere, offered Dyson, was the discovery of the "Monster group," resuming the hunt for sporadic finite groups, first discovered by Frenchman Emile Mathieu in the nineteenth century, but then abandoned.

> [R]ather suddenly, in the last twenty years, a magnificent zoo of new sporadic groups has been discovered by a variety of mathematicians working with a variety of methods . . . The only thing these various discoveries had in common was a concrete, empirical, experimental, accidental quality, directly antithetical to the spirit of Bourbaki. . . .
>
> What has all this to do with physics? Probably nothing. Probably the sporadic groups are merely a pleasant backwater in the history of mathematics, an odd little episode, far from the mainstream of progress. We have never seen the slightest hint that the symmetries of the physical universe are in any way connected with the symmetries of the sporadic groups. . . . But we should not be too sure that there is no connection. Absence of evidence is not the same thing as evidence of absence. Stranger things have happened in the history of physics than the unexpected appearance of sporadic groups. We should always be prepared for surprises. I have to confess to you that I have a sneaking hope, a hope unsupported by any facts or any evidence, that sometime in the twenty-first century the physicists will stumble upon the Monster group, built in some unsuspected way into the structure of the universe. This is of course only a wild speculation, almost certainly wrong. The only argument I can provide in its favor is a theological one. We have strong evidence that the creator of the universe loves symmetry, and if he loves symmetry, what lovelier symmetry could he find than the symmetry of the Monster?[3] The sporadic groups are only one example out of the treasure-house of weird and wonderful concepts which unfashionable mathematicians have created. I could mention others. Can you imagine a regular polyhedron, a body composed of perfectly symmetrical cells arranged in a perfectly symmetrical structure, having a total of eleven faces? Last year, my friend Donald Coxeter in Toronto discovered it . . .

Coxeter, to be sure, appreciated Dyson's paper and the moral of his story. He was still being made to feel very out of style. "Dear Freeman," Coxeter wrote in response. "Many thanks for your splendid lecture on 'Unfashionable Pursuits,' which seemed particularly relevant as it arrived at the same time as a letter from the Editor of the LMS [London Mathematical Society] enclosing the referee's report on 'My Graph.'" That report began as follows: 'I recommend the paper be accepted, subject to the modification suggested below. It is difficult to evaluate, since its subject-matter and style are so unfashionable . . . I think it is up to the standard of things like his "Regular Polytopes," though of course much more limited in scope. However, because of the very special nature of the subject, my recommendation cannot be a very strong one—papers on topics of greater generality must take precedence' . . . I still take pleasure in your remark that Plato would have been delighted if he had known about $_5\{3,5,3\}_5$,"[4]—this was Coxeter's symbol for the eleven-faced object described by Dyson: *Can you imagine a regular polyhedron, a body composed of perfectly symmetrical cells arranged in a perfectly symmetrical structure, having a total of eleven faces?*[5]

APPENDIX 7

CRYSTALLOGRAPHY AND PENROSE
TOILET PAPER

Around the same time Coxeter and like-minded mathematicians happened upon M. C. Escher, the crystallographers did as well. Escher's symmetry drawings, his tessellations of the plane, are often used in teaching crystallography—and in fact, Escher's works anticipated crystallographic research by decades.[1]

In 1891, Russian crystallographer E. S. Fedorov proved that all periodic tilings of the plane belong to one of seventeen symmetry groups (later rediscovered by Coxeter's Princeton friend, George Pólya, together with P. Niggli).[2] Finding tiles that would produce only nonperiodic tilings (featuring shapes in patterns that do not repeat by translation) was an unsolved geometric puzzle, until Robert Berger developed the first set of tiles in 1966. His set consisted of 20,426 tiles, which he soon reduced to 104. In the early 1970s, Raphael Robinson created a set of six tiles, with various notches and extensions to prevent a periodic pattern.[3]

Sir Roger Penrose discovered the lowest limit as yet, a two-tile set of rhombs, essentially skinny and fat diamonds that interconnect in a field of five-pointed stars (a tiling of darts and kites can also be derived from the same tiles). "I tend to doodle and tilings were one of the things I used to play around with," said Sir Roger, who attended the ICM in 1954 and was taken with Escher's work, as was Coxeter, but less by his periodic tilings than by his impossible pictures. "I came away from the meeting feeling I'd like to do something similar," he said, "something impossible." With his father, the psychiatrist Lionel S. Penrose, he created the impossible "tribar" and "stairs." They published their creations in the *British Journal of Psychology* and sent them to Escher, who later made use of the ideas in his prints *Waterfall* and *Ascending and Descending*. "My interest in the non-periodic tiles I suppose was partly stimulated by my interest in physics—I was looking for something which was simple, on the small scale, something that had rules but produced complicated structures, because one sees that sort of thing in the universe. One hopes that the laws are ultimately simple."[4]

Sir Roger had been playing around with six tiles for a while, and wondered whether he could do better. Within a few hours, drawing in a note-

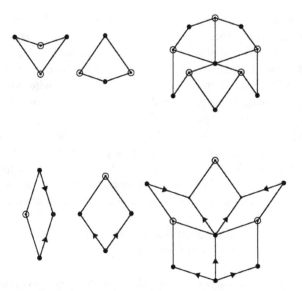

Penrose aperiodic tiles: the kites and darts, and the skinny and fat rhombs. When adjoining tiles to produce a Penrose tiling, all tile markings must match. Shown are legitimate vertex configurations (there is a "split and glue" process that turns either configuration into the other, nicely explained in *Quasicrystals and Geometry*, by Marjorie Senechal).

book, he got them down to two. "I thought, 'This is too easy, somebody must have thought of it before.' But it was the first time anybody had got it down to two. Unless you can find it somewhere in some Moorish design. But they haven't found them yet."[5]

It may seem like a trivial discovery. But whenever he lectured about his Penrose tilings, as they are known, somebody would always ask: "Does this raise an issue of a new type of crystallography where you could have these forbidden symmetries, five-fold symmetries in crystals?" "And my response to such a question tended to be 'Yes, indeed, in principle what you say is true. But I can't see how nature would produce such things.'" Nature didn't. Science did. In 1982 Dan Shechtman, then at Johns Hopkins University, observed quasicrystals (short for quasi-periodic crystals)—a form of solid in which the atoms are configured in an apparently regular but yet nonrepeating structure. And as with the study of crystallography, Coxeter groups and Coxeter diagrams find application in the investigation of quasicrystals.[6]

Any material, such as aluminum, is characterized by the pattern of its crystals. Quasicrystals alter the properties of the substance, often making it harder. Quasicrystals have found application, for example, as a nonstick coating for cookware named Cybernox. Upon the discovery of quasicrystals, the International Union of Crystallographers redefined the term crystal to allow for their acceptance. "It was a surprise to me," said Sir Roger. "But they were manufactured objects. Nobody has found such things in a cave."[7]

A less pleasant surprise came when Sir Roger discovered his Penrose tilings on a "Kleenex quilted" brand of toilet paper his wife brought home from the supermarket. Sir Roger and Pentaplex Ltd., the company that has licensing rights to the image of the Penrose tiling, took legal action against the manufacturer, Kimberly-Clark Corporation, the British division of the Dallas-based Kleenex. "When it comes to the population of Great Britain being invited by a multinational to wipe their bottoms on what appears to be the work of a Knight of the Realm without his permission, then a last stand must be taken," said a director of Pentaplex Ltd., as reported in the *Wall Street Journal*.[8]

The newspaper also reported that just as Penrose tilings applied to quasicrystals provided for a better frying pan . . .

> The same logic also makes for better toilet paper. A premium brand launched in the United Kingdom in 1993, Kimberly-Clark's Kleenex quilted toilet tissue is embossed with a pattern to fluff up the tissue, making it "thicker and softer," according to company literature. Sir Roger's writ argues that making the tissue fluffier allows manufacturers to reduce the amount of paper used on each roll. But if the pattern repeats itself, the tissue would likely bunch up, looking unattractive. That can be corrected using a Penrose-type pattern which lets the paper sit evenly on the roll, the suit contends. If the plaintiffs win, they can claim damages under British law equal to the Kleenex brand's U.K. profits. They can also demand that all remaining examples of the toilet paper be destroyed.

The parties ultimately reached an out-of-court settlement. One of the settlement conditions prevents Sir Roger from discussing the matter further (clearly, Kimberly-Clark's butt was at least somewhat kicked). He does make clear, however, that as a mathematical concept, Penrose tilings are free for the taking. As a graphic image, however, those skinny and fat diamonds are patented. The implication being that the chances were slim that a toilet paper company was borrowing a mathematical idea.[9]

The Penrose toilet paper is now a collector's item. At a recent Bridges conference ("Mathematical Connections in Art, Music, Science"), an annual gathering held in 2005 at the Banff Centre, in Alberta, Marion Walter, professor emeritus of mathematics education at the University of Oregon, brought a roll of the notorious Penrose toilet paper, and dolled out single squares to worthy participants.[10]

APPENDIX 8

THE MATHEMATICAL PUBLICATIONS
OF H. S. M. COXETER

1926   Mathematical Notes. *Math. Gazette* 13:205.

1928   The pure Archimedean polytopes in six and seven dimensions. *Proc. Camb. Phil. Soc.* 5 (24): 1–9.

1930   The polytopes with regular-prismatic vertex figures, part 1. *Phil. Trans. Royal Soc.* A 229:329–425.

1931   Groups whose fundamental regions are simplexes. *Journal of Math. and Phys.* 12:334–45.

        The densities of the regular polytopes. *Proc. Camb. Phil. Soc.* 27:201–11.

1932   The polytopes with regular-prismatic vertex figures, part 2. *Proc. London Math. Soc.* 6:132–36.

        The densities of the regular polytopes, part 2. *Proc. Camb. Phil. Soc.* 28:509–21.

1933   The densities of the regular polytopes, part 3. *Proc. Camb. Phil. Soc.* 29:1–22.

        Regular compound polytopes in more than four dimensions. *Journal of Math. and Phys.* 12:334–45.

1934   Discrete groups generated by reflections. *Annals of Math.* 35:588–621.

        On simple isomorphism between abstract groups. *Journal London Math. Soc.* 9:211–12.

        Abstract groups of the form $V_1^k = V_j^3 = (V_i V_j)^2 = 1$. *Journal London Math. Soc.* 9:213–19.

        (With J. A. Todd) On points with arbitrarily assigned mutual distances. *Proc. Camb. Phil. Soc.* 30:1–3.

1935   Finite groups generated by reflections, and their subgroups generated by reflections. *Proc. Camb. Phil. Soc.* 30:446–82.

        The functions of Schläfli and Lobatschefsky, *Quarterly Journal of Math.* 6:13–29.

        (With P. S. Donchian) An *n*-dimensional extension of Pythagoras' theorem. *Math. Gazette* 19:206.

        The complete enumeration of finite groups $R_i^2 = (R_i R_j)^{kij} = 1$. *Journal of London Math. Soc.* 10:21–25.

1936   Wythoff's construction for uniform polytopes. *Proc. London Math. Soc.* 2 (38): 327–39.

The representation of conformal space on a quadric. *Annals of Math.* 37:416–26.

The groups determined by the relations Sl = Tm = (S-ɪ T-ɪ ST) p = ɪ. *Duke Math. Journal* 2:61–73.

An abstract definition for the alternating group in terms of two generators. *Journal London Math. Soc.* 2 (11): 150–56.

The abstract groups Rm = Sm = (RjSj) pj = ɪ. *Proc. London Math. Soc.* 2 (41): 278–301.

(With J. A. Todd) A practical method of enumerating cosets of a finite abstract group. *Proc. Edinburgh Math. Soc.* 2 (5): 26–34.

On Schläfli's generalization of Napier's Pentagramma Mirificum. *Bull. Calcutta Math. Soc.* 28:123–44.

(With J. A. Todd) Abstract definitions for the symmetry groups of the regular polytopes in terms of two generators. *Proc. Camb. Phil. Soc.* 32:194–200.

1937   (With J. A. Todd) Abstract definitions for the symmetry groups of the regular polytopes in terms of two generators. part 2. *Proc. Camb. Phil. Soc.* 33:315–24.

Regular skew polyhedra in three and four dimensions and their topological analogues. *Proc. London Math. Soc.* 2 (43): 33–62.

1938   An easy method for constructing polyhedral group-pictures. *Amer. Math. Monthly* 45:522–25.

(With P. Du Val, H. T. Flather, and J. F. Petrie), *The Fifty-nine Icosahedra*, University of Toronto Studies (Math. Series, No. 6), 28.

1939   Polyhedra. In *Mathematical Recreations and Essays*, by W. W. Rouse Ball, 129–60. London: Macmillan.

The groups G m, n, p, Trans. *Amer. Math. Soc.* 45:73–150.

The regular sponges, or skew polyhedra. *Scripta Mathematica* 6:240–44.

1940   Regular and semi-regular polytopes, part ɪ. *Math. Zeitschrift* 46:380–407.

A method for proving certain abstract groups to be finite. *Bull. Amer. Math. Soc.* 446:246–51.

(With R. Brauer) A generalization of theorems of Schönhardt and Mehmke on polytopes. *Trans. Royal Soc. of Canada* 3 (34): 29–34.

The polytope 2₂₁, whose 27 vertices correspond to the lines on the general cubic surface. *Amer. Journal of Math.* 62:457–86.

The binary polyhedral groups, and other generalizations of the quaternion group. *Duke Math. Journal* 7:367–79.

1942   *Non-Euclidean Geometry* (Mathematical Expositions, No. 2). Toronto: University of Toronto Press.

1943   The map-coloring of unorientable surfaces. *Duke Math. Journal* 10:293–304.

A geometrical background for de Sitter's world. *Amer. Math. Monthly* 50:217–27.

1946   Quaternions and reflections. *Amer. Math Monthly* 53:136–46.

Integral Cayley numbers. *Duke Math. Journal* 13:561–78.

1947   The nine regular solids. *Proc. Canadian Math. Congress.* 1:252–64.

The product of three reflections. *Quarterly Applied Math.* 5:217–22.

1948   A problem of collinear points, *Amer. Math. Monthly* 55:26–28, 247.

Regular Polytopes. London: Methuen.

1949   *The Real Projective Plane.* McGraw-Hill, New York.

Projective geometry. *Math. Magazine* 23:79–97.

1950   Self-dual configurations and regular graphs. *Bull. Amer. Math. Soc.* 56:413–55.

(With A. J. Whitrow) World structure and non-Euclidean honeycombs. *Proc. Royal Soc.* A 201:417–37.

Extreme forms. *Proc. Internat. Congress of Mathematicians, Harvard* 1:294–95.

1951   Extreme forms. *Can. J. Math.* 3:391–441.

The product of the generators of a finite group generated by reflections. *Duke Math. J.* 18:765–82.

1952   Interlocked rings of spheres. *Scripta Math.* 18:113–21.

1953   (With J. A. Todd) An extreme duodenary form. *Can. J. Math.* 5:384–92.

The golden section, phyllotaxis, and Wythoff's game. *Scripta Math.* 19:135–43.

1954   (With M. S. Longuet-Higgins and J. C. P. Miller) Uniform polyhedra. *Phil. Trans. Royal Soc.* A 246:401–50.

Regular honeycombs in elliptic space. *Proc. London Math. Soc.* 3 (4): 471–501.

Six uniform polyhedra. *Scripta Math.* 20:227.

An extension of Pascal's Theorem. *Amer. Math. Monthly* 61:723.

Arrangements of equal spheres in non-Euclidean spaces. *Acta Math. Acad. Sci. Hungaricæa* 5:263–76.

Regular honeycombs in hyperbolic space. *Proc. Internat. Congress of Mathematicians*, Amsterdam 3:155–69.

1955   On Laves' graph of girth ten. *Can. J. Math.* 7:18–23.

The area of a hyperbolic sector. *Math. Gazette* 39:318.

The affine plane. *Scripta Math.* 21:5–14.

The Real Projective Plane, 2nd ed. New York: Cambridge University Press.

Reele Projektive Geometrie der Ebene [German translation of (66)]. Munich: Oldenbourg.

Hyperbolic triangles. *Scripta Math.* 22:5–13.

1956   The collineation groups of the finite affine and projective planes with four lines through each point. *Abh. Math. Sem. Univ. Hamburg* 20:165–77.

1957   Groups generated by unitary reflections of period two. *Can. J. Math.* 9: 263–72.

Map-coloring problems. *Scripta Math.* 23:11–25.

Projective geometry. In *The Tree of Mathematics*, ed. Glenn James. Pacoima, California: Digest Press, 173–194.

(With W. O. J. Moser) *Generators and Relations for Discrete Groups* (Ergeb. Math. 14). Berlin and New York: Springer.

Crystal symmetry and its generalizations [presidential address], *Trans. Royal Society of Canada* 3 (51): 1–13.

1958   The chords of the non-ruled quadric in PG (3, 3). *Can. J. Math.* 10:484–88.

Twelve points in PG (5, 3) with 95040 self-transformations. *Proc. Royal Soc. London* A 247:279–93.

On subgroups of the modular group. *J. de Math. Pures. Appl.* 37:317–19.

Close packing and froth. *Illinois J. Math.* 2:746–58.

Lebesgue's minimal problem. *Eureka* 21:13.

1959   Factor groups of the braid group. *Proc. 4th Canad. Math. Congress* [Banff, 1957] 95–122. Toronto: Toronto Univ. Press.

The four-color map problem, 1840–1890. *Math. Teacher* 52(4):283–89.

Symmetrical definitions for the binary polyhedral groups. *Proc. of Symposia in Pure Mathematics*, Amer. Math. Soc. 1:54–87.

Polytopes over GF(2) and their relevance for the cubic surface group. *Can. J. Math.* 11:646–50.

(With L. Few and C. A. Rogers) Covering space with equal spheres. *Mathematika* 6:147–57.

1961   On Wigner's problem of reflected light signals in de Sitter space. *Proc. of the Royal Society of London* A 261:435–42.

Similarities and conformal transformations. *Annali di Mat. Pura ed Appl.* 53:165–72.

Introduction to Geometry. New York: Wiley.

1962   Music and mathematics. *Canadian Music Journal* 6:13–24.

The problem of packing a number of equal nonoverlapping circles on a sphere. *Trans. New York Acad. Sci.* 2 (24): 320–31.

The classification of zonohedra by means of projective diagrams. *J. de Math. Pures Appl.* 41:137–56.

The total length of the edges of a non-Euclidean polyhedron. In *Studies in Mathematical Analysis and Related Topics* [Essays in honor of George Pólya, ed. G. Szegö, G. Loewner, et al.], 62–69. Stanford, CT: Stanford University Press.

The symmetry groups of the regular complex polygons. *Archiv der Math.* 13:86–97.

The abstract group G3,7,16. *Proc. Edinburgh Math. Soc.* 2 (13): 47–61, 189.

Projective line geometry. *Mathematicæ Notæ*, Universidad Naçional del Litoral, Rosario, Argentina 1:197–216.

1963   An upper bound for the number of equal nonoverlapping spheres that can touch another of the same size. *Proc. Symposia in Pure Mathematics* 7:53–71.

(With L. Fejes Tóth) The total length of the edges of a non-Euclidean polyhedron with triangular faces. *Quarterly J. Math.* 14:273–84.

(With B. L. Chilton) Polar zonohedra. *Amer. Math. Monthly* 70:946–51.

(With S. L. Greitzer) L'hexagramme de Pascal, un essai pour reconstituer cette découverte. *Le Jeune Scientifique* 2:70–72.

Unvergängliche Geometrie [German translation of *Introduction to Geometry*] Basel: Birkhäuser.

Regular Polytopes, 2nd ed. New York: Macmillan.

1964   *Projective Geometry*. New York: Blaisdell.

Regular compound tessellations of the hyperbolic plane. *Proc. Royal Soc.* A 278:147–67.

1965   *Non-Euclidean Geometry*, 5th ed. Toronto: University of Toronto Press.

(With W. O. J. Moser) *Generators and Relations for Discrete Groups*, 2nd ed., Berlin and New York: Springer.

Geometry. In *Lectures on Modern Mathematics*, by T. L. Saaty, 58–94. New York: Wiley.

Introduction to Geometry [Japanese translation]. Tokyo: Charles E. Tuttle Co.

1966   *Introduction to Geometry* [Russian translation]. Moscow: Nauka.

Reflected light signals. In *Perspectives in Geometry and Relativity*, by B. Hoffman, 58–70. Bloomington: Indiana University Press.

Achievement in maths. *Varsity Graduate*, 15–18.

The inversive plane and hyperbolic space. *Abh. Math. Sem. Univ. Hamburg* 29:217–42.

Inversive distance. *Annali di Mathematica* 4 (71): 73–83.

1967   The Lorentz group and the group of homographies. *Proc. Internat. Conf. on the Theory of Groups*, Australian National University–Canberra, ed. L. G. Kovacs and B. H. Neumann, 73–77.

Non-Euclidean geometry. In *Collier's Merit Students' Encyclopedia*.

Finite groups generated by unitary reflections. *Abh. Math. Sem. Univ. Hamburg* 31:125–35.

The Ontario K-13 geometry report. *Ontario Mathematics Gazette* 5(3):12–16.

(With S. L. Greitzer) *Geometry Revisited*, 193. Washington, D.C.: Mathematical Association of America.

Wstep do Geometrii dawnej i nowej [Polish translation of *Introduction to Geometry*] Warsaw: Panstowowe Wydawnictwo Naukowe. Analytic Geometry. In *Encyclopedia Britannica*, 857–860.

Solids, Geometric. In *Encyclopaedia Britannica*, 860–862.

(With G. W. Cu.) Mathematical Models. In *Encyclopaedia Britannica*, 1087–91.

1968   *Music and mathematics* [reprinted from the *Canadian Music Journal*], *Mathematics Teacher* 61:312–20.

The problem of Apollonius. *Amer. Math. Monthly* 75:5–15.

Mid-circles and loxodromes. *Math. Gazette* 52:1–8.

Loxodromic sequences of tangent spheres. *Aequationes Mathematic* 1:104–21.

Twelve Geometric Essays. Carbondale, IL: Southern Illinois University Press.

The mathematical implications of Escher's prints. *M.C. Escher, 8 June–21 July 1968* (catalog of a retrospective exhibition), 87–89. The Hague: Haags Gemeentemuseum.

Some examples of hyperdimensional awareness. *The Journal for the Study of Consciousness* 1(2):84–85.

1969   *Introduction to Geometry*, 2nd ed. New York: Wiley.

Affinities and their fixed points. *The New Zealand Mathematics Magazine* 6:114–17.

Affinely regular polygons. *Abhandlungen aus dem Mathematischen Seminar der Universität Hamburg* 34:38–58.

Helices and concho-spirals. *Proceedings of the 11ᵗʰ Nobel Symposium, Symmetry and Functions of Biological Systems at the Macromolecular Level*. Stockholm: Amqvist & Wiksell, reprint New York: Wiley & Sons).

1970   Inversive geometry. *Vinculum* 7:72–76.

Non-Euclidean geometry. In *Encyclopaedia Britannica*.

Products of shears in an affine Pappian plane. *Rendiconti di Mathematica* 6 (3): 1–6.

Solids, geometric. In *Encyclopaedia Britannica*.

Twisted honeycombs (Regional Conference Series in Mathematics No. 4). Providence, RI: American Mathematical Society.

1971   Cyclic sequences and frieze patterns. *Vinculum* 8:4–7.

An ancient tragedy, *Mathematical Gazette*. 55:312.

The mathematical implications of Escher's prints. In *The World of M. C. Escher*, ed. J. L. Locher, 49–52. New York: Harry N. Abrams.

The mathematics of map coloring. *Leonardo* 4:273–77.

Frieze patterns. *Acta Arithmetica* 18:297–310.

Virus macromolecules and geodesic domes. In *A Spectrum of Mathematics*, Essays presented to H. G. Forder, ed. J. C. Butcher, 98–107. Auckland: Auckland Oxford University Press.

The finite inversive plane with four points on each circle. In *Studies in Pure Mathematics*, ed. L. Mirsky, 39–52. London: Academic Press

Inversive geometry. In *The Teaching of Geometry at the Pre-College Level*, ed. H. G. Steiner, 34–45. Dordrecht, Holland: D. Reidel Publishing Company

[Spanish translation of *Introduction to Geometry*]. Mexico City: Limusa-Wiley S.A.

(With S. L.Greitzer) Redécouvrons la Géométrie [French translation of *Geometry Revisited*]. Paris: Dunod.

1972   Analytical geometry, mathematical models, and geometric solids.
        In *Encyclopaedia Britannica*.

        Cases of Hyperdimensional Awareness. In *Consciousness and
        Reality*, eds. Charles Muses and Arthur M. Young, 95–101.
        New York: Overbridge & Lazard.

        (With W. O. J. Moser) *Generators and Relations for Discrete
        Groups*, 3rd ed. Berlin and New York: Springer.

        The role of intermediate convergents in Tait's explanation for
        phyllotaxis. *Journal of Algebra* 20:168–75.

        (With J. H. Conway and G. C. Shephard) The centre of finitely
        generated group. *Tensor* 25:405–18; 26:477.

1973   (With J. H. Conway) Triangulated polygons and frieze patterns.
        Mathematical Gazette 57:87–94 and 175–86.

        The Dirac matrix group and other generalizations of the
        quaternion group. *Communications on Pure and Applied
        Mathematics* 26:693–98.

        Regular Polytopes, 3rd ed. New York: Dover Publications.

        [Hungarian translation of *Introduction to Geometry*]. Budapest:
        Müszaki Könyvkiadók.

        Projective Geometry, 2nd ed. Toronto: University of Toronto Press.

        Cayley diagrams and regular complex polygons. In *A Survey of
        Combinatorial Theory*, ed. J. N. Srivastava, 85–93.
        Amsterdam: North-Holland Publishing Co.

1974   (With W. W. Rouse Ball) *Mathematical Recreations and Essays*,
        12th ed. Toronto: University of Toronto Press.

        Regular Complex Polytopes, 185. London and New York:
        Cambridge University Press.

        Polyhedral numbers. *For Dirk Struik*, eds R. S. Cohen et al.
        Dordrecht, Holland: D. Reidel Publishing Company, 25–35.

        Kepler and mathematics. *Vistas Astron.* 18:585–93.

        The equianharmonic surface and the Hessian polyhedron. *Annali
        di Matematica* 4 (98): 77–92.

        Geometry, Non-Euclidean. In *Encyclopaedia Britannica*,
        1112–1120.

1975   Desargues configurations and their collineation groups. *Math.
        Proc. Camb. Phil. Soc.* 78:227–46.

        The space-time continuum. *Historia Mathematica* 2:289–98.

1977   (With G. C. Shephard) Regular 3-complexes with toroidal cells.
        Journal of Combinatorial Theory 22:131–38.

        The Erlangen Program. *Math. Intelligencer* 0:22.

The Pappus configuration and its groups. *Pi Mu Epsilon J.*
6:331–36.

The Pappus configuration and the self-inscribed octagon. *Proc.
Kongl. Nederlandse Akad. van Wetenschappen* A
80:256–300.

(With C. M. Campbell and E. F. Robertson) Some families of
finite groups having two generators and two relations. *Proc.
Royal Soc. London* A 357:423–38.

Gauss as a geometer. *Historia Mathematica* 4:379–96.

Projektivna geometrija. [Yugoslavian translation of *Projective
Geometry*] Zagreb: Skalska knjiga.

(With S. L. Greitzer) *Az urja felfedezett geometria* [Hungarian
translation of *Geometry Revisited*] Budapest: Gondolat.

1978   Polytopes in the Netherlands. *Nieuw Archief voor Wiskunde* 3
(36): 116–41.

(With A. F. Wells) Three-dimensional nets and polyhedra. *Bull.
Amer. Math. Soc.* 84:466–70.

The amplitude of a Petrie polygon. *C. R. Math. Rep. Acad. Sci.
Can.* 1:9–12.

Parallel lines. *Can. Math. Bull.* 21(4):385–97.

1979   On R. M. Foster's regular maps with large faces. *AMS Proc.
Symp. Pure Math.* 34:117–28.

(With R. W. Frucht) A new trivalent symmetrical graph with 110
vertices. *Annals New York Acad. Sci.* 319:114–52.

The non-Euclidean symmetry of Escher's picture *Circle Limit III.
Leonardo* 12:19–25, 32.

The derivation of Schoenberg's star polytopes from Schoute's
simplex nets. *C. R. Math. Rep. Acad. Sci., Can.* 1:195.

(With Peter Huybers) A new approach to the chiral Archimedea
solids. *C. R. Math. Rep. Acad. Sci. Can.* 1:269–74.

1980   Angles and arcs in the hyperbolic plane. *Math. Chronicle*
9:17–33.

(With W. O. J. Moser) *Generators and Relations for Discrete
Groups*, 4th ed. Berlin and New York: Springer.

Higher-dimensional analogues of the tetrahedrite crystal twin.
Match (9) 67–72.

1981   Angels and devils. In *The Mathematical Gardner*, ed. David
Klarner, 197–200. Belmont, CA: Wadsworth.

The derivation of Schoenberg's star-polytopes from Schoute's
simplex nets. In *The Geometric Vein: The Coxeter Festschrift*,

eds. C. Davis, B. Grünbaum, F. A. Scherk, 149–164. Berlin and New York: Springer-Verlag.

Unvergängliche Geometrie, 2nd ed. Basel: Birkhäuser.

(With R. Frucht and S. L. Powers) *Zero-symmetric Graphs.* New York: Academic Press.

1982   Rational spherical triangles. *Math. Gazette* 66:145–47.

The Fifty-nine Icosahedra, 2nd ed. Berlin and New York: Springer-Verlag.

Regular polytopes. In *McGraw-Hill Encyclopedia of Science and Technology.* New York: McGraw-Hill.

Ten toroids and fifty-seven hemi-dodecahedra. *Geom. Dedicata* 13:87–99.

The group of genus two. *Rend. Sem. Mat. di Brescia* 7:219–48.

1983   My graph. *Proc. London Math. Soc.* 3 (46): 117–36.

(With W. L. Edge) The simple groups PSL (2,7) and PSL (2,11). *C. R. Math. Rep. Acad. Sci. Can.* 5:201–6.

The affine aspect of Yaglom's Galilean Feuerbach. *Nieuw Archief voor Wiskunde* 4 (1) 212–23.

A symmetrical arrangement of eleven hemi-icosahedra. *Annals of Discrete Mathematics* 20:103–14.

The twenty-seven lines on the cubic surface. In *Convexity and Its Applications*, eds. P. Gruber and J. M. Wills, 111–19. Basel: Birkhäuser.

(With S. L. Greitzer) *Zeitlose Geometrie* [German translation of *Geometry Revisited*]. Stuttgart: Klett Verlag.

1984   Surprising relationships among unitary reflection groups. *Proc. Edinburgh Math. Soc.* 27:185–94.

(With Asia Ivic Weiss) Twisted honeycombs {3, 5, 3} and their groups. *Geom. Dedicata* 17:160–79.

1985   The Lehmus inequality. *Aequationes Mathematicæ* 28:4–8.

The simplicial helix and the equation tan nq = n tanq. *Can. Math. Bull.* 28:385–393.

A special book review: M.C. Escher, His Life and Complete Graphic Work, *Math. Intelligencer* 7 (1):59–69.

Polytopes, kaleidoscopes, Pythagoras and the future. *C. R. Math. Rep. Acad. Sci. Can.* 7:107–14.

The seventeen black and white frieze types. *C. R. Math. Rep. Acad. Sci. Can.* 7:327–31.

Regular and semi-regular polytopes, part 2, *Math. Z.* 188:559–91.

1986   (With G. F. D. Duff and L. Havercamp) Variations on
        Pythagoras, *James Cook Mathematical Notes* 4: 4,208–12.
        (With W. W. Rouse Ball) *Mathematical Recreations and Essays*,
        12th ed. [Russian Translation]. Moscow: Mir.
        The generalized Petersen graph. In *Symmetry Unifying Human
        Understanding*, ed. Istvan Hargittai, 579–84. New York:
        Pergamon.
        Coloured Symmetry, in *M.C. Escher: Art and Science*, eds. H. S.
        M. Coxeter, M. Emmer, R. Penrose, and M. L. Teuber, 15–33.
        Amsterdam: North-Holland Publishing Co.
1987   Alicia Boole Stott. In *Women in Mathematics*, ed. L. S. Grinstein,
        220–24. New York: Greenwood Press.
        Introduction for *The Non-Euclidean Revolution*, by Richard J.
        Trudeau. Stuttgart: Birkhäuser.
        (With W. W. Rouse Ball) *Mathematical Recreations and Essays*,
        13th ed. New York: Dover Publications.
        Projective Geometry, 2nd ed. Berlin and New York: Springer.
        A packing of 840 balls of radius 9°0'19″ on the 3-sphere. In
        Intuitive Geometry, eds. K. Böröczky and G. Fejes Tóth. New
        York: Elsevier Science Publishers, 127–37.
        A simple introduction to coloured symmetry. *Internat. J.
        Quantum Chemistry* 31:455–61.
        On Miller's generalized dihedral group. *C. R. Math. Rep. Acad.
        Sci. Can.* 9:265–69.
1988   A challenging definite integral. *Amer. Math. Monthly* 95:330.
        Foreword for *The Foster Census*, by R. M. Foster. Winnipeg:
        Charles Babbage Research Centre.
        Regular and semi-regular polytopes, part 3, *Math. Z.* 200:3–45.
        Regular and semi-regular polyhedra. In *Shaping Space*, eds. M.
        Senechal and G. Fleck, 67–79. Boston and Basel: Birkhäuser.
1989   Star polytopes and the Schläfli function É(a, b, g). *Elemente der
        Math.* 44:25–36.
        Escher's lizards. *Structural Topology* 15:23–30.
        Trisecting an orthoscheme. *Computer Math. Appl.* 17:59–71.
        Review of "Sphere-packings, Lattices and Groups," by J. H.
        Conway and N. J. A. Sloane. *Amer. Math. Monthly*
        96:538–44.
1991   Orthogonal trees. *Bull. Inst. Combin. Appl.* 3:83–91.
        Evolution of Coxeter-Dynkin diagrams. *Nieuw Archief voor
        Wiskunde* 9:233–48.

Foreword for *Journey into Geometries*, by M. Sved. Washington: MAA Spectrum, Mathematical Association of America.

Regular Complex Polytopes, 2nd ed. [reprinted with corrections and a new chapter 14]. London and New York: Cambridge University Press.

(With G. C. Shephard) Some regular maps and their polyhedral realizations. In *Applied Geometry and Discrete Mathematics, The Victor Klee Festschrift*, eds. Peter Gritzmann and Bernd Sturmfels, 157–74. Providence, RI: American Mathematical Society.

1992   (With G. C. Shephard) Portraits of a family of complex polytopes. *Leonardo* 25:239–44.

Affine regularity. *Abh. Math. Sem. Univ. Hamburg* 62:249–53.

1993   Cyclotomic integers, nondiscrete tessellations, and quasicrystals. *Indagationes Mathematicæ*, 4:27–38.

(With Jan van de Craats) Philon lines in non-Euclidean planes. *J. Geometry* 48:26–55.

Cubic curves related to a quadrangle. *C. R. Math. Rep. Acad. Sci. Can.* 15:237–42.

The Real Projective Plane, 3rd ed. [reprinted with corrections and an appendix for Mathematica by George Beck, including a diskette]. Berlin and New York: Springer.

(With G. C. Shephard) Portraits of a family of complex polytopes. In *The Visual Mind*, ed. Michele Emmer. Cambridge, MA: MIT Press, 19–26.

1994   (With J. F. Rigby) Frieze patterns, triangulated polygons and dichromatic symmetry. In *The Lighter Side of Mathematics*, eds. R. K. Guy and R. E. Woodrow, 15–27. Washington, DC: Mathematical Association of America,

The evolution of Coxeter-Dynkin diagrams. In *Polytopes: Abstract, Convex and Computational*, eds. T. Bisztriczky, Peter McMullen, R. Schneider, and Asia Ivic Weiss, 21–42. NATO ASI Series, Dordrecht, Holland: Kluwer.

Symmetrical combinations of three or four hollow triangles. *Math. Intelligencer* 16 (3): 25–30.

Projective Geometry, rev reprint of 2nd ed. Berlin and New York: Springer-Verlag.

1995   *Kaleidoscopes*, eds. F. A. Sherk, Peter McMullen, A. C. Thompson, and Asia Ivic Weiss), 439. New York: Wiley.

Symmetrical combinations of triangles. *Math. Intelligencer* 17 (1): Postscript.

Some applications of trilinear coordinates. *Linear Algebra* 228:375–88.

1996   Close-packing and froth [reprint of *Illinois J. Math.* 2: 746–58]. *Forma* 11 (3): 271–85.

(With István Hargittai) Lifelong symmetry: a conversation with H. S. M. Coxeter. *Math. Intelligencer* 18 (4): 35–41.

The trigonometry of Escher's woodcut *Circle Limit III*, *Math. Intelligencer* 18 (4): 42–46.

1997   (With J. C. Fisher and J. B. Wilker) Coordinates for the regular complex polygons, *J. London Math. Soc.* 2 (55): 527–48.

Numerical distances among the spheres in a loxodromic sequence. *Math. Intelligencer* 19 (4): 41–47.

Reciprocating the regular polytopes. *J. London Math. Soc.* 2 (55): 549–57.

(With S. L. *Greitzer*) *Redécouvrons la géométrie* [French translation of *Geometry Revisited*] Paris: Éditions Jacques Gabay.

The trigonometry of hyperbolic tessellations. *Can. Math. Bull.* 2 (40): 158–68.

Erratum: The trigonometry of Escher's woodcut *Circle Limit III*. *Math. Intelligencer* 19 (1): 79; and revised in *Hyper Space* 6 (2): 53–57 (original article, *Math. Intelligencer* 18 (4): 42–46).

Numerical distances among the spheres in a loxodromic sequence. *Math. Intelligencer* 19 (4) 41–47.

1998   (With B. Grünbaum) Face-transitive polyhedra with rectangular faces. *C. R. Math. Rep. Acad. Sci. Can.* 20:16–21.

Whence does an ellipse look like a circle? *C. R. Math. Rep. Acad. Sci. Can.* 20 (1), 16–21.

Numerical distances among the circles in a loxodromic sequence. *Nieuw Archief voor Wiskunde* 16:1–9.

Seven cubes and ten 24-cells. *Discrete Comput. Geom.* 19:151–57.

Non-Euclidean Geometry, 6th ed. [reprinted with corrections and an appendix, Angles and Arcs in the Hyperbolic Plane]. Washington DC: Mathematical Association of America,

1999   *The Beauty of Geometry: Twelve Essays*. Mineola, NY: Dover Publications.

(With P. Du Val, T. H. Flather, and J. F. Petrie) *The Fifty-nine Icosahedra*, 3rd ed., Stadbroke, Norfolk: Tarquin Publications.

(With *Kharchenko, A. V.*) Frieze patterns for regular star polytopes and statistical honeycombs: Discrete geometry and rigidity. *Period. Math. Hungar.* 39 (1–3): 51–63.

2000   Five spheres in mutual contact. *J. Geometry and Graphics* 4(2): 109–14.

2001   (With B. Grünbaum) Face-transitive polyhedra with rectangular faces and icosahedral symmetry. *Discrete Comput. Geom.* 25:163–72.

2005   The Descartes circle theorem. In *The Changing Shape of Geometry*, ed. Chris Pritchard, 189–192. New York: Cambridge University Press.

An Absolute Property of Four Mutually Tangent Circles. In *Non-Euclidean Geometries, János Bolyai Memorial Volume, Mathematics and its Applications*, eds. András Prékopa and Emil Molnár, 109–14. Berlin, NY: Springer-Verlag.

ENDNOTES

CHAPTER 0—INTRODUCING DONALD COXETER

1. Donald Coxeter, interview, Toronto, January 27, 2003.

2. Robert Connelly (professor of mathematics, Cornell University), interview, Budapest, July 10, 2002.

3. Susan Coxeter Thomas, interview, Banff, Alberta, August 28, 2001.

4. Michel Broué (director, L'Institut Henri Poincaré), interview, Toronto, May 15, 2004.

5. Coxeter, "Whence Does an Ellipse Look like a Circle?" *Comptes Rendus, Mathematical Reports of the Academy of Science, Royal Society of Canada* 20, no. 1 (1998), 16–21.

6. A more formal definition is found in the 1911 *Encyclopaedia Britannica*, a good point of reference for the sum total of all knowledge at the beginning of the twentieth century (and the edition Coxeter would have consulted when he was a precocious young student). The sixty-three pages on geometry begin with this rather old-fashioned overview: "GEOMETRY, the general term for the branch of mathematics which has for its province the study of the properties of space. From experience, or possibly intuitively, we characterize existent space by certain fundamental qualities, termed axioms, which are insusceptible of proof; and these axioms, in conjunction with the mathematical entities of the point, straight line, curve, surface and solid, appropriately defined, are the premises from which the geometer draws conclusions . . . The origin of geometry (Gr. [geos], earth, [metron], a measure), is, according to Herodotus, to be found in the etymology of the word. Its birthplace was Egypt, and it arose from the need of surveying the lands inundated by the Nile floods. In its infancy it therefore consisted of a few rules, very rough and approximate, for computing the areas of triangles and quadrilaterals; and, with the Egyptians, it proceeded no further, the geometrical entities—the point, line, surface and solid—being only discussed in so far as they were involved in practical affairs. The point was realized as a mark or position, a straight line as a stretched string or the tracing of a pole, a surface as an area; but these units were not abstracted; and for the Egyptians geometry was only an art—an auxiliary to surveying." Alfred North Whitehead, "Geometry," in *Encyclopaedia Britannica*, 11th ed. (1910), 11:675.

7. Coxeter, interview, Banff, Alberta, August 28, 2001; Lister Sinclair, "Math and Aftermath," CBC *Ideas*, May 13 and 14, 1997.

8. Marjorie Senechal, Coxeter Legacy Conference, Toronto, May 14, 2004, "Donald and the Golden Rhombohedra," in *The Coxeter Legacy: Reflections and Projections*, 159–177.

9. Coxeter, interviews, August 2001–March 2003.

10. Buckminster Fuller, *Synergetics: Explorations in the Geometry of Thinking*, ix.

11. Irving Kaplansky (director emeritus, Mathematicial Sciences Research Institute, University of California, Berkeley), interview, July 15, 2004; Robert Moody (professor of mathematical sciences, University of Edmonton), interview, Banff, Alberta, August 27, 2001.

12. David Logothetti, "An Interview with H. S. M. Coxeter, the King of Geometry," *The Two-Year College Mathematics Journal*, 11, no. 1 (January 1980), 2.

13. Doris Schattschneider (professor of mathematics emeritus, Moravian College),

interview, November 2005; John Conway (John von Neumann Professor of Mathematics, Princeton University), Princeton, interview, November 27, 2005.

14. William Moser (emeritus professor of mathematics, McGill University), interview, Toronto, March 28, 2003; András Prékopa (professor of mathematics, statistics, and operations research, Rutgers University; professor emeritus, Eötvös Loránd Tudományegyetem), interviews, Budapest, Hungary, July 7,10, 2002; Imre Toth (emeritus professor of the history and philosophy of mathematics, College International de Philosophie), interview, Budapest, July 12, 2002 and October 2004; John Ratcliffe (professor of mathematics, Vanderbilt University), personal interview, Budapest, Hungary, July 8, 2002 and January 25, 2005; Glenn Smith (geometry aficionado), personal interviews, Budapest, Hungary, July 7, 2002, and Toronto, May 13, 2004.

15. Walter Whiteley (director of applied mathematics, York University, Toronto), interviews, May 13, 2004, June 3, 2005, and numerous interviews by phone and e-mail 2002–2005.

16. László Lovász (mathematician in residence, Microsoft) interview, September 6, 2002.

17. Whitelely, interviews; Moody, interview; Sir Harry Kroto (professor of chemistry, Florida State University), interview, December 21, 2004; Gord Lang (communications scientist, Ferranti), e-mail interview, June 1–15, 2005; Neil Sloane (AT&T Fellow), interviews, February 3, 2005, October 24, 2005.

18. Leon Lederman and Christopher Hill, *Symmetry and the Beautiful Universe*.

19. Nima Geffen (associate professor of mathematics, Tel-Aviv University), Aspects of Symmetry Conference, Banff, Alberta, interview, August 27, 2001.

20. Ravi Vakil (assistant professor of mathematics, Stanford University), interview, January 29, 2005, October 30, 2005.

21. Broué.

22. Brian Greene, *The Elegant Universe*, 173–83.

23. Edward Witten (professor of physics, Princeton Institute for Advanced Study), e-mail correspondence "Re: Strings and Geometry," May 26, 2005.

24. Coxeter, *Introduction to Geometry*, ix.

25. Conway, interviews, Princeton, April 10–17, 2005.

26. Hermann Weyl, *Symmetry*; Lederman and Hill; István Hargittai and Magdolna Hargittai, *Symmetry: A Unifying Concept*; Mario Livio, *The Equation That Couldn't be Solved: How Mathematical Genius Discovered the Language of Symmetry*.

27. Whiteley.

28. Ibid.

29. Whiteley; Brad L. Neiger, "The Re-emergence of Thalidomide: Results of a Scientific Conference"; J. N. Gordon and P. M. Goggin, "Thalidomide and Its Derivatives: Emerging from the Wilderness"; Tommy Eriksson, Sven Björkman, Bodil Roth, Åsa Gyge, Peter Höglund, "Enantiomers of Thalidomide: Blood Distribution and the Influence of Serum Albumin on Chiral Inversion and Hydrolysis"; Marianne Reist, Pierre-Alain Carrupt, Eric Francotte, and Bernard Testa, "Chiral Inversion and Hydrolysis of Thalidomide: Mechanisms and Catalysis by Bases and Serum Albumin, and Chiral Stability of Teratogenic Metabolites."

30. Schattschneider.

31. Lederman and Hill.

32. Lederman and Hill; Schattschneider; Conway.

33. Vakil.

34. Chandler Davis (emeritus professor of mathematics, University of Toronto), interviews, August 2004, January 20, 2005, June 10, 2005, September 22, 2005, and e-mail correspondence June 2004–December 2005.

35. Coxeter, "Polytopes, Kaleidoscopes, Pythagoras and the Future," 107.

36. Conway, April, November 23–27, 2005.

37. Ibid.

38. Coxeter, *Regular Polytopes*, 1.

39. Conway.

40. Coxeter, *Regular Polytopes*, 118–41, 237–58.

41. Coxeter, interviews, August 2001–March 2003; David Pantalongy, "H. S. M. Coxeter's Unusual Collection of Geometric Models," 11–19; University of Toronto Museum of Scientific Instruments, http://www.chass.utoronto.ca/cgi-bin/utmusi/display?field—mathematics.

42. Ratcliffe.

43. E. T. Bell, *The Development of Mathematics*, 323.

44. Whiteley; Jeremy Gray (professor of the History of Mathematics and director of the Centre for the History of the Mathematical Sciences, Open University), interviews, July 10, 2002 and April 6, 2005; Philip Davis (emeritus professor of applied mathematics, Brown University), interview, April 27, 2004, and "The Rise, Fall, and Possible Transfiguration of Triangle Geometry: A Mini-history"; Chandler Davis, and e-mails "Re: history of geometry" and "What is geometry?" October 2004; I. M. Yaglom, "Elementary Geometry, Then and Now," in *The Geometric Vein: The Coxeter Festschrift*, 253–69.

45. Hans Freudenthal, "Geometry Between the Devil and the Deep Sea," 418–19.

46. Coxeter, *Introduction to Geometry*, pp. 161–162.

47. Whiteley, "The Decline and Rise of Geometry in 20th Century North America," 1999 CMESG conference paper, available at http://www.math.yorku.ca/Who/Faculty/Whiteley/menu.html; Whiteley, "Working with Visuals in the Mathematics Classroom: Why It Is Needed, Hard and Possible," (excerpt: "Learning to See Like a Mathematician"), May 2005, OAME conference paper; and Whiteley, "Introduction to Geometries," Math 3050.06, York University, Toronto, 2004–2005.

48. Toth; and Toth, *Palimpseste: Propos avant un triangle*, 40; Pierre Cartier (emeritus director, Institut des Hautes Études Scientifique), interviews, Montreal, April 12, 13, 2003; Yaglom.

49. Toth; Cartier; Marjorie Senechal, "An Interview with Pierre Cartier," Bob Moon. *The New Maths Curriculum Controversy: An International Story*," 99; Jean Dieudonné, "New Thinking in School Mathematics," in *New Thinking in School Mathematics*," OEEC, 31–49.

50. Coxeter, interview, Budapest, July 10, 2002.

51. Marjorie Senechal (professor of mathematics, Smith College), interviews, Toronto, May 12, 13, 2004; Seymour Schuster (emeritus professor of mathematics, Carleton College Minnesota), interview, Toronto, May 13, 2004; Asia Ivić Weiss (professor of mathematics, York University), interview, Toronto, May 16, 2004, and Budapest, July 11, 2002; Branko Grünbaum (emeritus professor of mathematics, University of Washington), e-mail correspondence, January 25, 2005, "Re: The Man Who Saved Geometry."

52. Coxeter has a shadow in Ignatius J. Reilly, from the Pulitzer prize–winning novel, *A Confederacy of Dunces*, by John Kennedy Toole. The protagonist Reilly, a modern Don Quixote, bemoans the lack of a proper geometry and theology in the world (Reilly: "What I want is a good, strong monarchy, with a tasteful and decent king, who has some knowledge of geometry and theology and to cultivate a Rich Inner Life"). The title is borrowed from Jonathan Swift: "When a true genius appears in the world, you may know him by this sign, that the dunces are all in a confederacy against him." John Kennedy Toole, *A Confederacy of Dunces*, 213; Jonathan Swift, *A Modest Proposal and Other Satires*, 188.

53. Witnessing Coxeter's all-consuming passion, and being consumed by it oneself, brings

to mind writer Lawrence Weschler's description of his "passion pieces": "There is something both marvelous and hilarious in watching the humdrum suddenly take flight . . . [people] just moseying down the street one day, minding their own business, when suddenly and almost spontaneously they caught fire, they became obsessed, the became intensely focused and intensely alive—ending up, by day's end, somewhere altogether different from where they'd imagined they were setting out that morning." Weschler, *A Wanderer in the Perfect City: Selected Passion Pieces*, xii–xiii.

CHAPTER 1—MR. POLYTOPE GOES TO BUDAPEST

1.   Donald Coxeter, "An Absolute Property of Four Mutually Tangent Circles," János Bolyai Conference on Hyperbolic Geometry, July 8, 2002.

2.   E. T. Bell, *Men of Mathematics*, 19–34.

3.   Coxeter, "An Absolute Property of Four Mutually Tangent Circles," in *Non-Euclidean Geometries, János Bolyai Memorial Volume—Commemorating the 200th Anniversary of His Birthday*, 109–14.

4.   Coxeter, interview, Budapest, July 8, 2002; Coxeter, "The Descartes Circle Theorem," in *The Changing Shape of Geometry*, 189–192.

5.   Bolyai Conference, Budapest, July 8, 2002.

6.   Coxeter, Budapest, "An Absolute Property of Four Mutually Tangent Circles."

7.   Ibid.

8.   Coxeter, interviews, August 2001–July 2002.

9.   Ibid.

10.   Carolyn Abraham, *Possessing Genius: The True Account of the Bizarre Odyssey of Einstein's Brain*.

11.   Dr. Sandra Witelson recalled the course of events somewhat differently. Coxeter's daughter, having read about the brain research at McMaster, called Dr. Witelson (unbeknownst to Coxeter) with the information that her father was a geometrical genius who might be worthy of study. Witelson then pursued procurement. Sandra Witelson (neuroscientist, McMaster University), interviews, March and April 2003; Coxeter, interview, August 2001.

12.   Coxeter, interview, July 5, 2002.

13.   Ibid.

14.   Prékopa.

15.   Smith.

16.   Ibid.

17.   Ibid.

18.   Coxeter, interview, Banff, August 23, 2001.

19.   Coxeter, interview, Budapest, July 7, 2002.

20.   As Gray described: "All historians of this period are essentially lawyers. They come on as: 'My client: spotlessly clean, invented everything in sight, only person around at the time . . . absolutely a load of rubbish that anybody could allege that *your* client did something similar.' They are very partisan people." Gray.

21.   Eric W. Weisstein, "Platonic Solid." From *MathWorld*—A Wolfram Web Resource. http://mathworld.wolfram.com/PlatonicSolid.html (accessed October 6, 2005).

22.   Hart is a polyhedron enthusiast, a research professor in the department of computer science at Stony Brook University, New York, and a sculptor who makes impressionistic polyhedral sculptures—sometimes with plastic knives and forks—that are so massive they require a "barn-raising" to erect. "My sculpture would not exist if Coxeter did not exist," Hart said. "Coxeter's book *Regular Polytopes* opened my eyes to what geometry could be." George Hart,

interview, January 18, 2005; Hart's Online Encyclopedia of Polyhedra, http://www.georgehart. com/virtual-polyhedra/roman_dodecahedra.html (accessed October 6, 2005).

23. Ibid.

24. Michele Emmer, "Art and Mathematics: The Platonic Solids," in *The Visual Mind*; Coxeter, *Introduction to Geometry*, 149.

25. In *Regular Polytopes*, Coxeter continued to say, "The tetrahedron, cube, and octahedron occur in nature as crystals (of various substances, such as sodium suphantimoniate, common salt, and chrome alum, respectively). The two more complicated regular solids cannot form crystals, but need the spark of life for their natural occurrence. Haeckel observed them as skeletons of microscopic sea animals called radiolaria, the most perfect examples being *Circogonia icosahedra* and *Circorrhegma dodecahedra*." The statement concerning "the impossibility of any inorganic occurrence" Coxeter later regretted and retracted in the preface to the third edition: "That statement must now be taken with a grain of borax, for the element boron forms a molecule B_{12} whose twelve atoms are arranged like the vertices of an icosahedron." In 1984, crystals were discovered with a dodecahedral form—"quasicrystals"—which pleased Coxeter immensely. Coxeter, *Regular Polytopes*, 13.

26. Whitehead.

27. Conway.

28. W. W. R. Ball, *A Short Account of the History of Mathematics*, 14–28.

29. Ibid., 19.

30. Coxeter once noticed a proof for the Pythagorean theorem while glancing beneath his feet at an expanse of interlocking patio stones—a tessellation of the plane by squares of two different sizes. Roger Penrose, *The Road to Reality*, 25–28; and Penrose, interview, March 23, 2005.

31. John J. O'Connor and Edmund F. Robertson, The MacTutor History of Mathematics Archive. "Pythagoras of Samos." http://www-groups.dcs.st-and.ac.uk/~history/Mathematicians/ Pythagoras.html (accessed October 6, 2005).

32. Ball, 21.

33. Ball, 43.

34. Penrose, 7–23, and interview.

35. A. E. Taylor, *Plato, The Man and His Work*, 441.

36. Christopher Green (York University), "Classics in the History of Psychology." "Timaeus." http://psychclassics.yorku.ca/Plato/Timaeus/timaeus2.htm; and Benjamin Jowett, "Selections from Plato's *Timaeus*," in *The Collected Works of Plato*, edited by Huntington and Cairns, Princeton University Press, 1980.

37. Coxeter, *Regular Polytopes*, 5–6, 15–17.

38. Green.

39. Ibid.; and Conway, interviews, April 10–17, 2005.

40. Green.

41. Green; "The Platonic Solids," http://www.math.dartmouth.edu/~matc/math5.geometry/ unit6/unit6.html; and "The Matter of Timaeus," http://www.anunnaki.org/library/codex/ timaeus.php.

42. Coxeter, *Introduction to Geometry*, 149.

43. And while Plato hypothesized that the universe has the shape of a dodecahedron, the Pythagoreans were the first to suggest that the shape of the Earth was spherical. Historians doubt there was any hard evidence behind this belief; the Pythagoreans likely based it on mystical, philosophical, and aesthetic influences—the sphere being the most perfect shape, the Earth and all the celestial bodies simply had to be spherical. Legend has it that some two centuries

later, Aristotle (384–22 BC) proved the Earth to be spherical (though generally he is not noted for his accuracy on astronomical matters). He noticed while traveling north to south that new constellations appeared and disappeared over the horizon. But he gathered his most convincing evidence during a lunar eclipse, when he observed the Earth's shadow on the moon—it was always round, and the only object that always has a circular shadow is a sphere. Ibid.; "The Platonic Solids" and "The Matter of *Timaeus*"; and O'Connor and Robertson, "Pythagoras of Samos," http://www-groups.dcs.st-and.ac.uk/~history/Mathematicians/Pythagoras.html (accessed January 19, 2006); and Thomas Fowler, "Aristotle's Astronomy," http://www.perseus .tufts.edu/GreekScience/Students/Tom/AristotleAstro.html.

44. Jeffrey Weeks (freelance geometer), interview, November 28, 2003; and Kroto.

45. Coxeter, *Regular Polytopes*, pp. 1–24; Conway; and Daud Sutton, pp. 14–17.

46. Coxeter, *Introduction to Geometry*, 149.

47. Here are Coxeter's instructions for an elasticized dodecahedron: "From a sheet of cardboard, cut out two nets, one for each bowl. Run a blunt knife along the five sides of the central pentagon so as to make them into hinged edges. Place one net crosswise on the other, with the scored edges outward, and bind them by running an elastic band alternately above and below the corners of the double star, holding the model flat with one hand. Removing the hand so as to allow the central pentagons to move away from each other, we see the dodecahedron rising as a perfect model." Barry Monson (professor of mathematics, University of New Brunswick), interview, Toronto, January 25, 2005, February 14, 2005, and June 29, 2005; and Ibid.

48. Coxeter, *Regular Polytopes*, 1. See pp. 5–6 for a version of Euclid's proof that there "cannot be more than five regular solids, and a simple construction to show that each of the five actually exists."

49. Coxeter, *Regular Polytopes*, chap. 1, 2; Sutton, 14–15; Conway; Senechal, numerous telephone and e-mail interviews, 2004–2005; Schattschneider, numerous telephone and e-mail interviews, 2004–2005.

50. Coxeter, *Regular Polytopes*, 1, 13–14; Ball, 52–62; O'Connor and Robertson, "Euclid of Alexandria," http://www-groups.dcs.st-and.ac.uk/~history/Mathematicians/Euclid.html (accessed January 19, 2006); Weisstein, "Elements," http://mathworld.wolfram.com/Elements.html (accessed January 19, 2006).

51. "Euclides," http://www.archaeonia.com/science/mathematics/euclid.htm (accessed January 19, 2006).

52. Jeremy Gray, *János Bolyai, Non-Euclidean Geometry, and the Nature of Space*, 28–29, 107.

53. In his book *Envisioning Information*, Edward Tufte contrasted Byrne's colorful visual approach with "an orthodox march through the Pythagorean theorem," which he said demands that, "too much time must be spent puzzling over an alphabetic macaroni of 63 encoded links between diagram and proof." Tufte also mentioned another edition of Euclid's *Elements* (published in London in 1570; translated to English by Henry Billingsley, with a preface by John Dee), in which nets of geometric models were pasted to the pages, allowing the reader an interactive experience, folding models on the spot as they read the corresponding text (and making it a pop-up geometry book for subsequent readers, that is, if the models were not removed after folded). Edward R. Tufte, *Envisioning Information*, 16, 84–87; and Ibid., p. 19.

54. University of British Columbia, Sun SITE Digital Math Archive, http://www.sunsite .ubc.ca/DigitalMathArchive/Euclid/byrne.html (accessed January 19, 2006).

55. Coxeter, *Introduction to Geometry*, 4.

56. Coxeter, *Non-Euclidean Geometry*, 2.

57. Coxeter, *Introduction to Geometry*, 5.

58. Ibid.

59. Gray, interview; and Gray, *János Bolyai*.

60. The most elaborate attempts, Coxeter noted, were those of Italian Jesuit Giovanni Saccheri (1667–1733), followed some fifty years later by Frenchman Johan Lambert (1728–77). In 1763, a German doctoral thesis recorded a total of twenty-eight attempts to prove the parallel postulate. "The very fact of the procession indicates that something was amiss," said Gray in his book *János Bolyai, Non-Euclidean Geometry, and the Nature of Space*, "else one of these attempts would surely have commanded assent thereafter." Even the most formidable mathematicians of their day could not resist temptation. Andrien-Marie Legendre (1752–1833), Pierre-Simon Laplace (1749–1827), and Joseph-Louis Lagrange (1736–1813) all made ill-fated attempts. Lagrange, Gray recounted, "the greatest mathematician in the world at the time, had the embarrassment of reading a paper on the parallel postulate to the Institut de France which so palpably assumed what it wanted to prove that the only response was a painful silence while Lagrange put his paper back in his pocket and the meeting moved on to next business." Coxeter, *Non-Euclidean Geometry*, 1–13; and Gray, *János Bolyai*, 40 and 32.

61. Gray, interviews.

62. Gray, *János Bolyai*, 49–52.

63. Ibid., 51–52.

64. Ibid., 51.

65. Ibid., 52.

66. Bolyai's comment that "certain things ripen at the same time" proved prophetic, as he was in good company with his discovery of non-Euclidean geometry. The term was actually invented by German Carl Friedrich Gauss (1777–1855) who, even before Bolyai's discovery, noticed, "The assumption that (in a triangle) the sum of three angles is less than 180 degrees leads to a curious geometry, quite different from our own, but thoroughly consistent, which I have developed to my satisfaction." Gauss did not publish his work because he feared no one would believe him—the laws of Euclidean geometry governed the world and anything else was heresy. Later, Russian Nikolai Lobachevsky (1792–1856), German Bernhard Riemann (1826–66), and Ludwig Schläfli (1814–95), in Switzerland, also developed new geometries by assuming the fifth postulate false. They all approached it by different tacks, and thus each produced subtly different geometries that now fall under the umbrella of non-Euclidean (Bolyai called his "absolute geometry," and Lobachevsky came up with "imaginary geometry"). As Coxeter recounted, "The subject was unified in 1871 by [Felix] Klein, who gave the names parabolic, hyperbolic, and elliptic to the respective systems of Euclid, Bolyai-Lobachevsky, and Riemann-Schläfli." Klein (1849–1925) published his famous Erlangen Program, proposing a unification by employing symmetry as the underlying principle. The different geometries were compatible and coexisted because shapes in their respective spaces were all governed by properties of symmetry and transformations. The symmetries Klein proposed could be united on an abstract level, collected and interconnected through the algebra of group theory. Ibid., 53.

67. Gray, 54–55.

68. Bolyai's geometry also allowed him to accomplish a feat that truly is impossible in Euclidean geometry: Bolyai squared the circle. Coxeter also devised a method for squaring the circle, which he demonstrated in the opening of a lecture he delivered at Pacific Institute of Mathematical Sciences in Vancouver in 2000. "You probably have heard of the ancient problem of squaring the circle. You see, here's the circle," he said, holding in both hands a large circle made from a half-inch band of metal. "But if you want to square it this is all you have to do": he snapped his fingers with a twist around the edge, popping the circle with a resounding metallic echo into a perfect

square, and garnering appreciative laughter and applause. The problem of squaring the circle asks: given a circle, how can a square be constructed whose enclosed area is precisely equal to that of the circle? One approach dating to ancient times is to approximate the circle by a regular polygon with a suitably large number of sides—Chinese geometer Liu Hui, circa AD 264, used a polygon of 3,072 sides. Ibid., 67–75; Coxeter, public lecture, "The Mathematics in the Art of M.C. Escher," Pacific Institute for the Mathematical Sciences, Changing the Culture 2000: Visualizing Mathematics, video recording, http://www.pims.math.ca/education/2000/CtC/coxeter/ (accessed January 21, 2006); O'Connor and Robertson, "Liu Hui."

69. Gray, 53–81.

70. Sir Michael Atiyah (honorary professor of mathematics, Edinburgh University), interviews, August 26, 2003, May 3, 2005, May 31, 2005; and Chandler Davis (emeritus professor of mathematics, University of Toronto), e-mail correspondence, "Re: Euclidean and Non Euclidean," September 23, 2005.

71. Atiyah.

72. Atiyah; and C. Davis.

73. Atiyah.

74. Atiyah; C. Davis; and Conway interview, September 23, 2005.

75. Coxeter, "Geometry," in *Lectures on Modern Mathematics*, 58.

76. Alexander Bogomolny (a New Jersey mathematician), e-mail interview, "Re: H. S. M. Coxeter," July 22, 2004; and A. Bogomolny, Non-Euclidean Geometries, 1998, http://www.cut-the-knot.org/triangle/pythpar/NonEuclid.shtml (accessed July 2004).

77. CBC Television, "Explorations: Space, Time and the Atom Bomb," with Lister Sinclair, February 26, 1957.

78. Ibid.

79. Coxeter, *Non-Euclidean Geometry*, 4.

80. Gray, *János Bolyai*, 52.

81. Bolyai Conference, July 8, 2002.

82. Gray, interview, Budapest, July 10, 2002.

83. Ibid.

84. Karoly Bezdek (director of the Centre of Computational and Discrete Geometry, University of Calgary), interview, Budapest, July 8, 2002, and September 16, 2002.

85. Prékopa, interview, July 10, 2002.

86. G. H. Hardy, *A Mathematician's Apology*, 1.

87. Paul Hoffman, *The Man Who Loved Only Numbers*.

88. Ron Graham, interview, April 20, 2005.

89. Coxeter once revealed his secret for longevity during a question-and-answer session at his ninetieth-birthday conference. "His talk didn't really reveal the energy of the young Coxeter—and by young I mean eighty-five," said John Coleman, one of Coxeter's first students. The question came from a middle-aged gentleman, "somewhat swarthy, slightly round, ample," who asked in an ingratiating voice, "Oh, Dr. Coxeter, have you any suggestion for us younger mathematicians as to how we may retain the acumen of mind that you have shown as you get older?" "Coxeter was a modest Englishman," recalled Coleman, "and he was embarrassed that someone was suggesting that he had acumen of mind. So he sort of stood there, thinking. Moving from one foot to the other, in that diffident English way. And then he came out with, 'Well, perhaps if you followed my pattern of life, were a vegetarian and did a hundred push-ups every morning before breakfast, that would help.'" The questioner hadn't really received the advice he wanted. "The poor guy, who was obviously someone who loved a twelve- or sixteen-ounce steak with potatoes, was very nonplussed, very disappointed that there wasn't some intellectual silver bullet that would

keep him alert while he indulged in his beer and steak." Coxeter's intellectual magic bullet may have been his adoption of his father's habit of beginning each morning with a headstand, which on another occasion he cited as a possible explanation for his intellectual endurance. John Coleman (emeritus professor of mathematics, Queen's University), interviews, Kingston, Ontario May 23, 2003, and March 8, 2005.

90. Coxeter, interview, Budapest, July 2002.

91. Bruce Schechter, *My Brain Is Open*, 179.

92. Schechter, *My Brain Is Open*.

93. Graham.

94. Schechter, 178–79.

95. Coxeter diaries, 10 and 12 November 1935, H. M. S. Coxeter Fonds, University of Toronto Archives, B2004–0024.

96. Coxeter diaries, 27 June 1965.

97. Erdös correspondence, Coxeter Fonds, University of Toronto Archives.

98. Ibid.

99. There are two links making Coxeter an Erdös 2: John Horton Conway, the John von Neumann professor of mathematics at Princeton, is an Erdös 1. With Erdös (and M. J. T. Guy and H. T. Croft) he published on the "Distribution of Values of Angles Determined by Coplanar Points," in 1979; and with Coxeter he published two papers—on "Triangulated Polygons and Frieze Patterns" in 1973, and "The Centre of a Finitely Generated Group," in 1972. C. Ambrose Rogers, professor emeritus at the University of London, is another link. Rogers wrote papers with Erdös in 1953 on "The Covering of *n*-Dimensional Space by Spheres," and in 1961 on "Covering Space with Convex Bodies," and wrote with Coxeter in 1959 on a similar topic, "Covering Space with Equal Spheres." Graham, citing MathSciNet; and Conway, interviews, November 2005.

100. Susan Thomas, interview, Budapest, July 8, 2002.

101. Ernest Vinberg (professor of mathematics, Moscow State University), interviews, Budapest, July 8, 11, 2002.

102. Daina Taimina (senior research associate, Cornell University), interviews, Budapest, July 8, 11, 2002.

103. John Ratcliffe (professor of mathematics, Vanderbilt University), interviews, Budapest, July 8, 10, 2002 and January 2005.

104. Coxeter, interviews, Budapest, July 8 and 9–13, 2002.

CHAPTER 2—YOUNG DONALD IN WONDERLAND

1. Coxeter, interviews, August 2001–March 2003.

2. Ibid.

3. Caroline Dakers, *The Holland Park Circle: Artists and Victorian Society*, 238.

4. "H. S. M. Coxeter, University of Toronto Library Oral History Project," University of Toronto Archives, B1986–0088.

5. Coxeter interviews; Susan Thomas, family papers.

6. Coxeter, being a vegetarian and lover of animals, was unhappy to discover the provenance of his surname: "I wasn't very pleased to hear that the origin was 'cock setter,' meaning someone who arranged cock fights. It was a disappointment. I had thought it might have been 'clock setter,' that one of my ancestors had been making clocks, but it wasn't so." Coxeter interviews; "University of Toronto Oral History Project."

7. Coxeter, interviews; Timothy Prus (resident at 34 Holland Park Road), interview, September 3, 2003.

8. Coxeter, interviews.

9.   Coxeter, CBC *Ideas*, "Math and Aftermath," May 13, 14, 1997.

10.   Coxeter, "Mathematics and Music," *The Canadian Music Journal* (1962): 15.

11.   Ibid., 13–14.

12.   Historically speaking, Coxeter observed that great periods of musical development loosely coincided with great periods of mathematical advancement. "As a counterpart for the music of the Italians from Palestrina to Respighi we have the mathematics of Leonardo of Pisa (alias Fibonacci), Piero della Francesca, Saccheri . . . Beltrami, and many great geometers of our own century. Germany's Bach, Beethoven, Schubert, Brahms, and Wagner are matched by Leibniz, Gauss, Riemann, Klein, and Hilbert; France's Saint-Saëns, Fauré, Debussy, and Ravel by Cauchy, Galois, Poincaré, and Elie Cartan; Russia's Tschaikowsky and Shostakovitch by Lobachevsky and Kolmogorov; Norway's Grieg by Abel; England's Purcell by Newton; and so on. A notable exception is Victorian England, which produced no musical counterpart for the great mathematican Cayley, whose collected works fill a dozen huge volumes." Ibid., 14.

13.   Ibid., 17.

14.   Ibid.

15.   Ibid., 18–19.

16.   Ibid., 15–16.

17.   Coxeter, "Math and Aftermath."

18.   Recently Coxeter's compositions were evaluated for performance (a celebration of Coxeter's life at the Fields Institute in Toronto in 2004 featured a string quartet concert of Coxeter's music). About 20 percent were judged suitable for arrangement, and in fact quite accomplished and complex for such a young composer. "He had clearly absorbed the turn-of-the-century musical vocabulary of his time," noted Charles Small, a former mathematician and lifelong chamber music enthusiast in Moffat, Ontario, who scored the music for performance. "There are Wagnerian flourishes, traces of Schumann and Mendelssohn, and especially echoes of Elgar and the English school. But his music (like his mathematics!) reveals also a strong original voice and a rich and playful imagination." Another performer of Coxeter's music, Robert Craig, a freelance musician, musicologist and theorist in Kansas City, commented that Coxeter's immature ambitions were evident. "His music doesn't always fall into the hands well," said Craig. He also noted that the style of music Coxeter attempted was characterized by "harmonic surprises," but Coxeter's surprises were often awkward or a little too unexpected. His neat and nuanced manuscripts indicate he was very detail-oriented: "his notation is excellent, and he's very conscious of tempo and dynamics." Coxeter, interviews; "University of Toronto Oral History Project"; Coxeter Legacy Conference, Charles Small, interview, Toronto, May 12, 2004, and e-mail "Re: Coxeter," January 23, 2006; and Bridges: Mathematical Connections in Art, Music, and Science, "Coxeter Day," Robert T. Craig, interview, Banff, August 3, 2005, and e-mail correspondence, "Re: Coxeter Music," October 17, 2005.

19.   Coxeter interviews; Eve Coxeter, interviews, Liverpool, September 2003; Nesta Coxeter, interviews, London, September 2003; and Joan Coxeter, interview, May 20, 2004.

20.   C. Sheridan Jones, "Under the Zepps!" *London in War-Time*, 106–121.

21.   Coxeter, interviews; and Thomas, interviews.

22.   Coxeter, interviews; and Coxeter, "Amellaibian," Coxeter Fonds, University of Toronto Archives.

23.   Coxeter, interviews.

24.   Harold Coxeter to Katie Gabler, 16 September 1920, Nesta Coxeter, family papers.

25.   Coxeter, interview, March 14, 2003.

26.   Ibid.

27.   One of Donald's half sisters remembered a different story: Harold hired a lady to meet

him at a hotel, tipping off an inspector who burst in on a fabricated tryst, thus providing the public proof of infidelity necessary for divorce. Coxeter, interviews; Thomas, interviews; and Eve Coxeter, interviews.

28. Coxeter, interviews; Thomas, interviews; Eve Coxeter, interviews; Nesta Coxeter, interviews; and Joan Coxeter, interview.

29. Harold Coxeter to Rosalie Gabler, 12 September 1924, Nesta Coxeter, family papers.

30. Coxeter, interviews.

31. Coxeter, interviews; John Wilker, interview with Coxeter, CMS Fiftieth Anniversary, 1995; and "University of Toronto Oral History Project."

32. Yaglom; Gray; P. Davis; and C. Davis.

33. The impact of the Pythagorean theorem, for example, cannot be disputed, and its ingenuity gave it legs. In 1907, an Ohio high school mathematics teacher, Elisha Scott Loomis, compiled over 350 demonstrations devised over the years for the Pythagorean theorem—very pictorial proofs with names like "Bride's Chair," "Peacock's Tail," Windmill," "Franciscan's Cowl," and "Pythagorean Pants"—which he published under the title *The Pythagorean Proposition*. In the years since Pythagoras, classical geometry had ebbed and flowed. It was on the rise, bubbling to a slow boil, up to and including Newton and Descartes. Those were the days when it was a designation of privilege and status among scientists and mathematicians to address one another as "Cher Géomètre." Then geometry fell into a shallow valley, followed by another peak, a renaissance spanning the nineteenth century, with the pinnacle of Einstein's geometry of space and time. Near the end of this golden era, many books like Loomis's appeared, chronicling the discovery of striking and unexpected classical theorems. Two such works were D. Efremov's *New Geometry of the Triangle*, in 1902, and J. L. Coolidge's *A Treatise on the Circle and the Sphere*, in 1916. Not long into the new century, however, geometry—classical elementary Euclidean geometry, anyway—was again heading into another dip, another period of eclipse. Bogomolny; Gray; P. Davis; Yaglom.

34. Gray, interviews.

35. Coxeter's retort, borrowing as he often did from his friend and colleague J. L. Synge, would be: "Can it be that all the great scientists of the past were really playing a game, a game in which the rules are written not by man but by God? . . . When we play, we do not ask why we are playing—we just play. Play deserves no moral code except that strange code which, for some unknown reason, imposes itself on play . . . You will search in vain through scientific literature for hints of motivation. And as for the strange moral code observed by scientists, what could be stranger than an abstract regard for truth in a world which is full of concealment, deception, and taboos? . . . In submitting to your consideration the idea that the human mind is at its best when playing, I am myself playing, and that makes me feel that what I am saying may have in it an element of truth." Coxeter, *Regular Polytopes*, 77, quoting from *Hermathena* 19 (1958): 40.

36. Senechal, May 12, 2004; and J. L. Heilbron, in *Geometry Civilized*, vi.

37. Gray, interviews, and *Non-Euclidean Geometry and the Nature of Space*, 107–19.

38. W. W. Rouse Ball, in his *Mathematical Recreations and Essays*, restored the tradition for amusement in his chapter on "Geometrical Recreations," providing one fallacy with a brain-teasing diagram that impossibly proved every triangle to be isosceles. W. W. R. Ball, *Mathematical Recreations and Essays*, 44–51; and Gray.

39. Senechal.

40. Gray, 95–96; Bill Richardson (secretary, the Mathematical Association) e-mail correspondence, April 14, 2005, "Re: The Mathematical Gazette," and January 14, 2006, "Re: Question re the Association for the Improvement of Geometrical Teaching"; and J. V. Armitage,

"The Place of Geometry in a Mathematical Education," in *The Changing Shape of Geometry*, 515–26.

41. Italians Giuseppe Peano (1858–1932) and Mario Pieri (1860–1913) fostered this approach earlier. Gray, interview; and *János Bolyai*, 114.

42. Ioan James, *Remarkable Mathematicians: From Euler to von Neumann*; Gray, *The Hilbert Challenge*; O'Connor and Robertson, "David Hilbert," http://www-groups.dcs.st-and.ac.uk/~history/Mathematicians/Hilbert.html (accessed January 21, 2006); and re: "beer mugs," Peter Galison (professor of the history of science and physics, Harvard University), interview, October 20, 2005; and Gray, *The Hilbert Challenge*, 49.

43. Gray, interviews.

44. Ibid.

45. Edwin A. Abbott, *Flatland*.

46. Other mystical titles on the topic included *The Unseen Universe*, 1875, by P. G. Tait and Balfour Stewart, which quickly went through seventeen editions. P. D. Ouspensky published *Tertium Organum* in 1911; the American edition was called *The Third Canon of Thought, a Key to the Enigmas of the World*, and translated by the architect Claude Bragdon, "perhaps the foremost American proponent of the fourth dimension," who developed an architectural style that "employed three dimensional sections of four dimensional hyperspaces as the unifying motifs and structural elements of his buildings. These 'shadows of the fourth dimension' were meant to serve as embodied reminders . . . of a higher spiritual reality," according to David Pacchioli, "Deflating Hyperspace," *Research/Penn State*. The books that Coxeter thought "expounded admirably" included: Duncan Sommerville's *An Introduction to the Geometry of N Dimensions*, 1929; and E. H. Neville's *The Fourth Dimension*, 1921. Coxeter, *Regular Polytopes*, 118; see also Hugh Shearman, "The Problem of the Fourth Dimension," in *The Theosophist*.

47. While at a restaurant during the Bolyai conference in Budapest, one of Coxeter's dinner companions, Glenn Smith, happened to have a multicolored cube in the pocket of his trousers, which led to a dinner lesson on Hinton's method of how to think in four dimensions. Jeff Weeks, a freelance geometer from Canton, New York, who was also there at the table, provided this explanation: "First, think about Flatlanders, how 2-dimensional creatures living on a flat surface would imagine a 3D cube. They can see the breadth and the depth but they can't see the height directly. One way they might imagine the height is by using color-coding. Say the bottom face of the cube is red and the top face is violet, and other colors come in the middle—so it would go red, orange, yellow, green, blue, violet, like a rainbow. Then even if you flatten that cube down into their plane, but keep track of the colors, it would provide the information on the height. And that would give them a way of starting to think about an extra dimension. We 3-dimensional people can do the same thing. If we start with a cube, a solid cube, say that is entirely red in its interior, and then take a violet cube in exactly the same position, the color difference represents being higher or lower in a fourth dimension. By thinking of these colors you construct a four dimensional cube in your mind." Bolyai Conference, Budapest, July 10, 2002; and Charles H. Hinton, "Fourth Dimensional Writings," http://www.ibiblio.org/eldritch/chh/hinton.html (accessed January 21, 2006); and "Charles Howard Hinton," http://en.wikipedia.org/wiki/Charles_Howard_Hinton (accessed January 21, 2006).

48. Coxeter, interview, March 14, 2003.

49. H. G. Wells, *The Time Machine*, 3–4.

50. Alan Lightman, *The Discoveries*, 60–83, 240; and Peter Coles, "The Eclipse that Changed the Universe," http://www.firstscience.com/site/articles/coles.asp, January 10, 2006.

51. O'Connor and Robertson, "Albert Einstein," http://www-groups.dcs.st-and.ac.uk/~history/Mathematicians/Einstein.html (accessed January 21, 2006); and Alan Lightman, "Relativity and the Cosmos," at http://www.pbs.org/wgbh/nova/einstein/relativity/ (accessed January 21, 2006).

52. Barry Shell, Coxeter videotaped interview, May 1994; Wilker; University of Toronto Oral History Project; and Coxeter, interviews.

53. Conway, interviews, Princeton, June 21–25, 2004.

54. Ibid.; Coxeter, interviews.

55. Conway, interviews.

56. *King Henry IV, Complete Works of William Shakespeare*, act 3, scene 1, 440.

57. C. Davis, interviews.

58. Conway to Coxeter, March 7, 1957, Coxeter Fonds, University of Toronto Archives.

59. Ibid.

60. Coxeter, in *Regular Polytopes*, warned: "Little, if anything is gained by representing the fourth Euclidean dimension as time. In fact, this idea, so attractively developed by H. G. Wells in *The Time Machine*, has led such authors as J. W. Dunne (*An Experiment with Time*) into a serious misconception of Relativity." Coxeter, *Regular Polytopes*, 119.

61. Conway, interviews, April and November 2005.

62. Ibid.

63. Conway, interviews.

64. After a lot of thinking and pages and pages of calculations, Conway figured out the Monster group had 196,884 dimensions by inputting certain conditions into his programmable calculator, setting the machine to find the solution, and letting it run overnight. "In the morning, there it was," he said. "It was as much a surprise to me that I found it as it was to anyone else." Conway also calculated the group's order—this number has fifty-four digits, which he can recite off the top of his head in the following groupings: "8080 17424 79451 28758 86459 90496 17107 57005 75436 80000 00000." (See appendix 6 and endnotes for more on the Monster group, as well as the Monster Moonshine conjecture.) Conway, interviews, April 2005; see also John Horton Conway and Neil J. A. Sloane, *Sphere Packings, Lattices and Groups*.

65. Coxeter, *Regular Polytopes*, 118.

66. Ibid. Coxeter quoted the French: *"Un homme qui y consacrerait son existence arriverait peut-être à se peindre la quatrième dimension."*

67. Ibid., 119.

68. Ibid.

69. Ibid.

70. Coxeter's "Dimensional Analogy" treatise continued through parts 1–5, filling five notebooks, and ran the gamut: regular figure, angles, vertex angles, coordinates, reciprocal figures, the circle-sphere series, the square-octahedron series, absolute properties generalized, all regular figures determined, truncating the regular figures, the Archimedean solids, the Archimedean hyper-solids, the pure Archimedean series, regular concave figure, vertex-figures and truncation, complete with appendixes, summary of results, and a legend of notations used. "Dimensional Analogy: An Essay by D. Coxeter, February, 1923," Coxeter Fonds, University of Toronto Archives.

71. Coxeter, 236.

72. Ibid.

73. Ibid., 236–62.

74. Ibid., 119.

75. James Gleick, *Isaac Newton*, 36.

76. René Descartes, *Discourse on Method, The Harvard Classics*, at http://www.bartleby.com/34/1/1.html (accessed January 21, 2006).

77. E. T. Bell, *Men of Mathematics*, 19–55.

78. Ibid.

79. C. Davis, interviews.

80. Bell.

81. Ibid.; and Conway, interviews.

82. Bell, 40.

83. The dictum that mathematical truths preexist and are discovered rather than being invented is perhaps supported by the birth of analytical geometry (as it was with the simultaneous discoveries of non-Euclidean geometry). Within the same century, three mathematicians found this mathematical treasure independent of one another. Descartes' rival, fellow Frenchman Pierre de Fermat (1601–65), made the same discovery but lost out in the glory because he wasn't in the habit of publishing his work. Instead, he is famous for his last theorem, which took more than three and a half centuries to prove (this was done finally by Andrew Wiles, at Princeton, in 1994). And both Fermat and Descartes were preceded in their discovery by Frenchman Nicole d'Oresme (1323–82), who produced a similar form of coordinate geometry. Ibid., 47.

84. Ibid., 53–54.

85. Coxeter, interviews; Thomas, interviews; Eve Coxeter, Nesta Coxeter, and Joan Coxeter, interviews.

86. O'Connor and Robertson, "Bertrand Arthur William Russell," http://www-groups.dcs.st-and.ac.uk/~history/Mathematicians/Russell.html (accessed January 21, 2006).

87. Coxeter, interviews; Eve Coxeter, Nesta Coxeter, and Joan Coxeter, interviews; and Robert Kanigel, *The Man Who Knew Infinity: A Life of the Genius Ramanujan*, 186–191.

88. "Reading's Great People," http://www.readinglibraries.org.uk/services/local/morley.htm (accessed October 7, 2005).

89. Edith Morley to E. H. Neville, 11 September 1923, Coxeter Fonds, University of Toronto Archives.

90. Donald Coxeter to E. H. Neville, 11 September 1923, Coxeter Fonds, University of Toronto Archives.

91. Coxeter, to E. H. Neville, 11 October 1923, Coxeter Fonds, University of Toronto Archives.

92. Coxeter, interview; "University of Toronto Oral History Project."

93. Coxeter, interviews.

94. Ibid.

95. Ibid.

96. Ibid.; and Richardson.

97. Ibid.

98. An integral, a fundamental concept in calculus, is a mathematical object that can be interpreted as an area or a generalization of area. A definite integral is an integral with an upper and lower limit and is a way of defining certain regions in the plane. Weisstein, "Definite Integral," http://mathworld.wolfram.com/DefiniteIntegral.html.

99. Coxeter, "Mathematical Notes," *Mathematical Gazette*, 205.

100. Coxeter, interviews; Coxeter, interview, Shell.

101. Coxeter, interviews.

102. István Hargittai, "Lifelong Symmetry: A Conversation with H. S. M. Coxeter," *The Mathematical Intelligencer*, 37.

103. G. H. Hardy to "D. Coxeter Esq.," November 1926, Wren Library, Trinity College, Cambridge, File: Add. Ms. a. 275, 47–51.

CHAPTER 3—AUNT ALICE, AND THE CAMBRIDGE CLOISTER

1. David Blackburn and Jonathan Rosenhead, "Cambridge Mathematics," *The Eagle,* 46–47.

2. Ibid.

3. Coxeter, Personal Records of Fellows of the Royal Society, Coxeter Fonds, University of Toronto Archives.

4. O'Connor and Robertson, "Quotations by J. E. Littlewood," http://www-groups.dcs .st-and.ac.uk/~history/Quotations/Littlewood.html (accessed January 21, 2006).

5. Coxeter's obtaining a PhD in 1931 was more an aberration than the norm. His supervisor, H. F. Baker, enjoyed a distinguished mathematical career based only on a first degree, which remained an adequate demonstration of one's credentials well into the mid–twentieth century. In creating the PhD degree, British universities were responding to strong political pressure. "Imperial considerations were important. There had long been anxiety that students from the dominion would go to Germany or the United States and be weaned away from the mother country." John Aldrich, "The Mathematics PhD in the United Kingdom."

6. Dame Mary Cartwright (1900–98) had studied with G. H. Hardy at Oxford, working toward her PhD in 1928. In 1930, Cartwright received a fellowship that brought her to Girton College, Cambridge, where she continued working on her doctoral thesis and produced Cartwright's theorem. June Barrow-Green and Jeremy Gray, "Geometry at Cambridge, 1863–1940," in *Historia Mathematica*, 27 (prepress copy); O'Connor and Robertson, "Dame Mary Lucy Cartwright," http://www-groups.dcs.st-and.ac.uk/~history/Mathematicians/ Cartwright.html (accessed January 21, 2006).

7. Gray and Barrow-Green (both at the Open University), chronicle the ancestry of illustrious practitioners who built Cambridge's reputation as a place to study geometry, especially the classically based projective geometry: Cayley with his "botanist's eye," William Clifford, "a spellbinding lecturer," as well as Francis Macaulay, Bertrand Russell, A. F. Whitehead, and H. F. Baker—the latter more than any of his predecessors vaulted geometry at Cambridge to high status. However, in the early 1960s, "when the pressure built up to revise the syllabus and install key features of Bourbaki's vision of mathematics—much of it in response to the work of Hilbert and Emmy Noether in any case—projective geometry was a natural candidate making way for the new mathematics. And that is what happened: projective geometry . . . moved into the area of specialist options and out of the core, mainstream provision. From there, it fell somewhat into disrepute, criticized for its various imprecisions . . . The status of the subject has fluctuated ever since, as has its claim on the school syllabus, because the subject can seem less than rigorous and, even if rigorous, baroque." Ibid.

8. Coxeter's teachers were P. W. Wood (analytic geometry), Herbert Richmond and Thomas Room (projective geometry), Frank Ramsey (differential geometry) and his father, Arthur Ramsey (electricity), Max Newman (topology), Philip Hall (theory of groups), Albert Ingham (theory of numbers), S. Pollard, Abram Besicovitch, and Littlewood (analysis), and George Birtwistle (mechanics). A. A. Robb delivered the lecture on the geometry of time and space, which stressed, Coxeter recalled, "that the special theory of relativity becomes most neutral in terms of a real affine 4-space with a Minkowski metric determined by a cone." In this class, Coxeter said that his friend Patrick Du Val sat beside him, and was way ahead of the game—"he already understood the analogous geometric interpretation of [the Dutch mathematician and Einstein collaborator] Willem de Sitter's world [proposing that at a four-

dimensional space-time would fit in with cosmological models based on general relativity] . . . a model that provides a most convincing explanation for the expanding universe in terms of a zig-zag of world lines." Coxeter, Personal Records of Fellows of the Royal Society.

9.   June Barrow-Green, interview, London, September 12, 2003; and Gray, interviews.

10.   Coxeter was known to misplace his bicycle, and once reported it stolen to the police; somewhat later, "Walking along one day I saw it nicely parked by the side of the road, exactly where I had left it." Coxeter was also absent-minded about his appearance: "I should like to know who has been grumbling about my clothes this time," he groused in a letter home. "All right, I will try to remember about the flannel trousers and sports jacket when I next see Forbes." And he was too distracted to keep track of his keys: "Thought for three hours I had lost my keys. At last found them in trouser pocket." Coxeter, interviews; Michael Longuet-Higgins (geophysicist and geometer, UCSD, and Trinity Fellow), interviews, Cambridge, August–September 2003, and Toronto, May 2004; and Coxeter to Katie Coxeter, 13 February 1932, Coxeter Fonds, University of Toronto Archives; and Coxeter diaries, 11 November 1961, Coxeter Fonds, University of Toronto Archives.

11.   Longuet-Higgins.

12.   "A Simple Story," *The Trinity Magazine*, December 1926, 18.

13.   Coxeter, interviews.

14.   Longuet-Higgins.

15.   Gleick.

16.   Gleick, 56.

17.   Gleick.

18.   O'Connor and Robertson, "Sir Isaac Newton," http://www-groups.dcs.st-and.ac.uk/~history/Mathematicians/Newton.html (accessed November, 2005).

19.   Ibid.

20.   Gleick.

21.   Gleick, 130; and Gleick, interview, November 19, 2004.

22.   O'Connor and Robertson; Gleick, 15.

23.   Coxeter indicated that Kepler followed an illustrious line of polyhedron buffs—Pythagoras, Plato, Euclid, Archimedes, Archbishop Thomas Bradwardine, Albrecht Dürer, and Leonardo da Vinci. Dürer (1471–1528), the German Renaissance painter whose woodcut *Melancholia I* depicts an exasperated-looking thinker sitting staring at an unusual polyhedron, made a significant contribution to the literature on polyhedra with his book *Painter's Manual*. Leonardo da Vinci (1452–1519) made his contribution in Luca Pacioli's book *The Divine Proportion* (though many of Leonardo's sketches show polyhedra). In *The Divine Proportion*, Leonardo employs his deft artist's, mathematician's, and engineer's hand by presenting a new view of polyhedra, as hollow cages with solid edges, giving a view of both the front and back of the polyhedra at the same time. Coxeter said approvingly: "[Leonardo] made skeletal models of polyhedra, using strips of wood for their edges and leaving the faces to be imagined." Coxeter, "Kepler and Mathematics," in *Vistas in Astronomy*, 661–64.

24.   Of Kepler's attempts, Coxeter commented, "He sought a connection between the five solids and the six planets that were known in his time. Although he later discovered two new semi-regular polyhedra, there is no evidence that he predicted the discovery of Uranus and Neptune. In 1810 Louis Poinsot discovered another pair of regular polyhedra, and the astronomers responded with the asteroids . . . and Pluto . . . but we must resist these numerological temptations!" Coxeter, "Kepler and Mathematics," 661–70; and Kitty Ferguson, *Tycho and Kepler*, 181–99.

25.   Ferguson, 191.

26. Coxeter, "Kepler and Mathematics," 665, quoted from Koestler's *The Sleepwalkers (A History of Man's Changing Vision of the Universe)*.

27. Ferguson, 197.

28. O'Connor and Robertson, "Tycho Brahe," http://www-groups.dcs.st-and.ac.uk/~history/Mathematicians/Brahe.html (accessed January 21, 2006); and http://www.learningmatters.co.uk/education/a_to_z/maths/b.html (accessed January 21, 2006).

29. Ferguson, 337–51.

30. The solids attributed to Archimedes (287–212 BC) can be constructed from the Platonic solids in very specific ways—by truncating (cutting off their corners) or exploding them in such a way that when the interstices are filled in with regular polygons they meet the "semiregular" criteria. The truncated tetrahedron, for example, is comprised of hexagons and equilateral triangles; the truncated icosahedron is a patchwork of hexagons and pentagons—the shape on which a soccer ball is modeled. Conway, interviews; Coxeter, *Regular Polytopes*, 30; and Sutton, 32–33.

31. The two-dimensional cousins of the star polyhedra are the star polygons. Just as there are an infinite number of ordinary regular polygons, {3}, {4}, {5}, {6} . . . {n}, there are also infinitely many star polygons, {5/2}, {7/2}, {7/3}, {8/3}. The symbol for the star polygon {n/d} indicates that a star polygon is formed inside an ordinary regular n-gon by connecting one vertex to another that is d vertices away (counting clockwise), then repeating the process from that vertex, and so on. Or, arrange n number of dots in a circular configuration, equally spaced apart, and join the dots at the interval of d—so 5/2 requires five dots arranged equidistant in a circular pattern, every second dot to be joined until the polygon is closed; and 7/3 requires seven dots, every third dot joined until the polygon is closed. Also, just as there are only finitely many regular polyhedra—the Platonic solids: {3,3}, {4,3}, {3,4}, {5,3}, {3,5}—there are also only finitely many regular star-polyhedra, namely, {5/2,3} and {5/2,5}. Coxeter, *Regular Polytopes*, 93–97. Conway, interviews; Sutton, 28–29.

32. Coxeter (with P. Du Val, H. T. Flather, and J. F. Petrie), *The 59 Icosahedra*.

33. Wilker, Coxeter interview.

34. *The 59 Icosahedra* was coauthored by Flather as well as Coxeter's classmate at Trinity, Patrick Du Val, with illustrations done by John Petrie, and published in 1938 by the University of Toronto Press. They tried to get it published by the Royal Society of Edinburgh (RSE), but plans fell through, pursuant to this 1936 letter from RSE's president, Sir D'Arcy Thompson: "My dear Coxeter, I can't help telling you that you have given us a great deal of trouble with your paper, and that trouble is not over yet . . . Your paper will cost us a very considerable sum, a large sum, to print; and it is only worth that expenditure if it is to be, in its way, a standard work, as sort of locus classicus for the Icosahedron. The text is not written up with anything like sufficient care to make it so; and the Bibliography, which Du Val has appended to it is simply childish—simply making fools of us! . . . The Paper will come before us in October, and what the Council then says or does will, of course, depend entirely on what the Reporters tell them. I expect they will advise postponement, and revision. You may be quite sure they will not accept the paper, in its present state, with open arms. I am really astonished, my dear Coxeter, that among you, you have not spent more care and pains upon it; you have not treated it seriously." To this day, one set of Flather's models resides at Cambridge, and another at York University in Toronto. During the First World War, Flather sent his second set to Coxeter in Toronto for safekeeping. Coxeter subsequently gave it to his last PhD student, Asia Ivić Weiss. Conway, interviews; correspondence Thompson to Coxeter, Coxeter Fonds, University of Toronto Archives.

35. Barrow-Green; and Trinity College student records.

36. Coxeter to Katie Coxeter, 1 November 1931, Coxeter Fonds, University of Toronto Archives.

37. Coxeter to Katie Coxeter, 21 July 1932, Coxeter Fonds, University of Toronto Archives.

38. Coxeter to Katie Coxeter, 2 August 1932, Coxeter Fonds, University of Toronto Archives.

39. "Magpie & Stump," *The Trinity Magazine*, December 1928.

40. Coxeter, interviews.

41. Coxeter, interview, March 14, 2003.

42. Coxeter, interview, March 14, 2003.

43. Ibid.

44. Coxeter, Personal Records of Fellows of the Royal Society.

45. Coxeter, interviews.

46. Ibid.

47. In another relevant passage, Stekel addressed disguises of "love urges," including shame, anxiety and disgust, and said: "How many disorders of the stomach are only nervous and can be traced back to unconscious feelings of disgust! Many girls who suddenly cease to eat meat have within them a disgust of the "flesh," which they transfer to the process of nutrition. This manifestation is also to be seen in men; and especially in some fanatical vegetarians one can, here and there, detect these origins. Naturally, the disgust at meat (flesh) is then rationalized with social and humanitarian motives. Such people attach themselves all the more easily to a great movement, because they seek social explanations for their individual struggles and conflicts." Wilhelm Stekel, *Disguises of Love*, 32, 41.

48. Coxeter, dream diary, 20 June 1928–7 August 1928, dreams IV, VI, LXI, VIII, XXVI, VLII, XXXVII, XLVIII, Coxeter Fonds, University of Toronto Archives.

49. Ibid., dreams IX, VII, X, XV, XLVII.

50. Coxeter, interviews.

51. Coxeter, Personal Records of Fellows of the Royal Society.

52. Smith, interview.

53. The others he mentioned, in *Regular Polytopes*, were British Cambridge professor Arthur Cayley (1821–95), Polish Hermann Günter Grassmann (1809–77), and German August Möbius (1780–1868). Coxeter, *Regular Polytopes*, 141.

54. Ibid., 118; and Conway, interviews.

55. Coxeter, *Regular Polytopes*, 118–64, 292–95; and Conway, interviews.

56. Coxeter; Conway; and Schattschneider, e-mail correspondence "Re: diagrams, and a question," December 29, 2005.

57. Schläfli called the hypercube the "measure polytope," because it was considered the unit of measurement; and the generalized octahedron the "cross polytope," because it can be constructed by drawing a cross (of either two or three lines, depending on the dimension), and joining the ends. Ibid.

58. Nearly thirty years after Schläfli's inventions, some of the same ideas were rediscovered by an American. "The result was," said Coxeter, "that many people imagined W. Stringham to be the discoverer of the regular polytopes." And incidentally, Coxeter noted, the time he spent sitting reading Schläfli's work in the University Library "gave me the proper method for treating those integrals which had aroused Hardy's interest." Coxeter, *Regular Polytopes*, 143. Coxeter, Personal Records of Fellows of the Royal Society.

59. Schläfli's work book-ended Coxeter's oeuvre, figuring into both his PhD and his last paper delivered at the Budapest conference. In concluding his lecture "An Absolute Property of

Four Mutually Tangent Circles," Coxeter gave a theorem by Schläfli, and then asked, "Is there, for this elementary theorem, a demonstration more simple than the one derived from the theory of cubic forms?"—almost as if Coxeter was co-opting Schläfli's query to provide himself with one final question to leave to the world of mathematics. Harold Scott MacDonald Coxeter, PhD diss., "Some Contributions to the Study of Regular Polytopes," December 18, 1931; and Coxeter, "An Absolute Property of Four Mutually Tangent Circles."

60. Coxeter, interview, March 14, 2003; "University of Toronto Oral History Project."

61. Barrow-Green and Gray, 21–22.

62. Ibid.

63. Ibid., 21.

64. In his day, Coxeter attended with Patrick Du Val, John A. Todd, Jack G. Semple, William V. D. Hodge, William L. Edge, T. G. Room, and Gilbert de Beauregard Robinson, who was studying abroad from Toronto. Most of Baker's followers were keen, though some found these inescapable meetings rather tiring (this came out in one of Baker's obituaries). "Baker was not very inspiring as a lecturer," remembered Coxeter. "He went steadily on. I was apt to find it a little bit dry. One time I was asleep and then suddenly awake and heard Baker say, 'I see Coxeter is asleep.' That was a little embarrassing." Coxeter, Personal Records of Fellows of the Royal Society; and Barrow-Green and Gray.

65. "Henry Frederick Baker," obituary, *The Eagle*, 6.

66. Gray, interviews.

67. A distinction must be made between the liking for recondite but elementary geometry of triangles, conics, etc., elevated to n dimensions, and the serious geometry of algebraic curves, surfaces, and higher dimensional varieties in the Italian tradition. Coxeter liked the elementary branch, as later demonstrated in his book *Introduction to Geometry*, and this was also a British taste among the higher reaches of the mathematics profession. As Gray noted, to boil it down, there were three strains of geometry going on in Cambridge: (1) work in the Italian tradition of algebraic geometry, which was solid enough but the practitioners themselves knew it wasn't truly brilliant; (2) Coxeter's new departure, which has proved to be much more lasting; and (3) fun with elementary geometry. There was an affinity, if not an overlap, among these strains; certainly as far as Coxeter was concerned there was an intersection between the latter two branches. Ibid.

68. Coxeter, Personal Records of Fellows of the Royal Society.

69. When Coxeter independently rediscovered these Archimedean polytopes he had stumbled upon the same intellectual revelation that struck hobby geometer Thorold Gosset (1869–1962) a quarter century before. Gosset was a lawyer by training, but as Coxeter recounted, "Having no clients, he amused himself by trying to find out what regular figures might exist in n dimensions." After rediscovering all of them he proceeded to enumerate the "semi-regular figures." Following this path he discovered three polytopes in six, seven, and eight dimensions that are analogs of the Archimedean polyhedra. He recorded the results in an essay "On the Regular and Semi-regular Figures in the Space of n Dimensions," and it was sent to three reputable mathematicians at Cambridge to be evaluated for publication. One of them said "the author's method, a sort of geometrical intuition," failed to appeal to him; he found the ideas "fanciful." It was published with only the barest outline in the *Mathematical Messenger* in 1900. "That published statement remained unnoticed until after its results had been rediscovered by E. L. Elte and myself," wrote Coxeter. "As [Gosset] was a modest man, [he] let the subject drop, and pursued his career as a lawyer." Ibid.; Coxeter, *Regular Polytopes*, 162–64.

70. A Del Pezzo surface is a complex two-dimensional algebraic surface, named after Italian Pasquale de Pezzo (1859–1936), which Coxeter described in terms of Cayley numbers (named

after Arthur Cayley). Weisstein, "Del Pezzo Surface," http://mathworld.wolfram.com/DelPezzo-Surface.html (accessed January 22, 2006).

71. Coxeter, Personal Records of Fellows of the Royal Society.

72. This was one of a triumvirate of papers on the topic Coxeter published in *The Proceedings of the Cambridge Philosophical Society*. In a paper in 1934, he acknowledged that a chance question by Du Val had launched him on the study of groups generated by reflections—what became known as Coxeter groups. Coxeter stated: "In connection with his work on singularities of surfaces, Du Val asked me to enumerate certain subgroups in the symmetry groups of the 'pure Archimedean' polytopes n_{21} ($n < 5$), namely those subgroups which are generated by reflections . . . The work involved being somewhat intricate, several slips would have been overlooked but for the information that Du Val was able to supply from the (apparently remote) theory of surfaces." Barrow-Green and Gray; and see appendix 8.

73. Coxeter, Personal Records of Fellows of the Royal Society.

74. An 8-D polytope, which Coxeter called the "Gosset polytope," can be represented by Coxeter's notation as 4_{21}. In the context of the arrangement of kaleidoscopic mirrors, which Coxeter used to investigate polytopes, and the components of the Coxeter diagram he invented (see chapter 4), the notation reads as such: the 4 pertains to the "trunk" of four nodes, or mirrors, of a Coxeter diagram, while the 2–1 pertains to two separate "tails" of two nodes and one node each (the central eighth node adjoins the trunk and tails). The node on which the blob is indicated is said to be the "root." This polytope belongs to the Coxeter group E_8—the E standing for "exceptional." Using Coxeter's notation, the other Gosset polytopes would be represented by 2_{21} (in 6-D), and 3_{21} (in 7-D). Ibid; Coxeter, *Regular Polytopes*, 150–53, 162–64, 202–4; and Conway, interviews.

75. See appendix 8.

76. Coxeter to Katie Coxeter, 1 November 1931, Coxeter Fonds, University of Toronto Archives.

77. Coxeter, interviews; Conway, interviews; and Coxeter, "Some Examples of Hyperdimensional Awareness," *The Journal for the Study of Consciousness*, 84–85; and Coxeter, *Regular Polytopes*, vi, 258–59.

78. Coxeter, *Regular Polytopes*, 258.

79. Ibid.

80. Ibid.

81. Coxeter, *Regular Polytopes*, 258–59.

82. Coxeter, Personal Records of Fellows of the Royal Society.

83. Coxeter, "Some Examples of Hyperdimensional Awareness," 85.

84. Coxeter diaries, 20 October 1933.

85. Coxeter, PhD diss.; and O'Connor and Robertson, "Godfrey Harold Hardy," http://www-groups.dcs.st-and.ac.uk/~history/Mathematicians/Hardy.html (accessed January 22, 2006).

86. Barrow-Green, interview.

87. *Trinity College, Cambridge, Ordinances*, 1931, 107–35.

88. O'Connor and Robertson.

89. Mirrors are the object of many ancient myths—they reflect our souls or foretell the future (before the days of mirrors, gods and goddesses peered into the still water to catch a glimmering of their fates), and when broken they bring seven years of bad luck. These mysterious properties make them rich material for literature and analysis: they are prominent in Scottish anthropologist Sir James Frazier's book *The Golden Bough: A Study in Magic and Religion*, which Coxeter owned a copy of; *The Documents in the Case* by Dorothy L. Sayers tells a mur-

der mystery whose solution turns on understanding "optical activity" and mirror-image molecules, which Coxeter referenced in *Introduction to Geometry*; and Jorge Luis Borges's "Death
and the Compass" is a story wherein a detective is led to a gruesome end by his obsession with
symmetry. Coxeter, *Introduction to Geometry*. Coxeter, interviews; Asia Ivić Weiss, interview,
September 11, 2002.

90. Sir David Brewster, *The Kaleidoscope, Its History, Theory, and Construction*, 6–8.

91. Coxeter, interviews; Marjorie Senechal, interviews, November 23, 2004, and follow-
up; and Eli Moar, "The Magic World of Mirrors," *To Infinity and Beyond*, 149–54.

92. Senechal; and Conway, interviews; Schattschneider, e-mail correspondence "Re: Diagrams of Kaleidoscopes," December 16–21, 2005; Roe Goodman (professor of mathematics,
Rutgers), interview, January 12, 2006. Roe Goodman, "Alice through Looking Glass after
Looking Glass: The Mathematics of Mirrors and Kaleidoscopes," in *The MAA Monthly*,
281–98.

93. Ibid.

94. There are three arrangements of kaleidoscopes that generate all five Platonic solids. The
kaleidoscope that generates the icosahedron is called, as you might expect, the icosahedral kaleidoscope; it also generates the dual of the icosahedron, the dodecahedron. Similarly, the octahedral kaleidoscope, with its precise arrangement of mirrors and correctly positioned props,
generates both the octahedron and its dual, the cube. The tetrahedral kaleidoscope produces
only a solid tetrahedron, which is self-dual. Several United States patents have been granted for
three-dimensional kaleidoscopes: Patent #5,651,679, titled "Virtual polyhedra models," to inventor Frederick Altman in 1997; and Patent #5,475,532, titled "Infinite space kaleidoscope"
(and described as a "kaleidoscopic housing" with "triangular viewing windows"), issued to in-
ventors Juan Sandoval and Javier Bracho in 1995. Ibid; and http://www.uspto.gov/patft (accessed January 12, 2006).

95. Senechal; Conway; and Goodman.

96. Coxeter, *Regular Polytopes*, 92; and Goodman, 283.

97. Senechal; Conway; Goodman.

98. Instead of lugging kaleidoscopes around as Coxeter did, geometers often replace the
physical mirrors by geometric abstractions of mirrors. This practice is usually restricted to two
dimensions, where a mirror becomes a line called the axis of reflection (also called the mirror,
axis or symmetry, axis). Coxeter, *Regular Polytopes*, 196–204; and Conway.

99. Roe Goodman constructed a set of kaleidoscopes to help his graduate students (and
the undergraduate math club) understand reflection groups in the theory of Lie algebras. In a
recent paper, he explored the mechanics of the kaleidoscopes by drawing an analogy to
Through the Looking Glass, imagining Alice's trip through a peculiar cone-shaped arrangement of three looking glasses: "She steps through one of the looking glasses and finds herself
in a new *virtual chamber* that looks almost like her own. On closer examination she discovers
that she is now left-handed and her books are all written backward. There are also *virtual mirrors* in this chamber. Stepping through one of them, she continues her exploration and passes
through many virtual chambers until, to her great relief, she suddenly finds herself back in her
own real chamber, just in time for tea. Eager to have new adventures, Alice wonders how
many different ways the mirrors could be arranged so that she could have other trips through
the looking glasses and still return the same day for tea . . . Alice's problem was solved (for all
dimensions) by H. S. M. Coxeter, who classified all possible systems of *n* mirrors in *n*-
dimensional Euclidean space whose reflections generate a finite group of orthogonal matrices."
Goodman, "Alice Through Looking Glass After Looking Glass," 281; Senechal; Conway; and
Coxeter, *Regular Polytopes*.

100. Coxeter diaries, 23 March 1935, 3 December 1935, 6 December 1935, 20 January 1936.

101. Coxeter, interviews.

102. Alicia Boole Stott to Coxeter, November 1935, Coxeter Fonds, University of Toronto Archives.

103. G. H. Hardy, *A Mathematician's Apology*, 5, 24–25.

104. Coxeter, PhD diss., iii.

105. Coxeter, Personal Records of Fellows of the Royal Society.

106. Blackburn and Rosenhead, 46.

CHAPTER 4—COMING OF AGE AT PRINCETON WITH THE GODS OF SYMMETRY

1. *The Princeton Mathematics Community in the 1930s: An Oral History Project*, Transcript Number 3, George W. Brown and Alexander M. Mood, 4

2. Coxeter, Personal Records of Fellows of the Royal Society.

3. Coxeter, Rockefeller Personal Record Card.

4. "Kingwood College Library, American Cultural History, 1930–1939," http://kclibrary.nhmccd.edu/decade30.html (accessed January 22, 2006).

5. Robert Shaplen, *Toward the Well-Being of Mankind: Fifty Years of the Rockefeller Foundation*, foreword by George Harrar, v.

6. Shaplen, 5–8.

7. Raymond B. Fosdick, *The Story of the Rockefeller Foundation*, 140.

8. One of Rose's associates commented that his scheme might appropriately be subtitled "Tactics in the campaign against human ignorance." George Gray, *Education on an International Scale: A History of the International Education Board*, 1923–1938, 8.

9. Fosdick, 141; and Gray, 10.

10. Gray, v.

11. Gray, v–vi.

12. Hilbert, it should be noted, had a foot on each side of the axiomatic/formalist–visual/intuitive geometry divide. He had a soft spot for intuitive and visual geometry, as expressed in his book coauthored with S. Cohn-Vossen, published in German in 1932 (the German title was *Anschauliche Geometrie*, translated as *Descriptive Geometry*—*anschauliche* also translates to "visual," "clear," and "illustrative"—whereas the English edition was published in 1952 under the title *Geometry and the Imagination*). The considerable surface area of its pages covered in diagrams shows Hilbert's pleasure in visual questions. And Hilbert stated in the preface: "In mathematics, as in any scientific research, we find two tendencies present. On the one hand, the tendency toward abstraction seeks to crystallize the logical relations inherent in the maze of material that is being studied, and to correlate the material in a systematic and orderly manner. On the other hand, the tendency toward intuitive understanding fosters a more immediate grasp of the objects one studies, a live rapport with them, so to speak, which stresses the concrete meaning of their relations." But this book was an exercise in nostalgia for Hilbert—he appreciated the importance of diagrams in learning mathematics, and in creating it, but he insisted that it be recast in abstract form. Notwithstanding his support for his friend Hermann Minkowski, who expressed the unvisualizable number theory in terms of visual geometry and similarly showed that Einstein's relativity theory was better comprehended in geometric-pictorial terms, in practice Hilbert believed that the real power of geometry lay in formalism, rigor, and logic; he didn't work much on classical geometry; he was much more algebraic in temperament; his preference for analysis trumped pictures. Jeremy Gray, interviews; and David Hilbert, *Geometry and the Imagination*, iii–v.

13. George Gray, vii.

14. Ibid.

15. Ibid.

16. Other mathematicians crisscrossing the globe as Rockefeller Fellows that year included Gerrit Bol (from Germany to Princeton), Lars V. Ahlfors (Finland to France), Marie Charpentier (France to America), Erich E. Kahler (Germany to Italy), Philon M. Vassiliou (Greece to Germany), Julljusz Pawet Schauder (Poland to Germany), Stanislaus Saks (Poland to America), Grigoire Moisil (Romania to Italy and France), Max Gut (Switzerland to America), and Chinlai Shen (China to America). Rockefeller Archive Center, "Natural Science Fellowships, 1932" (RF RG 1.2, Series 100E, box 37, folder 279); Wilker, interview with Coxeter.

17. Harold Coxeter to Lucy Coxeter, 19 August 1932, Coxeter Fonds, University of Toronto Archives.

18. Ibid.

19. Coxeter to Harold Coxeter, 25 August 1932, Coxeter Fonds, University of Toronto Archives; and Coxeter to Katie Coxeter, 1 September 1932, Coxeter Fonds, University of Toronto Archives.

20. Coxeter to Katie Coxeter.

21. Coxeter to Harold Coxeter, 1 November 1932, Coxeter Fonds, University of Toronto Archives.

22. Coxeter to Katie Coxeter, 2 October 1932, Coxeter Fonds, University of Toronto Archives.

23. Alice Calaprice, ed., *The New Quotable Einstein*, 48.

24. Leopold Infeld, *Quest*, 299.

25. Infeld, 243.

26. The tragic death in a bicycle accident of mathematician Henry Burchard Fine (1858–1928), the dean of the mathematics faculty who has orchestrated Princeton's rise as a world power in mathematics and physics, prompted a donation from the family of Thomas D. Jones, a Princeton alumnus, lawyer, and lifelong friend of Fine's, which financed the construction of Fine Hall. Sylvia Nasar, *A Beautiful Mind*, 52–53.

27. Infeld, 294.

28. Simon Kochen (Henry Burchard Fine Professor of Mathematics, Princeton University), interview, Princeton, June 22, 2004; Conway, interviews.

29. Coxeter, correspondence to Katie Coxeter, 2 October 1932.

30. The Italian school included Guido Castelnuovo, Federigo Enriques, and Francesco Severi. From 1880 to 1914, the Italian school was known for the Enriques-Castelnuovo classification of algebraic surfaces. After the First World War, Francesco Severi extended this work but with less success, and, as Jeremy Gray described, "they acquired a reputation for being vague. Oscar Zariski, who passed through in the 1930s, wrote the final account of their work on surfaces, but spoke of leaving paradise thereafter, and the rest of his work was in a much more abstract tradition responsive to contemporary developments in commutative algebra. The Italians worked exclusively over the complex field, and used transcendental methods inapplicable in abstract algebraic geometry over an arbitrary (algebraically closed) field; these, too, contributed to their eventual partial eclipse. Lefschetz differed from the Italians in being driven by the desire to work topology into the subject, but shared a degree of visionary imprecision with them." *The Princeton Mathematics Community in the 1930s: An Oral History Project*, "Leon W. Cohen," Transcript Number 6, 3; and O'Connor and Robertson, "Solomon Lefschetz," http://www-groups.dcs.st-and.ac.uk/~history/Mathematicians/Lefschetz.html (accessed January 21, 2006); and Gray, e-mail "Re: Lefschetz and the Italian School," October 30, 2005.

31. *The Princeton Mathematics Community Oral History Project*, "Leon W. Cohen."

32. O'Connor and Robertson.

33. Ibid.; Conway, interviews; and Kochen, interview.

34. Veblen was putting final touches on a paper titled "The Foundations of Differential Geometry"—the study of geometry using calculus, invented in the early nineteenth century in the context of making maps and surveying a region of the earth—but "What Is Geometry?" was a main issue addressed. O'Connor and Robertson, "Oswald Veblen," http://www-groups.dcs.st-and.ac.uk/~history/Mathematicians/Veblen.html (accessed January 23, 2006); Yaglom, 253–54; *The Princeton Mathematics Community Oral History Project*; and Deane Montgomery, "Oswald Veblen," in *A Century of Mathematics in America*, part 1.

35. Yaglom, 254.

36. O'Connor and Robertson, "John von Neumann," http://www-groups.dcs.st-and.ac.uk/~history/Mathematicians/Von_Neumann.html (accessed January 23, 2006); *Princeton Mathematics Community Oral History Project*.

37. Wigner had studied the effects of the symmetry of molecular configurations on their vibrational spectrum, and in the following year introduced time-reversal symmetry into quantum theory. With his affection for symmetry, Coxeter would have been intrigued by the overlap. O'Connor and Robertson, "Eugene Paul Wigner," http://www-groups.dcs.st-and.ac.uk/~history/Mathematicians/Wigner.html (accessed January 23, 2006); "George Pólya," http://www-groups.dcs.st-and.ac.uk/~history/Mathematicians/Polya.html (accessed January 23, 2006); and *The Princeton Mathematics Community Oral History Project*.

38. O'Connor and Robertson, "Eugene Paul Wigner."

39. Ibid.; Coxeter to Harold Coxeter, 16 November 1932, and Coxeter diaries, 9 January 1932, Coxeter Fonds, University of Toronto Archives.

40. See appendix 8.

41. C. Davis, interviews.

42. Coxeter to Harold Coxeter.

43. Coxeter diaries, 20 March 1933, 20 April 1933, Coxeter Fonds, University of Toronto Archives.

44. Prior to Roosevelt's election to the office of president in a landslide on November 8, a campus publication reported the results of the student presidential poll—incumbent Herbert Hoover 1,392, Franklin Roosevelt 425, and 283 for Norman Thomas, the Socialist candidate and a Princeton graduate, whom Coxeter would have favored. The reporter noted, "It is not expected, however, that these tidings will throw much of a fright into the Democratic National Committee, especially after the successful way in which Miss Mary McCarthy and other election board pugilists repulsed the student attempts at registering last spring." "Poll on Campus," *Princeton Alumni Weekly*, November 4, 1932.

45. Coxeter to Harold Coxeter.

46. Coxeter attended lectures by James Alexander, though Alexander and his wife, Natalia, were best known for their wild parties—"humdingers"—with dancing outside their house on Princeton Avenue late into the night. Graduate students marveled at how Alexander and von Neumann could whoop it up until the early hours of the morning and then deliver perfectly lucid lectures at 8:30 a.m. the next day. Alexander was a man of independent means and did not accept a full salary from the university. This meant, however, that he was not a reliable lecturer. He began the term with enthusiasm for one of his latest ideas, but after a few weeks, when his attention waned (a mountain climber, he liked to scale the Graduate Tower during term for practice) or a solution to the posed problem was not readily emerging, a notice went up on the message board in Fine Hall announcing: "Professor Alexander will not meet his seminar this

week." A few meetings followed and then another notice: "Professor Alexander will not meet his seminar until further notice." Of which there was none. Alexander, in collaboration with Veblen, showed that the topology of manifolds could be extended to polyhedra, a result Coxeter would have appreciated. Ibid.; Coxeter Diaries, March 4, 1933, Coxeter Fonds, University of Toronto Archives; *The Princeton Mathematics Community Oral History Project*; and O'Connor and Robertson, "James Waddell Alexander," http://www-groups.dcs.st-and.ac.uk/~history/Mathematicians/Alexander.html (accessed January 23, 2006).

47. Coxeter to Harold Coxeter.

48. Coxeter to Katie Coxeter, 29 March 1933, Coxeter Fonds, University of Toronto Archives.

49. Coxeter to Harold Coxeter.

50. Coxeter, "University of Toronto Oral History Project."

51. Coxeter to Harold Coxeter.

52. Ibid.

53. Coxeter diaries, 16 January 1933, Coxeter Fonds, University of Toronto Archives.

54. Sloane, interview.

55. Ibid.

56. Coxeter, interview, March 14, 2003.

57. Ibid.

58. C. Davis, interviews; Conway, interviews; and Senechal, interviews.

59. Coxeter, *Regular Polytopes*, 202–4.

60. Peter McMullen (professor emeritus at University College London), interview, February 1, 2005; Conway; Schattschneider.

61. Conway, interview, April 2005.

62. Senechal, "Coxeter and Friends," *The Mathematical Intelligencer*, 16.

63. Coxeter diaries, 17 February 1933, Coxeter Fonds, University of Toronto Archives.

64. Coxeter diaries, 16 June 1933, Coxeter Fonds, University of Toronto Archives.

65. Coxeter diaries, 15 June 1933, Coxeter Fonds, University of Toronto Archives.

66. Coxeter to Katie Coxeter, 15 June 1933, Coxeter Fonds, University of Toronto Archives.

67. Ibid.

68. "Paul S. Donchian Opens Door to a Fairlyand . . ." in *The Hartford Daily Courant*, January 20, 1935, E3.

69. Coxeter recounted Donchian's life story in *Regular Polytopes* (Coxeter also used photographic plates of Donchian's models for this book and others), and in an article in the *Journal for the Study of Consciousness*—an odd journal in which to find Coxeter, alongside such musings as "Electroencephalographic Analysis of Changes of Consciousness," complete with EEG brain wave analysis, and another article, "Evolution, Love and Mystical States" (then again, on occasion Coxeter's name is cited by astrologers to lend their work some credence). Coxeter's article, however, included straight-up biographical profiles of practitioners who, in his experience, had demonstrated a talent for hyperdimensional awareness—his usual chorus of hyperspacers: Alicia Boole Stott, Thorold Gosset, John Flinders Petrie, Ludwig Schläfli, and Donchian. Donchian's great-grandfather had been a jeweler at the court of the sultan of Turkey, and many of his ancestors were jewelers and handicraftsmen. Coxeter observed that it must have been this ancestral predisposition to intricate craftsmanship that made Donchian such a superb model-maker. "At about the age of thirty," wrote Coxeter, "Donchian suddenly began to experience a number of startling and challenging dreams of the previsionary type

soon to be described by Dunne in *An Experiment with Time.* In an attempt to solve the problems thus presented, Donchian determined to make a thorough analysis of the geometry of hyperspace." Donchian aimed to reduce the subject to its simplest terms, "so that anyone like himself with only elementary mathematical training could follow every step." On one visit, Donchian gave Coxeter a tutorial, explaining all he could of his methods in model making, saying that he wanted one person in the world to know his methods. His approach to three-dimensional rendering of polytopes was to treat them as plan and elevation, like an architect. "My system is to build first the central grouping, then the exterior shell, with the central grouping inserted at the last moment and suspended by temporary stay-cords," said Donchian. "The process of connecting the innermost and outmost portions proceeds by constant testing of the results and the plodding application of common sense. The models are fortunately foolproof, because if a mistake is made it is immediately apparent and further work is impossible. The final joining of the inner and outer portions carries something of the thrill experienced by two tunneling parties, piercing a mountain from opposite sides, when they finally break through and find that their diggings are exactly in line." Coxeter, *Regular Polytopes*, 260; and Coxeter, interviews.

70.   When Coxeter and Donchian and Patrick Du Val made a joint presentation at the Pittsburgh meeting of the American Association for the Advancement of Science, the *New York Times* picked up the story, running ten column-inches, of which this is an instructive excerpt:

"A new type of design for geometrical models, appealing alike to the scientist, the artist and the philosopher, has been constructed after many years of painstaking work by Paul S. Donchian of Hartford . . . Donchian has discovered a new method of projecting symmetrically into three-dimensional space the hypercube series to any number of dimensions [similar to the method of projection Coxeter espoused, discussed in chapter 2]. Beginning with the simplest, the cube, he progresses step by step, dimension by dimension, to the hyper-space world of four, six, seven, nine, ten, twelve, thirteen, fifteen, twenty-one, until he finally reaches the very rarefied mathematical atmosphere of the hyper-cube in twenty-four dimensions.

"An illustration of how the infinite may be given a certain measure of concreteness in the finite is the well-known example of two mirrors hung facing each other on opposite walls of a room. In this case the mirrors give an infinite number of reflections in a finite space. In the same manner Donchian builds his geometrical models of infinite dimensions in three-dimensional figures.

"Each succeeding dimension is . . . obtained by multiplying the number of points in the preceding dimension by two. In this manner a hyper-cube of, say, twenty-four dimensions would have the astounding number of 16,777,216 points.

"Obviously, the making of a model of such proportions would require the efforts of several generations of men working full time. Donchian, however, has discovered and perfected a method of symmetrically projecting these increasingly complicated hyper-cubes down into either three or two-dimensional space with perfect accuracy." *Hartford Daily Courant*; and William I. Laurence, "Projecting Hyper-Cubes," *New York Times*, July 21, 1935.

71.   Coxeter, Personal Records of Fellows of the Royal Society.

72.   Chandrasekhar won the 1983 Nobel Prize in Physics, shared with W. A. Fowler, for his studies of the physical processes of importance for the structure and evolution of stars, as described in his work, "The Mathematical Theory of Black Holes." O'Connor and Robertson, "Subrahmanyan Chandrasekhar," http://www-groups.dcs.st-and.ac.uk/~history/Mathematicians/Chandrasekhar.html (accessed January 23, 2006); and Ibid.

73.   Coxeter, personal records of Fellows of the Royal Society.

74.   *The Princeton Mathematics Community Oral History Project.*

75. Coxeter, Personal Records of Fellows of the Royal Society.

76. *The Princeton Mathematics Community Oral History Project*, Transcript Number 8, "William L. Duren, Nathan Jacobson, and Edward J. McShane," 13.

77. Einstein also made use of the tensor calculus of two turn-of-the-century Italian mathematicians, Tullio Levi-Civita (1873–1941) and Gregorio Ricci-Curbastro (1853–1925).

78. Conway, interviews.

79. Beatrice M. Stern, "Chapter IV, The School of Mathematics," in *A History of the Institute for Advanced Study*, 1930–1950, 121–96.

80. Calaprice, 228.

81. Calaprice, 230.

82. Lightman, 67.

83. Leon Lederman (director emeritus, Fermilab), interview, September 22, 2005.

84. *The Princeton Mathematics Community Oral History Project*, "Leon W. Cohen," Transcript Number 6, 8.

85. Ibid.

86. Lederman and Hill, 23, 97.

87. Lederman recently coauthored a book on this subject, *Symmetry and the Beauty of the Universe*, with Christopher Hill. They have now launched a campaign of sorts to increase symmetry's profile. They are working to convert the book into a teachers' manual, as symmetry is abysmally absent from most high school science curriculums. "It shouldn't be, because it is not difficult, and it is a charming idea," said Lederman. His symmetry crash course focuses on three key words: "system," "transformation," and "invariance." The system could be a round table, a Platonic solid, an atom, or a galaxy—"it's just what you decide, in your effort to understand the world, is going to be your subject of study." The transformation is a change in the system, or a change in the way you look at the system—the way Coxeter, attending the Royal Society party, changed his perspective on the table, hanging above it or viewing it from across the room. A system can be transformed by rotation, or reflection in a mirror, or a kaleidoscope of mirrors, or by variables of time, or energy. Finally, if the system does not change certain fundamental properties when subjected to this transformation, then the system is invariant, or symmetrical. Lederman, interview.

88. "Symmetry," CBC *Ideas*, November 20, 1972; Lederman and Hill, 67.

89. Some have speculated that one of Einstein's lectures planted a seed in Coxeter's brain, to germinate later along his "world line" (in Einstein's theory of relativity, time and space, according to the mathematician Herman Minkowski, "fade away into mere shadows." They are replaced by four-dimensional space-time in which a person's entire history becomes his world line). In 1966, Coxeter published a paper, "Reflected Light Signals," in the book *Perspectives in Geometry and Relativity*. But the inspiration also might have come from a colleague at the University of Toronto, J. L. Synge, and his book, *Relativity: The Special Theory*. Smith, interviews.

90. Einstein had abandoned quantum theory, to which he made a fundamental contribution. The original quantum theory was replaced with quantum mechanics in the mid-1920s. Einstein couldn't support quantum mechanics because it settled for an account of physical behavior that depended on probabilities. "Probabilities are philosophically loaded," said Hans Halvorson, a Princeton philosopher who specializes in quantum field theory. "Who knows what probability is? Probability could be someone's ignorance. What's the probability that— suppose I was in a coma during the election—what's the chance that Bush won?" Einstein wanted a more complete theory with fewer uncertainties. "Quantum mechanics is very worthy of regard," Einstein said in 1926. "But an inner voice tells me it is not yet the right track. The theory yields much, but it hardly brings us closer to the Old One's secrets. I, in any case, am

convinced that He does not play dice." Hans Halvorson (professor of philosophy, Princeton), interview, December 23, 2004; Calaprice, 231.

91. Moffat read all of Einstein's papers and his work on unified theory as a student in Copenhagen in the early 1950s. "I came up with some issues that I found problematic about what he was doing," he said. "So I wrote a manuscript and sent it to him and of course never expected to hear back." Moffat's comments were genuine and well-founded, criticisms, and Einstein replied, beginning a correspondence back and forth in the final years of Einstein's life, which was published in Einstein's collected correspondence. John Moffat (physicist, Perimeter Institute), interview, March 2005.

92. Coxeter, "University of Toronto Oral History Project."

93. *The Princeton Mathematics Community Oral History Project*, "George W. Brown and Alexander M. Mood," Transcript Number 3, 12.

94. Vakil, interview; Goodman, interview; and Conway, interviews.

95. Conway, interviews.

96. The history of the enumeration of symmetry groups begins with German mathematician Felix Klein (1849–1925) and Norwegian Sophus Lie (1842–99), who decided as students in the latter nineteenth century to partition the world of groups between them. Klein took the discrete groups (as later would Coxeter), while Lie undertook the continuous groups—now known as Lie groups. Lie developed a tool for exploring his continuous groups, called Lie algebras, much the way Coxeter developed his Coxeter diagrams and Coxeter groups. A major achievement of the twentieth century was the classification of the continuous groups, through the work of Lie, as well as Wilhelm Killing (1847–1923) and Emil Artin (1898–1962), both from Germany; Frenchman Elie Cartan (1869–1951); Hermann Weyl; and others. Remarkably, the problem in the end reduces to the classification of "root systems." They are essentially the same as Coxeter's later inventions, the Coxeter groups. In other words, the theory of Lie's continuous domain reduces to or encompasses the most interesting case of Klein's discrete domain—the realm in which Coxeter was making fundamental contributions with his kaleidoscopic groups generated by reflections. Coxeter learned about all this when he attended and recorded the notes for Weyl's course. Thus, the way in which Coxeter groups permeate much of mathematics is often through Lie groups—Lie groups are the continuous symmetry groups, and the finite Coxeter groups are essentially associated with the Lie groups. To venture into some technical details, one of Coxeter's great theorems (according to John Conway) is that the abstract groups generated by elements of order 2, whose only other relations give the orders of their products in pairs, are isomorphic to the finite spherical reflection groups, which Coxeter himself enumerated. The association of Lie groups and Coxeter groups took an interesting turn through the independent work of Victor Kac (then in Moscow; now at MIT) and Robert Moody, who was a graduate student at the University of Toronto in the mid-1960s (now professor emeritus, University of Alberta). Namely, the association could be turned on its head, yielding Lie algebras and Lie groups from Coxeter groups. This resulted in a whole new collection of Lie algebras and groups arising from the (infinite) Coxeter groups. These are now called Kac-Moody Lie algebras and they have found numerous applications in mathematics and mathematical physics. Fittingly, Moody's work on this arose directly from his simultaneous exposure to Lie theory in one course and to reflection groups in a course on regular polytopes, which was taught by Coxeter. The indispensability of Coxeter groups was demonstrated recently in the paper "Bad Upward Elements in Infinite Coxeter Groups," showing that there are no "evil elements" in an infinite Coxeter group, and the paper "Evil Elements in Coxeter Groups," showing that there are no evil elements in a finite Coxeter group. Said one of the authors, Sarah Perkins: "The two results together show that there are no evil elements in ANY Coxeter groups (I always like to vanquish evil, don't you?)." Conway, interviews; Vakil, interviews;

and Goodman, interview; Sarah Perkins (Lecturer in Mathematics, University of London), e-mail correspondence, January 2006; and Robert Moody, interview March 11, 2006.

97. Coxeter diaries, 11 April 1935 ("My first lecture to Weyl's seminar, 1 of 5"), Coxeter Fonds, University of Toronto Archives.

98. Hermann Weyl, *The Structure and Representation of Continuous Groups*, 186–210.

99. Hermann Weyl to Donald Coxeter, Coxeter Fonds, University of Toronto Archives.

100. The Princeton Oral History Project discusses in passing the difficulties experienced by a number of well-qualified mathematicians in trying to find a job. Of Sam Wilks, who was working in the new field of statistics, one interviewee commented: "I think if he had been Coxeter . . . he would have had the same problem." As it was, Veblen had to argue to get approval even for the refugee postings; the opinion of at least one faculty member was that, "If these distinguished people come and take the positions, the young American mathematicians will become hewers of wood and drawers of water."

101. That the book should be passed on to Coxeter made him heir to a mathematical classic. And the ancestry fit: one of Ball's tutees at Cambridge, in 1903, was J. E. Littlewood; and since Littlewood directed Coxeter's undergraduate studies, Coxeter could legitimately be regarded as Ball's "grand student." Coxeter to Harold Coxeter, 18 December 1934, Coxeter Fonds, University of Toronto Archives.

102. Coxeter to Harold Coxeter, 11 December 1934, Coxeter Fonds, University of Toronto Archives; and Coxeter, interviews.

103. Thorstein Veblen, "The Higher Learning in America," http://www.ditext.com/veblen/veblen.html (accessed January 23, 2006).

104. Coxeter to Harold Coxeter, 11 December 1934.

105. Coxeter to Harold Coxeter, 7 February 1935, Coxeter Fonds, University of Toronto Archives.

106. Coxeter diaries, 27 and 29 April 1935, 29 May 1935, 15 and 18 February 1935, Coxeter Fonds, University of Toronto Archives.

107. Coxeter diaries, 8 January 1935, 13 and 20 April 1935.

108. Coxeter diaries, 14 June 1935, 30 August 1935.

CHAPTER 5—LOVE, LOSS, AND LUDWIG WITTGENSTEIN

1. Coxeter diaries, 12 and 15 September 1935, 11 and 31 October 1935, Coxeter Fonds, University of Toronto Archives.

2. "H. S. M. Coxeter, A Biographical Sketch," in *Kaleidoscopes: Selected Writings of H. S. M. Coxeter*, xiii, edited by Arthur Sherk, Peter McMullen, Anthony Thompson, and Asia Ivić Weiss.

3. Harold Coxeter to Donald Coxeter, 6 January 1936, Coxeter Fonds, University of Toronto Archives.

4. Coxeter diaries, 15 October 1936, Coxeter Fonds, University of Toronto Archives.

5. Barrow-Green, interview.

6. Coxeter, interviews.

7. Coxeter diaries, 8 January 1936.

8. Coxeter diaries, 31 January 1936, 19 February 1936.

9. Coxeter diaries, 8 October 1935.

10. Ray Monk, *Ludwig Wittgenstein: The Duty of Genius*, 336–37; and Coxeter interview, Toronto, February 9, 2002.

11. Coxeter, interview; and Coxeter diaries, 13, 23, and 26 October 1933, 2 and 20 November 1933, Coxeter Fonds, University of Toronto Archives.

12. Coxeter, interview.

13. Barrow-Green and J. Gray; Barrow-Green, interview; and Barrow-Green, e-mail correspondence "Re: technical Trinity question," May 9, 2005.

14. Lefschetz had invited Hodge to Princeton and Weyl published a corrected account of Hodge's main theorem. A crude indicator of the level at which Hodge worked is the fact that one of the million-dollar Millennium Prize Problems—"important classic questions that have resisted solution over the years"—is the Hodge conjecture. There are seven problems in total, posed in 2000 by the Clay Mathematics Institute, based in Cambridge, Massachusetts. Atiyah, e-mail correspondence "Re: Hodge question," May 11, 2005; and J. Gray, interviews.

15. O'Connor and Robertson, "William Vallance Douglas Hodge," http://www-groups.dcs. st-and.ac.uk/~history/Mathematicians/Hodge.html (accessed January 23, 2006).

16. Barrow-Green e-mail, "Re: technical Trinity question."

17. Harold Coxeter to Donald Coxeter, 6 and 13 June 1936, Coxeter Fonds, University of Toronto Archives.

18. Coxeter diaries, 3 and 6 June 1936, Coxeter Fonds, University of Toronto Archives.

19. H. F. Baker correspondence to Donald Coxeter, 20 August 1936; and Coxeter diaries, 7 June 1936, Coxeter Fonds, University of Toronto Archives.

20. Coxeter diaries, 8 June 1936, Coxeter Fonds, University of Toronto Archives.

21. Coxeter diaries, 23 November 1936, Coxeter Fonds, University of Toronto Archives.

22. Harold Coxeter to Rosalie Gabler, 27 August 1934, Nesta Coxeter family papers.

23. Coxeter diaries, 7 March 1936, Coxeter Fonds, University of Toronto Archives.

24. Coxeter, interviews; and O'Connor and Robertson, "Luitzen Egbertus Jan Brouwer," http://www-groups.dcs.st-and.ac.uk/~history/Mathematicians/Brouwer.html (accessed January 23, 2006).

25. Coxeter to Rien Brouwer, 21 March 1936, Susan Thomas, family papers.

26. Harold Coxeter to Donald Coxeter, 15 May 1936, Coxeter Fonds, University of Toronto Archives.

27. Coxeter had proposed once before, to a long-term girlfriend at Princeton. He wrote home, "Dorothy Henry really does give me a feeling of companionship . . . being interested in the right sort of things. She is a vegetarian because she once saw the Chicago Stockyards. She is religious without being sectarian. And she actually seems to like me (which is such a new complication that I am not quite sure which way to turn.)" But when his father met Dorothy during a visit to America, Harold concluded his son had made a mess of the relationship, as he recounted in a letter to Katie: "Donald, now, has duly proposed to her (after waiting and waiting too long, in my opinion). She is fond of him and has enjoyed him as a friend but she is very sensible and understanding and realizes that he is not really in love with anyone but himself, nor likely to be. She sees his lack of development clearly enough and as she says to me, she does not want to mother him as well as children. I don't think there is much likelihood of her marrying him—unless by some miracle he should want her so much that he changes and becomes less self-centered. I must say he has been very charming and considerate during the whole long time we have been together—quite a test, I think—but he is not very considerate really of anyone else and very greedy and anxious to have the best possible, always." As Harold predicted, Dorothy rejected Coxeter's proposal. Coxeter diaries, 24 May 1936, Coxeter Fonds, University of Toronto Archives; and Coxeter correspondence to Katie Coxeter, 15 June 1933, Coxeter Fonds, University of Toronto Archives; and Harold Coxeter to Katie Coxeter, 9 August 1933, Nesta Coxeter, family papers.

28. Rien Brouwer, May/June 1936 diary translation notes, Susan Thomas, family papers.

29. Congratulations also arrived from Coxeter's psychoanalyst, Wilhelm Stekel. It read (as written in the original): "Dear Donald, I am very glad, you will be maaried and finish the peri-

ode of childhood, infantilisme. I am shur, you got the right life-compagnon and much better she does not understand mathematics—there can't be rivalry and contest in ambition. I wish I could be present at your wedding but in my phantasy I shall accompagny you the road to manhood and fathership. My advice: Give up horse riding only for a while. Perhaps after getting injured to the pleasure of a real man, you will laugh at the hobby-pleasure." Stekel concluded by advising that the secret of happiness was not to expect too much. Wilhelm Stekel to Donald Coxeter, 6 June 1936, Coxeter Fonds, University of Toronto Archives.

30. Harold Coxeter to Donald Coxeter, 3 and 11 June 1936, Coxeter Fonds, University of Toronto Archives.

31. Harold Coxeter to Donald Coxeter, 13 June 1936, Coxeter Fonds, University of Toronto Archives.

32. Coxeter, interviews; Eve Coxeter, interviews; Nesta Coxeter, interviews.

33. Coxeter to Rien Brouwer, 7 July 1936, Susan Thomas, family papers.

34. Coxeter diaries, 14–17 July 1936, Coxeter Fonds, University of Toronto Archives.

35. O'Connor and Robertson, "John Lighton Synge," http://www-groups.dcs.st-and .ac.uk/~history/Mathematicians/Synge.html (accessed January 23, 2006); and Coxeter, interviews.

36. Coxeter, *Introduction to Geometry*, 135.

37. "The Fields Medal," http://www.fields.utoronto.ca/aboutus/jcfields/fields_medal.html (accessed January 23, 2006); and "John Charles Fields," http://www.fields.utoronto.ca/ aboutus/jcfields/index.html (accessed January 23, 2006).

38. Ahlfors received the medal for "research on covering surfaces related to Riemann surfaces of inverse functions of entire and meromorphic functions" and Douglas for "important work on the Plateau problem which is concerned with finding minimal surfaces connecting and determined by some fixed boundary." Coxeter was not one of the many modern geometers to win the Fields Medal. However, "There is no shame in not getting the Fields medal," said Ken Davidson, former director of the Fields Institute. "You have to solve a really big problem when you are really young, and then you have to get lucky." Coxeter's luck was hindered by the fact that between 1936 and 1950—his prime qualifying years—only two Fields Medals were awarded due to the Second World War. Also, the Fields is given for a bombshell of a mathematical contribution, such as solving Fermat's last theorem (Andrew Wiles might have received it for doing exactly that, except he was over forty). Coxeter's contribution with polytopes did move the mathematical firmament, but years after the fact, and in a more organic, ever-evolving way. In 1995, Coxeter was awarded the first Fields-CRM Prize (given jointly by the Fields Institute and the Centre de Recherches Mathématiques in Montreal), an award acknowledging either overall career contribution or a startling piece of work. "There was a confidential discussion about who were possible candidates," said Davidson. "Once Coxeter's name came up there was general agreement that he had both [he met both criteria]. Coxeter was the first premier mathematician in this country. And you could argue that there aren't many in this country that have since achieved comparable stature. He was a remarkable man and certainly doing the work he did back in the 1940s, when there was almost no mathematical research here, he was centre stage in the world as a mathematician." Ibid.; "Mittag-Leffler and Nobel"; http://www.fields. utoronto.ca/aboutus/jcfields/fieldsnobel.html (accessed January 23, 2006); and Ken Davidson (professor of mathematics, University of Walterloo), interview, May 17, 2005.

39. Coxeter diaries, 1 August 1936, Coxeter Fonds, University of Toronto Archives.

40. Coxeter diaries, 15 August 1936, Coxeter Fonds, University of Toronto Archives.

41. Harold Coxeter to Donald Coxeter, 12 August 1936, Coxeter Fonds, University of Toronto Archives.

42. Eve Coxeter, interviews; and Nesta Coxeter, interviews.

43. Coxeter diaries, 15–18 August 1936, Coxeter Fonds, University of Toronto Archives; and letters of condolence, Susan Thomas, family papers.

44. Letters of condolence, Susan Thomas, family papers.

45. Thorold Gosset to Donald Coxeter, 15 August 1936, Coxeter Fonds, University of Toronto Archives.

46. Coxeter was familiar with Soddy's creation. "I was so impressed by Soddy's poem that I wrote to him to let me visit him and take him out to lunch," Coxeter recalled. They kept in touch over the years, and in one letter Soddy confessed: "Dear Professor Coxeter, Naturally I was very much interested in your letter of 21ˢᵗ , and your enclosed paper 'Interlocked Rings of Spheres'. As you know I'm no mathematician and my 'Kiss Precise' and 'The Hexlet' you quote were *tour de force* hammered out by sheer algebra and luck. They depended on the reduction of a biquadratic equation of I think 23 terms to a quadratic by a transformation I have never really understood." Coxeter credited Soddy with having influenced four of his papers—the above-mentioned "Interlocked Rings of Spheres" as well as "Loxodromic Sequences of Tangent Spheres," "Numerical Distances among the Spheres in a Loxodromic Sequence," and "Numerical Distances among the Circles in a Loxodromic Sequence." Frederick Soddy to Donald Coxeter, 23 February 1951, Coxeter Fonds, University of Toronto Archives; and Coxeter, "Descartes Circle Theorem," in *The Changing Shape of Geometry*, 191.

47. Gosset.

48. Letters of condolence, Thomas.

49. Alicia Boole Stott to Donald Coxeter, 17 August 1936, Coxeter Fonds, University of Toronto Archives.

50. Coxeter diaries, 2 September 1936, Coxeter Fonds, University of Toronto Archives.

51. Eve Coxeter, interviews; Nesta Coxeter, interviews.

52. Coxeter diaries, 3 September 1936, Coxeter Fonds, University of Toronto Archives.

CHAPTER 6—"DEATH TO TRIANGLES!"

1. Coxeter diaries, 10 September 1936.

2. Cathleen Synge Morawetz (emeritus professor of mathematics, Courant Institute of Mathematical Sciences, New York University), interview, January 25, 2006.

3. Coxeter diaries, 2 November 1936; Coxeter, interviews.

4. Coxeter diaries, 19 and 27 October 1936.

5. Coxeter, interviews.

6. Schattschneider, interviews.

7. Coxeter diaries; Coxeter, interviews; and Susan Thomas, interviews.

8. A paper with J. A. Todd, "An Extreme Duodenary Form," a form of twelve variables, in the *Canadian Journal of Mathematics*, was not on the topic of his gastrointestinal difficulties but with its pun perhaps a fine example of Coxeter's sense of humor (the intestinal duodenum is so called because it is 12 inches long). See appendix 8; and Conway, interview, April 2005.

9. Thomas, interviews.

10. Coleman, interviews.

11. Coxeter diaries, 5 March 1937.

12. Jeremy Gray, *The Hilbert Challenge*, 241–82.

13. Ibid.; and Gray, interviews.

14. Gray, interviews; and O'Connor and Robertson, "Frank Morely," http://www-groups.dcs.st-and.ac.uk/~history/Mathematicians/Morley.html (accessed January 29, 2006).

15.   Gray, *The Hilbert Challenge.*

16.   Galison, "Images Scatter into Data, Data Gather into Images," 302.

17.   The back and forth in mathematics and science regarding the power and sanctity of images was recently chronicled at the 2002 exhibit in Karlsruhe, Germany, "Iconoclash: Beyond the Image Wars in Science, Religion, and Art." As one of the contributors, Peter Galison articulated the constant push and pull on the use of images and pictures in science with his contribution, "Images Scatter into Data, Data Gather into Images." In the exhibit catalogue (edited by Bruno Latour and Peter Weibel), Galison summed up this phenomenon with the statement: "We must have images; we cannot have images." In elaborating, he wrote, "We *must* have scientific images because only images can teach us. Only pictures can develop within us the intuition needed to proceed further toward abstraction . . . We are human, and as such, we depend on specificity and materiality to learn and understand . . . What are we humans good at? We are good at recognizing and seizing upon visual patterns . . . Perhaps this is because the long process of evolution has left us with a pattern-recognition capability well matched to the world . . . And yet: we *cannot* have images because images deceive . . . We are human and as such are easily led astray by the siren call of material specificity. Logic, not imagery, is the acid test of truth that strips away the shoddy inferences that accompany the mis-seeing eye." The visual vs. antivisual played out in quantum mechanics. Werner Heisenberg was antivisual, while Erwin Schrodinger was "furiously pro-visual" and repelled by "transcendental algebra"; Heisenberg, in turn, found Schrodinger's visual theories "disgusting." Niels Bohr brought both approaches together in his "complementarity," allowing the practitioner to choose. As far as relativity went, J. A. Wheeler's classic textbook, *Gravitation* (coauthored with Charles Misner and Kip Thorne, 1973), decries the abstract and is full of images, rods perforating egg cartons triggering ringing bells, while Nobel laureate Steven Weinberg's textbook, *Gravitation and Cosmology* (1972), contains no pictures, in keeping with his belief that shapes delude logical understanding. Galison, interview; and Ibid.

18.   Yaglom, 258.

19.   Liliane Beaulieu, "A Parisian Café and Ten Proto-Bourbaki Meetings (1934–1935)," *The Mathematical Intelligencer*, 27–35; Beaulieu, "Dispelling a Myth: Questions and Answers about Bourbaki's Early Work, 1934–1944," in *The Intersection of History and Mathematics*, 241–52; Beaulieu, "Bourbaki's Art of Memory," *Osiris*, 219; and O'Connor and Robertson, "Bourbaki: The Pre-war Years," http://www-groups.dcs.st-and.ac.uk/~history/HistTopics/Bourbaki_1.html, 29 January 2006, and "Bourbaki: The Post-war Years," http://www-groups.dcs.st-and.ac.uk/~history/HistTopics/Bourbaki_2.html (accessed January 26, 2006).

20.   O'Connor and Roberston, "Bourbaki: The Pre-war Years," and "Bourbaki: The Post-war Years"; Armand Borel, "Twenty-Five Years with Nicolas Bourbaki, 1949–1973," *Notices of the AMS*, 374; and Henri Cartan, "Nicolas Bourbaki and Contemporary Mathematics," *The Mathematical Intelligencer*, 178.

21.   Paul R. Halmos, " 'Nicolas Bourbaki,' " *Scientific American*, 88.

22.   Coleman, interviews.

23.   Halmos, 88.

24.   The Pythagoreans, similarly, were a secret society. And some historians speculate that Euclid was in fact a pseudonym for a number of mathematicians who together compiled and wrote *Elements.* O'Connor and Robertson, "Euclid of Alexandria," http://www-groups.dcs.st-and.ac.uk/~history/Mathematicians/Euclid.html (accessed January 29, 2006).

25.   Henri Cartan, "Nicolas Bourbaki and Contemporary Mathematics," *The Mathematical Intelligencer*, 179.

26.   Beaulieu, "Dispelling a Myth," 243. Simone Weil, *Gravity and Grace*, p. 139; and

Weil, *Notebooks I*, p. 9; and Lawrence E. Schmidt, "A Paper Presented to the Annual Colloquy of the American Weil Society," p. 3.

27. Halmos, 89.

28. Beaulieu, "Bourbaki's Art of Memory."

29. S. K. Berberian, "Bourbaki, the Omnivorous Hedgehog: A Historical Note?" *The Mathematical Intelligencer*, 104–5.

30. Jean-Pierre Serre, e-mail correspondence, "Re: H. S. M. Coxeter," September 19, 2003.

31. Nicolas Bourbaki, "The Architecture of Mathematics," *American Mathematical Monthly*, 228.

32. Beaulieu, "Bourbaki's Art of Memory"; and O'Connor and Robertson, "Bourbaki."

33. Privately, Bourbakis invented terminology for every branch of mathematics they studied. For example, they called topological vector spaces "bornographic" space. They used "sexylinear" for sesquilinear forms. And stated comparatively: "the spectral sequence is like the mini-skirt; it shows what is interesting while hiding the essential." And they often used the word "member" to refer not only to any Bourbaki, but also the "membrum virile." Halmos, 94; and Beaulieu.

34. Toth, interviews.

35. Coxeter, interviews; Senechal, interviews.

36. Halmos, 93.

37. As the Bourbaki society evolved, its membership ranged from ten to twenty individuals at a time, with new members invited first as a "cobayes," or "guinea pigs," to determine whether they were worthy. Members formed a "close-knit, inbred company with shared social and intellectual roots as well as mathematical tastes." The privileged were bound by an unstated vow of secrecy forbidding the naming of members. Almost all members of this private coterie were French, one exception being Samuel Eilenberg—he was known as S^2P^2, short for Smart Sammy the Polish Prodigy. Denis Guedj (translated by Jeremy Gray), "Nicholas Bourbaki, Collective Mathematician: An Interview with Claude Chevalley," *The Mathematical Intelligencer*, 18; Senechal, "The Continuing Silence of Bourbaki: An Interview with Pierre Cartier," *The Mathematical Intelligencer*, 22; Beaulieu, "Bourbaki's Art of Memory"; and Halmos, 94.

38. Cartier, interviews, Montreal, April 12 and 13, 2003.

39. Jean Dieudonné, "The Work of Nicholas Bourbaki," in *American Mathematical Monthly*, 134–36.

40. Martin Jay, *Downcast Eyes: The Denigration of Vision in Twentieth-Century French Thought*; Jay (Sidney Hellman Ehrman Professor of History, University of California, Berkeley), interview, July 2004; and Cartier, interviews.

41. Jay, 212–13.

42. Cartier calls the Bourbaki endeavor that of a "mathematical surrealist," surrealism being the artistic movement pervading France in the interwar era. Cartier gave the example of the surrealist game "the exquisite corpse": A piece of paper is folded accordion style into four sections; one participant draws the feet, the other the torso, the other the arms, and finally the neck and the head. What results is an oddly disjointed creature. This, Cartier said, "is similar to the Bourbaki mandate," producing a beast whose head is too far removed from its tail. The surrealists' style was to create fantastic visual imagery from the subconscious mind, or, as its philosophical founder André Breton put it: "to bewilder the senses." Cartier, interviews.

43. Jay, *Downcast Eyes*, and interview.

44. Ibid.

45. Toth, interviews.

46. Pitting Coxeter versus Bourbaki works ideologically, but not on a personal level.

Bourbaki distinguished between the mathematics that did not fit into its grand edifice and the mathematician responsible for the work. This distinction may have been clearer to Bourbaki than to the mathematicians involved, as it is human nature to identify oneself with one's creations and take criticism to heart. But there was no Bourbaki animus toward Coxeter the man (and Coxeter seems never to have criticized the Bourbaki mandate in any lecture or publication). Fields Medal winner and Bourbaki member Jean-Pierre Serre is adamant that this portrayal of Bourbaki versus Coxeter not be misinterpreted. Bourbaki had no professional contempt for Coxeter, or any geometer. It is a mistake, he said, to interpret that Bourbaki had some fixed and prejudiced opinion of any given mathematician. "On the contrary, Bourbaki was very careful in writing his books: he spoke only of the ideas and of the theorems, but not of the persons," said Serre in an e-mail discussion of the issue. A telephone interview to discuss Bourbaki's perspective on geometry in general—and the public interpretation of Bourbaki as antigeometry—would be a waste of his time; the questions posed, muddied as they were by the Bourbaki myth, were not worth addressing, based as they were on a misunderstanding of Bourbaki. With Bourbaki's legend clouded by a myth, to this day exacerbated by secrecy and elitism, it is no wonder. Serre, e-mail correspondence September 19, 24, 2003, and June 25, 2004; Senechal, interviews.

47. Coxeter diaries, 7 July 1939.

48. When Coxeter came to Trinity in 1926, he visited Ball's house in Cambridge (just one year after Ball died, in 1925) and found his back garden littered with the remnants of a maze constructed from wooden posts and string. Coxeter, interviews.

49. Coxeter, *Mathematical Recreations and Essays*, 151–52.

50. See appendix 8.

51. Coxeter and S. L. Greitzer, *Geometry Revisited*, 68; Coxeter, *Regular Polytopes*, 143; C. Davis, interviews; and Conway, interviews.

52. Conway, interviews.

53. Lister Sinclair, "Math and Aftermath," CBC *Ideas*, May 13 and 14, 1997.

54. Coxeter diaries, 18 April 1945.

55. "University of Toronto Oral History Project."

56. Coxeter became good friends with Robinson, though ultimately their friendship ended with a falling-out spurred by professional rivalry. Thomas, interviews; Synge Morawetz, interview.

57. Mandelbrot did his part reviving geometry in the twentieth century, discovering the Mandelbrot set, and generating interest in fractal geometry by showing how fractals are ubiquitous in both mathematics and nature. *Fractal* derives from the Latin word *fractus*, "irregular or broken up"—fractals formed a new geometry of nature, finding order in chaotic shapes and processes, everywhere from the coastline of Britain, to the branches of trees, the flow of blood our veins, and the behavior of the stock market—Mandelbrot's latest book is *The (Mis)Behaviour of Markets: The Fractal View of Risk, Run and Reward*. The Bourbakists, ever the jokesters, made fun of Mandelbrot and his fractals—melding their obsession with mathematics and food, they pondered the "problem of confinement for fractal-like pancakes." Benoit Mandelbrot (Sterling professor of mathematics, Yale University, IBM Fellow), interviews, October 2004, March 2005; and Beaulieu, "Bourbaki's Art of Memory."

58. Mandelbrot.

59. Mandelbrot; and Anthony Barcellos, "Benoit Mandelbrot," in *Mathematical People*, 206–25.

60. Ibid.

61. Ibid.

62. Mandelbrot.

63. In 1944, Coxeter prefaced his diary with his horoscope: "Aquarius people achieve what others deem impossible, even though not always correct in judging practical matters." Coxeter diaries, 1944 end paper.

64. Coxeter, interviews; and Thomas, interviews.

65. Coxeter, interviews.

66. Monson, interviews; Chris Fisher (professor of mathematics, University of Regina), interview, January 26, 2005.

67. Monson, interviews.

68. Monson, interviews; and Weiss, interviews.

69. Sherk, McMullen, Thompson, and Weiss, *Kaleidoscopes*, xiv.

70. Coxeter, Personal Records of Fellows of the Royal Society.

71. Alan Robson to Donald Coxeter, Coxeter Fonds, University of Toronto Archives.

72. Coxeter, *Regular Polytopes*, vi.

73. Coxeter, vii, 191.

74. If a mathematician chooses not to use a graphical shorthand (perhaps due to a discomfort or lack of familiarity with visual-geometric presentation of information), then an algebraic shorthand would be similarly useful in making the whole message more compact than a disjointed series of statements. Conway, interviews; Monson, interviews; Senechal, interviews; and Weiss, interviews; C. Davis, interviews.

75. Although Lagrange initiated work in this field, Evariste Galois (1811–32), whose story has all the elements of a Shakespearean tragedy, developed group theory in 1830 and is considered its founder. Whenever Coxeter told the Galois tale he launched into it as if it were a yarn about an ill-fated friend from college. "Oh!" he'd gasp. "The terrible tragedy of Evariste Galois!" In his historical remarks section in *Regular Polytopes*, Coxeter recounted the foundations of groups with a chronology of its pioneers: Joseph-Louis Lagrange (1736–1813), Niels Henrik Abel (1802–29), Galois, Augustin Louis Cauchy (1789–1857), and Camille Jordan (1838–1922). However, he concluded by saying, "Galois made such important contributions to the subject that he eventually became recognized as the real founder of group theory; yet his contemporaries scorned him and he was murdered at the age of twenty." Galois was killed in a pistol duel, in an argument over a woman. The night before the duel, which he seemed to know he would lose, Galois spent his last hours desperately writing down all his mathematical ideas. "I have made some new discoveries," he said. He filled eleven pages, explaining some vagaries apologetically, "I have no time and my ideas are not sufficiently developed on that terrain—which is immense." Historian E. T. Bell observed: "What he wrote in those last desperate hours before the dawn will keep generations of mathematicians busy for hundreds of years. He had found, once and for all, the true solution of a riddle which had tormented mathematicians for centuries: under what conditions can an equation be solved?" Broadly speaking, the increasing abstraction in geometry in the nineteenth century—with the rising tides of n-dimensional and projective geometry, and non-Euclidean geometries, later united under the umbrella of symmetry by Klein's Erlangen Program—contributed to the rise of group theory (together with the burgeoning number theory and theory of algebraic equations). Coxeter, *Regular Polytopes*, 55–56, 141; and Bell, 362–77.

76. Gray, interviews.

77. Monson, interviews; Conway, interviews.

78. In their book *Symmetry*, Christopher Hill and Leon Lederman give a whimsical and illuminating account of a symmetry group, using the equilateral triangle as their guinea pig. "The equilateral triangle presents us with one of the simplest, yet nontrivial, examples of sym-

metry," they wrote. "All . . . equilateral triangles, irrespective of color, size, position, orientation, or whatever, share [as] a common abstract feature their unique symmetry—which is the defining symmetry of the equilateral triangle, or what it means to be an equilateral triangle. If we could somehow communicate the essence of the symmetry of an equilateral triangle to Martians, they could reconstruct what we are talking about, but they wouldn't know how big or what color or what position of equilateral triangle we are communicating to them. It doesn't matter—the particular symmetry is the essence of what it means to be an equilateral triangle." Lederman and Hill, 295–96.

79. Schattschneider, e-mail correspondence, "Re: Symmetry of the Square," December 2005, January 2006.

80. C. Davis, interviews.

81. Conway, interview, Princeton, April 2005.

82. Stephen Strauss, "Art Is Math Is Art for Professor Coxeter," *The Globe and Mail*, May 9, 1996.

83. Coxeter's main interest was symmetry groups generated by reflections, but in the broader study there are many kinds of symmetry groups. There are infinitely many groups—any set with an operation that combines elements so as to satisfy the group laws forms a group. For example, the set of integers, under the operation of addition, forms a group. Schattschneider, e-mail correspondence.

84. Coxeter, *Regular Polytopes*, 75.

85. Monson, interviews.

86. Ibid.

87. Ibid.

88. Ibid.

89. Goodman, interview.

90. Ratcliffe, interviews.

91. Simon Kochen, interview, August 2005.

92. Vakil, interview.

93. Atiyah, interview.

94. Coxeter acknowledged the transcendent nature of his oeuvre—even if it was going places he might not himself investigate any further—in the preface to *Regular Complex Polytopes*, his 1974 sequel to *Regular Polytopes*, "Its relationship to my earlier *Regular Polytopes*," he said, "resembles that of *Through the Looking-Glass* to *Alice's Adventure in Wonderland*. The sequel is more profound . . . I have made an attempt to construct it like a Bruckner symphony, with crescendos and climaxes, little foretastes of pleasure to come, and abundant cross-references. The geometric, algebraic and group-theoretic aspects of the subject are interwoven like different sections of the orchestra." These crosscurrents of the Coxeter oeuvre butted heads not long ago, when organizers of the Coxeter Legacy conference (held in Toronto in 2004) tried to decide whom to enlist as invited lecturers. One faction of the organizing committee insisted that geometers with a classically Coxeterian spirit, such as John Conway, and Branko Grünbaum, at the University of Washington, should be the heart of the legacy conference. On the other hand, the more mainstream Coxeterian legacy resides with mathematicians not quite of Coxeter's tradition—the more modern and algebraic geometers who work with Coxeter groups in a way that has transcended Coxeter's original intent, such as Ravi Vakil; Alexandre V. Borovic, from the University of Manchester Institute of Science; Bertram Kostant, from MIT; and Michel Broué. The desire among some of the organizers for a healthy representation of the latter spurred recrimination among the faithful. They argued that Coxeter would find such a

roster of lecturers "preposterous, it would be an abomination." Surely the Coxeter Legacy conference should be a conference that Coxeter himself would find interesting and want to attend. Ultimately, it was resolved amicably, with no great drama, achieving a fine balance of both the classical and modern Coxeterians. Weiss, interviews; C. Davis, interviews; Vakil, interviews; Coxeter, *Regular Complex Polytopes*, xi.

95. Vakil, interview.

CHAPTER 7—TANGENTS ON POLITICS AND FAMILY VALUES

1. Linear programming was invented in 1947 by Stanford professor George Dantzig, spurred by his work at the Pentagon during the Second World War, as mathematical advisor to the U.S. Air Force Comptroller. According to Walter Whiteley, director of applied research at York University, in Toronto, there is a funny story—perhaps true, perhaps apocryphal—that goes along with its genesis: "One of the early problems of linear programming was to find the cheapest diet for soldiers that meets all the minimum daily nutritional requirements. The claim was that when they first ran the problem with linear programming, the answer they got was a diet of 100 percent carrots because there was no variable or equation that stipulated the need for variety, or a maximum amount of anything." Linear programming is also used for scheduling airlines—with a variable for every plane, every departure and destination airport, every pilot and crew member. And its awesome power is applied in allocating resources, planning production, scheduling workers, managing investment portfolios, and devising marketing and military strategies. Linear programming pertains to Coxeter and his polytopes because polytopes provide 3-D models of the problems. As Whiteley explained, if there are three variables in a problem, then the model would be a 3-D polytope, but the beauty of this method is that it can be done in *n*-dimensions and thus can handle a massive number of variables. Each side of the polytope, or the intersecting dimensional planes, delineates the constraints. So all the possible solutions that satisfy the constraints lie within the convex shape. "The solution space is the convex polyhedron," said Whitely. "Then you add one more condition, you say I want the cheapest solution. The optimal result, then, is found at the uppermost corner, or the uppermost edge or face of the polytope. Certainly not the interior. The principle of linear programming is how do I find the top spot, the optimal solution to the given problem." Whiteley, interviews.

2. "University of Toronto Oral History Project"; and Coxeter diaries, 1947–1952.

3. Coxeter, *Regular Polytopes*, viii.

4. He added, prophetically: "On the other hand, a reader whose standpoint is more severely practical may take comfort in [Russian mathematician Nikolai] Lobatchevsky's assertion that 'there is no branch of mathematics, however abstract, which may not some day be applied to phenomena of the real world.'" Ibid., vi.

5. Hardy, *A Mathematician's Apology*, 80.

6. John Bryden, *Best-Kept Secret*, 47–48.

7. Ibid.

8. Sinkov was one of three people (Neil Sloane and John Leech, the latter best known for his Leech lattice, at the University of Stirling in Glasgow, being the other two) whom Coxeter called upon when his calculations required more might than mere pencil and paper provided and he had to defer to computers. Coxeter never used a computer himself. Coxeter, "University of Toronto Oral History Project."

9. Coxeter diaries, 30 April 1941.

10. Coxeter, interviews; "Coxeter, University of Toronto Oral History Project."

11. Ibid.

12. Coxeter always spoke his mind, and managed a pointed comment about the Second World War even in the preface to *Regular Polytopes*, published three years after the war ended: "The history of polytope-theory provides an instance of the essential unity of our Western civilization and the consequent absurdity of international strife. The Bibliography lists the names of thirty German mathematicians, twenty-seven British, twelve American, eleven French, eight Swiss, seven Dutch, four Italian, two Austrian, two Hungarian, two Polish, two Russian, one Norwegian, one Danish, and one Belgian. (In proportion to population the Swiss have contributed more than any other nation.)" He was also known to drop his opinions into his lectures—obiter dicta, like a judge letting loose a side remark not relevant to the case being heard. John Coleman recalls Coxeter finishing a lecture on the blackboard and concluding that the proof in question "leaves us with one." Thereupon Coxeter slipped in a comment about his latest religious dalliance, with the Unitarians, believers in the oneness of God (as opposed to the Christian Trinity) and known through the ages as free thinkers and dissenters, evolving their belief system toward freedom, tolerance, and humanism. It would seem a perfect fit for Coxeter, but his liaison with the Unitarians was brief. As his daughter explained, "He was more a naturalist, not a humanist, because humanists place themselves above animals, which he would never do." Coxeter, *Regular Polytopes*, vii; Coleman, interviews; Thomas, interviews; and Coxeter diaries, August 14 and 30, 1945.

13. Tim Rooney (retired professor of mathematics, University of Toronto), interview, Toronto, March 24, 2003.

14. Coxeter diaries, 21 January 1942.

15. Coxeter diaries, 2 January 1940.

16. Coxeter diaries, 11 May 1943.

17. Coxeter diaries, 28 October 1944.

18. Coxeter diaries, 18 June 1940.

19. Lee Lorch (emeritus professor of mathematics, York University, Toronto), interview, Toronto, March 3, 2005.

20. Coxeter diaries, 17 July 1942.

21. Coxeter diaries, 19 May 1978—"*Globe & Mail* questionnaire proves I agree with NDP [New Democratic Party]."

22. Eric Infeld, interview and e-mail correspondence, January, February, 2005; Helen Infeld to Donald Coxeter, 6 January 1975, University of Toronto Archives, Coxeter Fonds; "Mathematician Hard to Figure, Prof. Infeld Stays in Poland but Doubt He Took A-Secrets," *Globe and Mail*, August 22, 1950; and "Drew Demands Gov't Investigate UofT Professor," *Canadian Press*, March 17, 1950.

23. C. Davis, interviews; and Davis, "The Purge," in *A Century of Mathematics in America*, vol. 1, 413–28.

24. Steve Batterson, *Stephen Smale: The Mathematician Who Broke the Dimension Barrier*, 140.

25. Lee Lorch, a longtime activist against racism, sexism, and militarism, also personally felt Coxeter's support. Coxeter made an impression when he was a visiting professor to Columbia University, in New York, shortly after the war. Coxeter showed his solidarity with his Columbia colleagues' protesting Lorch's dismissal from nearby City College due to Lorch's activism against racism in the Stuyvesant Town housing project. Lorch was also grateful for Coxeter's backing when he was vice president of the American Mathematical Society. The National Convention of the Democratic Party National Convention was held in Chicago, in preparation for the 1968 elections. Many protestors demonstrated against the Vietnam war at the convention, and the Chicago police suppressed the demonstration with ferocious brutality. The American

Mathematical Society had scheduled a regional meeting in Chicago, to be held in 1969. At the AMS annual general meeting in 1968, Lorch introduced a motion to move the AMS meeting away from Chicago in protest of the violence. The AMS by-laws give little or no power to the general membership, placing all the power in the hands of the council. At that time Coxeter, as one of three AMS vice presidents, was automatically on council. "I spoke with him about this matter. He expressed the view that council would comply with our motion and promised without hesitation his own support," recalled Lorch. By a nearly unanimous vote, the pending Chicago meeting was moved to Cincinnati. Lorch, interview; C. Davis.

26. Coxeter diaries, 11 February 1960.

27. Coxeter diaries, 23 November 1959.

28. Seymour Schuster (emeritus professor of mathematics, Carleton College, Minnesota), interview, Toronto, May 13, 2004, and January 14, 2006; Schuster to Coxeter, 28 February 1967, and Coxeter to Schuster, 7 March 1967, Schuster's personal papers.

29. Marie-Jeanne Coleman, interview, Kingston, Ontario, March 8, 2005.

30. Thomas, interviews.

31. Coxeter, interview, Toronto, February 28, 2003.

32. Coxeter diaries, 13–15 July 1944.

33. Coxeter diaries, 24 December 1947.

34. Coxeter diaries, 4 June 1944.

35. Coxeter diaries, 16 February 1948.

36. Coxeter diaries, 15 November 1955, 16 February 1956.

37. Coxeter diaries, 19 December 1955.

38. Coxeter diaries, 22 June 1959; and Thomas, interviews.

39. H. H. Punke, "The Family and Juvenile Delinquency," *Peabody Journal of Education*, 98.

40. Coxeter to H. H. Punke, 7 February 1957, Coxeter Fonds, University of Toronto Archives.

41. According to Plutarch, Archimedes was infatuated with geometry with every ounce of his being: "Oftimes Archimedes' servants got him against his will to the baths, to wash and anoint him, and yet being there, he would ever be drawing out of the geometrical figures, even in the very embers of the chimney. And while they were anointing of him with oils and sweet savours, with his fingers he drew lines upon his naked body, so far was he taken from himself, and brought into ecstasy or trance, with the delight he had in the study of geometry." O'Connor and Robertson, "Archimedes of Syracuse," http://www-groups.dcs.st-and.ac.uk/~history/Mathematicians/Archimedes.html (accessed January 29, 2006); and Ibid.

42. Coxeter diaries, 17 June 1949.

43. They had explored commissioning a house designed by Coxeter's friend and university colleague, architect Eric Arthur, but in the end they found a fittingly cubic house, in the early modernist or international style with octagonal windows flanking the front door. Built in 1935, and designed by architect Mackenzie Waters, an associate architect of Maple Leaf Gardens, it is now a Toronto heritage property. City of Toronto Inventory of Heritage Properties, http://app.toronto.ca/heritage/browseLetter.do?letter=R (accessed January 29, 2006); Coxeter diaries, 8 April 1953.

44. Coxeter diaries, 8 December 1938–15 April 1941.

45. Coxeter diaries, 4 December 1942.

46. Thomas, interviews.

47. Coxeter diaries, 23 October 1941.

48. Coxeter diaries, 6 April 1956.

49. Coxeter diaries, 27 November 1939.
50. Coxeter diaries, 1 January 1946.
51. Coxeter diaries, 17 December 1941, 4 and 5 October 1946, 4 November 1946.
52. Marie-Jeanne Coleman, interview.
53. Coxeter diaries, 25 November 1944.
54. Coxeter diaries, 19 February 1943, 25 November 1944.
55. Senechal, "Coxeter and Friends."

CHAPTER 8—BOURBAKI PRINTS A DIAGRAM

1. Coxeter published one of the most interesting papers on polyhedra in 1952, in the *Philosophical Transactions of the Royal Society of London*. It numbered more than fifty pages, and its title read: "H. S. M. Coxeter and Others on Uniform Polyhedra." The "others" were Coxeter's former research student at Cambridge, J. C. P. Miller, and distinguished geophysicist and geometer Michael Longuet-Higgins. Longuet-Higgins connected with Coxeter in the early 1950s via a mutual friend—Freeman Dyson, a celebrated theoretical physicist at Princeton's Institute for Advanced Study. Dyson was in Toronto for a physics conference and paid a visit to his friend Coxeter, with whom he'd had leisurely interactions over the years (leisurely, as polyhedra, for the polymath Dyson, were one of those Sunday-afternoon activities that he enjoyed very much). Dyson drew Coxeter's attention to some work by Longuet-Higgins and his brother, Christopher Longuet-Higgins, the distinguished chemist and cognitive scientist. As high school students, the brothers and Dyson had been inspired by Coxeter's book *The 59 Icosahedra*. This has led the brothers to try to enumerate the class of uniform polyhedra. And Michael began constructing edge models using a novel material of galvanized iron wire, or garden wire (the construction was innovative, requiring no solder at the vertices; a kink in the wire allowed them to click into place). He produced models for each of their discoveries; though for the most complicated, with ten edges passing through some vertices, he devised an alternative method: he soldered a wire frame, painted it black, and then used the frame to string white cotton threads representing the edges. Little did the Longuet-Higgins brothers know that, not long before, Coxeter and Miller had embarked on a similar enumeration at Cambridge. Coxeter and Miller believed their list to be complete, but because they lacked a rigorous proof, they sat on their results. When Coxeter learned of the Longuet-Higginses' enterprise, he and Miller had given up hope for finding a proof and were at long last writing up their results for publication. Coxeter wrote to the brothers, enclosing his list of the uniform polyhedra, and asked how many they had found. They had all but one on the list, the one Coxeter had dubbed "Miller's monster." Coxeter and Miller went to see the models, and it proved fortuitous timing. Miller was just drawing all the polyhedra, and the model with the white threads allowed Miller "to detect an error before its consequences became serious," as Coxeter said. Coxeter invited Michael to become a coauthor of the paper, with photographs of his models providing additional illustrations. Coauthorship was also on the merit of Michael's discovery of some relevant theoretical relations between certain chains, or sequences, of the uniform polyhedra; and he discovered a uniform tessellation that Coxeter and Miller had missed. The paper was of particular interest because it surveyed the many approaches to these polyhedra through history, and then presented a unified method for their construction. The authors still could not claim to have proven completeness for their enumeration, but they expressed hope that it was indeed complete. This was proven to be true in 1975, by John Skilling—with the aid of a computer—and published in his paper, "The Complete Set of Uniform Polyhedra." Longuet-Higgins, interviews; Coxeter–Longuet–Higgins correspondence, Coxeter Fonds, University of Toronto Archives; Longuet-Higgins, personal papers; and Coxeter et al., "Uniform Polyhedra," *Philosophical Transactions of the Royal Society*, 401–50, see appendix 8.

2. Coxeter, *Introduction to Geometry*, x.

3. While the upper echelons of research mathematicians at the university level—those who populated the prestigious American Mathematical Society and concerned themselves with producing the next generation of mathematicians proper—appreciated Coxeter's pioneering work in group theory, they more often than not dismissed his predilection for the gems of classical geometry, the trivial tinkering with toys. Math educators, however—college and high school teachers who populated the MAA—recognized Coxeter's appeal as a wizard of a teacher, his ability to lure young students and engage them with mathematics through the playful, and yet still profound, magic of elementary geometry (of course, some mathematicians belonged to both the AMS and the MAA, and were concerned with the quality and continuity of mathematics education on all levels, but generally the distinction holds). With a solid grounding in classical geometry absent at the university level, teachers needed to be trained elsewhere on this subject, and there was no better geometry teacher for teachers than Coxeter. Schuster, interview; Senechal, interviews; Schattschneider, interviews; and Martin Gardner, "Math and Aftermath."

4. Otherwise a man of measured tempo, Coxeter was known among his students as a frightfully reckless driver who sped through traffic, zigzagging with scarcely an inch of space to spare. On a number of occasions he landed himself in traffic court. He had numerous accidents—"200 degree skid"—and his cars—Citroën, MG, Triumph—were forever in the shop. Then he traveled by bus, but one day a friend of Rien's spied him hitchhiking home on a busy downtown Toronto street (he lived no more than a thirty-minute walk away). Weiss, interviews; Monson, interviews; Fisher, interviews; Thomas, interviews; Coxeter diaries, 31 December 1938.

5. Willy Moser, whose brother was mathematician Leo Moser, in fact made breakthrough contributions to geometry in his own right. His first book, coauthored with Coxeter, was the highly regarded *Generators and Relations for Discrete Groups*. Known as a "definitive handbook in the area," for a time it was miscatalogued at the University of Toronto library and misshelved in the genealogy section. Moser, interviews.

6. Coxeter, *Introduction to Geometry*, p. ix.

7. Weiss, interviews; and Moser, interviews.

8. Martin Gardner, "Math and Aftermath."

9. Coxeter, "Math and Aftermath."

10. Martin Gardner, *Martin Gardner's New Mathematical Diversions from Scientific American*, 196.

11. To wit: polygons, circles, spheres, regular polytopes, complex numbers, the pseudosphere, the five Platonic solids, and a statistical honeycomb, the product of two reflections, isometry in the Euclidean plane, dominoes to illustrate the space groups of two-dimensional crystallography, the close packing of equal spheres, the golden section and phyllotaxis, tensor notation and reciprocal lattices, projective geometry and Desargues's theorem, geodesics and the Euler-Poincaré characteristic, topology of surfaces and the four-color problem, the correction of a "prevalent error" concerning the shape of the monkey saddle, absolute geometry and the polyhedral kaleidoscope, hyperbolic geometry and the finiteness of triangles, differential geometry of curves and the circular helix, the simplest construction of four-dimensional figures, ordered geometry and Sylvester's problem of collinear points. Coxeter, *Introduction to Geometry*, xi–xvi, 295.

12. Atiyah, interviews.

13. David Mumford (professor of mathematics, Brown University), interview November 18, 2005.

14. Mumford offered a recent ode of sorts to Coxeter, with this description of his role in twenty-first-century mathematics: "In my book, Coxeter has been one of the most important 20th century mathematicians—not because he started a new perspective, but because he deepened and extended so beautifully an older esthetic. The classical goal of geometry is the exploration and enumeration of geometric configurations of all kinds, their symmetries and the constructions relating them to each other. The goal is not especially to prove theorems but to discover these perfect objects and, in doing this, theorems are only a tool that imperfect humans need to reassure themselves that they have seen them correctly. This is a flower garden whose beauty has almost been forgotten in the 20th century rush to abstraction and generality. I share Coxeter's love of this perspective, which has deep roots in algebraic as well as Euclidean geometry. I always found the algebraic roots-and-weights approach to Lie groups arid and unsatisfying until I found Coxeter's work fleshing this out with a rich tapestry of examples." Ibid.; and Jonathan M. Borwein, *Mathematics by Experiment: Plausible Reasoning in the 21st Century*, 86.

15. Of Bourbaki's influence on the mathematical world, and his pride at the enormity of the Bourbaki project, Chevalley once commented: "I absolutely had the feeling of bringing light into the world—the mathematical world . . . It went hand in hand with the absolute certainty of our superiority over other mathematicians—a certainty that we held something of a higher level than the rest of mathematics of the day. For example, there is a word which was—which still is—in current usage, *to bourbakise* (*bourbachiser*). This means to take a text that one considers screwed up and to arrange it and improve it." Guedj and Gray, "An Interview with Claude Chevalley," 20; O'Connor and Robertson, "Claude Chevalley," http://www-groups.dcs.st-and.ac.uk/~history/Mathematicians/Chevalley.html (accessed January 30, 2006).

16. O'Connor and Robertson, "Jean Alexandre Eugène Dieudonné," http://www-groups.dcs.st-and.ac.uk/~history/Mathematicians/Dieudonne.html (accessed January 30, 2006).

17. R. P. Boas, "Bourbaki and Me," *The Mathematical Intelligencer*, 84.

18. Beaulieu, "Bourbaki's Art of Memory."

19. Beaulieu, "Bourbaki's Art of Memory"; and Halmos; Borel; Cartan; Senechal, "The Continuing Silence of Bourbaki," *The Mathematical Intelligencer*, 22–28; and Pierre Cartier, interviews.

20. Beaulieu, "Bourbaki's Art of Memory."

21. Ibid.

22. Dieudonné, "The Work of Nicholas Bourbaki," 142.

23. Cartier, interviews.

24. Senechal, "The Continuing Silence of Bourbaki."

25. Thomas, interviews.

26. Beaulieu; Halmos.

27. Dieudonné, "The Work of Bourbaki during the Last Thirty Years," *Notes of the AMS*, 620.

28. Senechal, "The Continuing Silence of Bourbaki."

29. Senechal, interviews.

30. "The Bourbakists play the sort of evil Darth-Vader figure here with their ultra formalism . . . It was mini Bourbakist," observed William Higginson, a professor in the Mathematics, Science and Technology Education Group at Queen's University, Kingston, Ontario.

31. William Higginson, interview, March 2005; and Bob Moon, *The "New Maths" Curriculum Controversy*.

32. Euclid and his geometry also met criticism as being pedagocially difficult. One of Euclid's theorems, stating that the angles at the base of an isosceles triangle are equal, was named "pons asinorum," meaning bridge of asses, which, as Coxeter explained, "probably arose from

the bridgelike appearance of Euclid's figure (with the construction lines required in his rather complicated proof) and from the notion that anyone unable to cross this bridge must be an ass." Coxeter, "The Ontario K-13 Geometry Report," *Ontario Mathematics Gazette*, 12–16; and *Geometry, Kindergarten to Grade Thirteen*, Report of the (K-13) Geometry Committee; and Coxeter, *Introduction to Geometry*, 6.

33. Ibid.

34. "The Mystique of Modern Mathematics," in *Black Paper Two*, as cited in *The "New Maths" Curriculum Controversy*, 146–47.

35. Moon.

36. Dieudonné, "New Thinking in School Mathematics;" Moon, 5, 99; and Toth.

37. Liliane Beaulieu, writing the unauthorized Bourbaki biography to be published by Springer-Verlag, has gathered her own observations, during her decade of research, on the interpretation of Bourbaki as antigeometry. "The reasons for Bourbaki being labeled as antigeometry are manifold," she said. "Yes, Dieudonné's cry had its effect especially since, to my knowledge, there was no official disclaimer from the group itself. This does not mean that the group as a whole endorsed his ideas, or that they were rejected. It just means that it [Bourbaki] didn't address the issue officially, which is a common way of dealing with various issues within the group, up to this day." Beaulieu continued to point out that there was some geometry in Bourbaki's *Elements*, though swallowed up by algebra and analysis. She also shed light on the lack of diagrams, pertaining to geometry or not. "The near absence of figures in the published treatise was unusual in 'Analysis' and thus rightly gave the impression that this team did not wish to foster their use," she said. "Some [diagrams] appear, nevertheless, in the unpublished drafts before they are left out, once the idea is conveyed and integrated into a more abstract or general conceptual setting." Beaulieu, e-mail correspondence, "Re: Coxeter and Bourbaki," March–May 2005; and Broué, interviews.

38. Cartier, interviews.

39. Cartier; Broué.

40. Moon; Ángel Ruiz and Hugo Barrantes, *The History of the Inter-American Committee on Mathematics Education*; and Á. Ruiz (professor of mathematics, University of Costa Rica and the National Unversity; director, Research Center for Mathematics and Meta-Mathematics), e-mail correspondence, November 18, 2005.

41. Moon, 49.

42. Ruiz and Barrantes.

43. Another Canadian mathematician, Irving Kaplansky, an algebraist and group theorist, discussed the second-class status of geometry with lively prose and mixed metaphors in the preface to his book *Linear Algebra and Geometry*: "Linear algebra, like motherhood, has become a sacred cow. It is taught everywhere; it is reaching down into the high schools and even the elementary schools; it is jostling calculus for the right to be taught first. Yet all is not well. The courses and books all too often stop short just as the going is beginning to get interesting. And classical geometry, linear algebra's twin sister, is a bridesmaid whose chance of getting near the altar becomes ever more remote. Generations of mathematicians are growing up who are on the whole splendidly trained, but suddenly find that, after all, they do need to know what the projective plane is." Kaplansky, who received his undergraduate and master's degrees in mathematics at the University of Toronto, appreciated Coxeter's "incredible geometric visualization— he could really see things." He took a few courses from Coxeter, who was then just starting out, a "genial, shy, diffident, somewhat bashful" lecturer. Nonetheless he answered questions— "even questions on the stupid side"—in a polite and illuminating way. Kaplansky recalled: "I

was never on a plane of equality." Kaplansky, interviews; Coleman, interviews; and Conway, interview, June 2004, discussing views expressed by William Thurston (a mathematician at Cornell University); and Ruiz and Barrantes.

44. Higginson, interview.

45. Moon, 58–59.

46. Other songs on that album included "Smut," in praise of pornography; "Send the Marines"; and "The Vatican Rag"—the New Math was in good company. Tom Lehrer, "New Math," http://www.lyricsfreak.com/t/tom-lehrer/ (accessed January 30, 2006).

47. The French mathematician and Fields medallist René Thom also took a stand in his article " 'Modern' Mathematics: An Educational and Philosophical Error?" The subtitle read: "A Distinguished French Mathematician Takes Issue with Recent Curricular Innovations in His Field." Thom began by saying, "In the minds of most of our contemporaries, so-called modern mathematics holds a place of high prestige lying somewhere between cybernetics and information theory in the bag of tricks promoted by deceptive publicity as the essentials of modern technology, the indispensable tools for the future development of all scientific knowledge." To the contrary, he said, ". . . the contemporary trend to replace geometry with algebra is educationally baneful and should be reversed." And he concluded with a thought on the "genetic importance of geometry," saying: "the geometric continuum is the primordial entity. If one has any consciousness at all, it's consciousness of time and space; geometric continuity is in some way inseparably bound to conscious thought." René Thom, " 'Modern' Mathematics: An Educational and Philosophic Error?" *American Scientist*, 695–99; and Moon, 97–119.

48. Atiyah, interviews.

49. Atiyah, "What Is Geometry?" in *The Changing Shape of Geometry*, 24–29.

50. Atiyah, interviews.

51. Whiteley, interviews; Conway, interviews; David Henderson (professor of mathematics, Cornell University), interview, Budapest, July 11, 2002.

52. Conway, interviews.

53. Emma Castelnuovo, interview, May 2, 2005; and Castelnuovo to Siobhan Roberts, May 9, 2005.

54. Coxeter, *Geometry, Kindergarten to Grade Thirteen*, 19.

55. Freudenthal.

56. Coxeter, *Geometry, Kindergarten to Grade Thirteen*, 10.

57. Freudenthal, 240, 429, 431.

58. Coxeter, interviews; Joerg Wills (professor of mathematics, University of Siegen), interview, Budapest, July 10, 2002.

59. Miyazaki is the author of thirty-five books (he sent almost all of them to Coxeter), such as *Polyhedra and Architecture* and *An Adventure in Multidimensional Space*. His most recent book is *Science of Higher-dimensional Shapes and Symmetry*. "In it I inserted Prof. Coxeter's portrait with Plato, Kant, Moebius, Schlafli, and Einstein," he said. He also issued the dedication to Coxeter, saying, ". . . this book has no originality about the basic theoretical contents. All of them are of Prof. Coxeter himself . . . The main basic theoretical contents in this book were taught by his so many letters from the starting point of the investigation around 1970. His letters were every time filled with easy understandable warmhearted contents including about manners to write English letters, as if it were sent to his grand-grand-son." In 1965, when Miyazaki was first introduced to Coxeter's work, he decided on a whim to write him, asking a question about four-dimensional curved hypersurfaces and regular polytopes.

"At that time I didn't know how to write English letter and I wrote his name, "Professor Dr. H. S. M. Coxeter Esq." I wrote the letter using over polite and arrogantly modest words and sentences about which I could learn from our English dictionary," said Miyazaki. "At that time I thought I could not get the return mail, but astonishingly enough, soon I could receive his letter which started from 'Dear Koji' But I could not write such as Dear Donald and every time I wrote Dear Prof. Coxeter. He kindly taught me indirectly how to write an English letter. I have learnt indirectly about the meaning of Sincerely yours, Best wishes, Kind regards, Yours truly, etc. etc. from Prof. Coxeter." According to Miyazaki, the impact of Coxeter's book in Japan was considerable. "Us Japanese originally don't like geometric logics and if there are no Coxeter's understandable geometry us Japanese have not become aware of morphologic contents about things. Japanese version of his book 'Introduction to Geometry' gave much impact to so many Japanese including me." Koji Miyazaki, e-mail correspondence, July 2003, March–April 2005.

60. Coxeter diaries, 19 and 20 January 1959, 5 and 20 February 1959, 2 April 1959, 19 February 1960; Coxeter, *Twelve Geometric Essays*, 180–81.

61. See appendix 8.

62. Schuster, interviews.

63. The films were produced by Sy Schuster and Allen Downs; the second senior mathematician on the production crew was Daniel Pedoe (an Englishman who overlapped with Coxeter at both Cambridge and Princeton; his career took him to university postings around the world, retiring from his position at the University of Minnesota; he authored several books, including *The Gentle Art of Mathematics* and *Geometry and the Visual Arts*; after retiring he collaborated with Hidetosi Fukagawa, a high school teacher in Japan, who sought to interest Japanese academics in San Gaku—centuries-old Japanese wooden tablets that hung in temples and shrines and contained geometric theorems; they coauthored *Japanese Temple Geometry Problems*, published in 1984 by the Charles Babbage Research Centre in Canada); Schuster.

64. Coxeter wasn't quite geometry's equivalent of Canadian literary maven Northrop Frye, who chaired every committee he could on English literature curriculum reform. But according to the author and critic Guy Davenport, the two were comparable in a more general sense: "When the dust settles, someone will notice that the Canadian genius is for the synoptic view of things—Douglas Bush, Northrop Frye, H. S. M. Coxeter, Herbert Marshall McLuhan, Barker Fairley, Hugh Kenner." Davenport, "The Kenner Era," *National Review*, December 31, 1985.

65. Rooney, interview.

66. Ibid.

67. The committee was chaired by Coxeter's colleague at the university, George Duff (later Coxeter's favorite head of the department) and also included Ontario high school mathematics teachers, and the notable W. W. Sawyer, a man with similar provenance to Coxeter's, being born in England in 1911. Sawyer spent his early years lecturing at British universities, and went to the University of Ghana as the head of the math department, followed by New Zealand and the United States, where he was a vocal critic of New Math. Sawyer, like Coxeter, was an amazing populizer of mathematics, writing the Penguin classics *Mathematician's Delight* and *Prelude to Mathematics*. In 1965, Sawyer assumed the appointment of joint professor to the mathematics and education departments at the University of Toronto, where he stayed until he retired back to England. *Geometry, Kindergarten to Grade Thirteen*, 1, 4, 18.

68. Ibid., 34–92.

69. In 1968, Coxeter noted in his diary, "Mark Kac says geometry is dead in USA." Kac, a Polish mathematician then at Rockefeller University, New York, who was famous for his paper

"Can One Hear the Shape of a Drum?" once remarked: "I am fascinated . . . particularly [with] the role of dimensionality: why certain things happen in 'from three dimensions on' and some others don't. I always feel that that is where the interface . . . of nature and mathematics is deepest. To know why only certain things observed in nature can happen in the space of a certain dimensionality. Whatever helps understand this riddle is significant." Coxeter diaries, December 14, 1968; and O'Connor and Robertson, "Mark Kac," http://www-groups.dcs.st-and.ac.uk/~history/Mathematicians/Kac.html (accessed February 3, 2006).

70. N. Bourbaki, *Groupes et Algèbres de Lie,* chaps. 4, 5, 6, 243; and Bourbaki, *Elements of the History of Mathematics,* 269–73; and Senechal, interviews.

71. In addition to Jacques Tits, two other mathematicians also stake claim, at various stages, to introducing Bourbaki to Coxeter's work. John Coleman suggested he may have provided the first point of contact. Coleman completed his master's under Bourbaki member Claude Chevalley at Princeton. In 1949, he met with his thesis advisor at a Bourbaki conference near Versailles. He learned Bourbaki was working on a volume about Lie groups, which prompted Coleman to tell Chevalley of Coxeter's work pertaining to this subject. "He was surprised to hear of it," said Coleman. "He wasn't aware of Coxeter's work until I told him." Pierre Cartier recounts a similar scenario. When Bourbaki was working on the volume on groups in the 1950s (each volume involved a long, drawn-out process), he proposed Coxeter's theories as the way to go. "I can tell you that one of the volumes of Bourbaki was thoroughly inspired by the work of Coxeter and that I was instrumental in convincing my colleagues within the Bourbaki group that this was the right approach," said Cartier. "There was one week of fighting to convince [them] that Coxeter was the way to do it. It was not without a fight." The issue was this: The Bourbakists were aware of another mathematician who had done some similar work in this area of groups, Russian-born Eugene Dynkin, then at Moscow State University and currently a professor of mathematics at Cornell University. Dynkin discovered a notation nearly identical to Coxeter's. This presented a problem, however, for the Bourbakists. In the volume about Lie groups, Bourbaki had planned to use and reference Dynkin's diagram, which was used widely among physicists. But learning of Coxeter's work, they were forced to reevaluate, since it is of the utmost importance in mathematics to credit the first practitioner of a discovery or invention. Coxeter's first published reference to the use of such a graph was in a paper he published in the *Journal of the London Mathematical Society,* submitted on his birthday, February 9, 1931; and the first published appearance of the graphs was in the *Annals of Mathematics* in 1934. Dynkin's notation dated to his studies at Moscow University between 1940 and 1945 (he was exempt from war service due to poor eyesight). Dynkin's discovery came from his attempt to understand the papers by Weyl on semisimple Lie groups. Coxeter's discovery came from his work on crystallographic groups. As a consequence of the two discoveries, so close in timing, some mathematicians call these diagrams "Coxeter-Dynkin diagrams." Coxeter greeted this news cordially rather than competitively. He was particularly pleased by the communication with Dynkin that came about as a result. Coxeter liked to recount details of Dynkin's letter dated April 3, 1984, in which Dynkin remarked: "It is striking that my notation turned out to be so similar to yours. This probably shows how natural these notations are." Coleman, interviews; Cartier, interviews; Eugen Dynkin, interview, May 2004; and Coxeter-Dynkin correspondence, Coxeter Fonds, University of Toronto Archives; Jacques Tits, "Groupes et Géométries de Coxeter," *Wolf Prize in Mathematics,* vol 2., 740–54; and Tits correspondence to Siobhan Roberts, July 28, 2004.

72. And since then, the Coxeter terms have reproduced. There is the Coxeter element, Coxeter matroid, the Todd-Coxeter coset enumeration algorithm and the Boerdijk-Coxeter helix—a linear necklacelike packing of regular tetrahedra, which provides an efficient solution to some

close-packing problems, with applications in biology, such as the helical structures of amino acids and proteins. The Coxeter terminology also includes: Coxeter theory, Coxeter tessellation method, Coxeter transformation, Coxeter scheme, Coxeter complex, Coxeter polynomial, Coxeter orbifold, Coxeter functor, Tutte-Coxeter graph, and Coxeter-Knuth insertion procedure.

73. Coxeter diaries, 25 and 26 March 1969.

74. Dieudonné, "The Work of Bourbaki in the Last Thirty Years," 620.

75. With Bourbaki having acknowledged the importance of Coxeter's work, his tools gained in popularity, spreading through the mathematical community. In 1974, Coxeter's colleague, Israel Halperin, said he had been hearing about "Coxeter groups," and, in 1976, at a conference at the University of Warwick, Coxeter discussed his namesakes with Roger Penrose and George Lusztig, the latter then a budding "modern" geometer, now at MIT (Lusztig's main interest is in theory of group representations and Coxeter groups, he said, have always played an important role in his work). The next day, Coxeter noted in his diary: "I looked at Bourbaki's three chapters." Senechal, "The Continuing Silence of Bourbaki"; George Lusztig, Norbert Wiener Professor of Mathematics, MIT, e-mail correspondence, 3 February 2006; and Coxeter diaries, 6 August 1974, 13 and 14 January 1976.

76. Dieudonné, "The Work of Nicholas Bourbaki," 144; Cartier, interviews; and Senechal, "The Continuing Silence of Bourbaki."

77. *The Two-Year College Mathematics Journal* 11, no. 1 (January 1980).

78. Ibid., 19–20.

79. Ibid., 18.

CHAPTER 9—BUCKY FULLER, AND BRIDGING THE "GEOMETRY GAP"

1. Coxeter noted of one find: "Preparing for my lecture I found a frightful error on p. 147, spent hours devising two new lines to occupy same space as two nonsense lines, sent them to Mr. Coles."

2. Coxeter diaries, 21 February 1974, 2 January 1973.

3. Coxeter diaries, 4 February 1975.

4. Buckminster Fuller, *Synergetics, Explorations in the Geometry of Thinking*, ix.

5. Smith, interviews; Conway, interviews; and Robert W. Marks, *The Dymaxion World of Buckminster Fuller.*

6. Coxeter, interviews; Coxeter diaries, 10 August 1967; Coxeter–Fuller correspondence, Coxeter Fonds, University of Toronto Archives; Buckminster Fuller Collection (M1090), Stanford University Libraries; Smith, interviews; Conway, interviews; Marks.

7. Coxeter, interviews; and Coxeter–Fuller correspondence.

8. Fuller, *4D Time Lock*; and see "Introduction to Geodesic Domes and Structure," at the Buckminster Fuller Institute Web site, http://www.bfi.org/domes/ (accessed February 3, 2006).

9. Ibid.; and Smith, interviews.

10. Smith was drawn to Fuller in 1970, when had had just returned from Vietnam, "intellectually in a coma and very discouraged about humanity. While wandering in a book store in Puerto Rico, looking for a science fiction book, by mistake I picked up RBF's *Utopia or Oblivion: The Prospects for Humanity*." In Fuller, Smith found hope for humanity that "snapped me out of my coma." He tried to get his hands on everything Fuller had written, and Fuller ultimately led him to Coxeter. Smith, interviews.

11. Ibid.

12. Coxeter diaries, 29 February and 1 March 1968.

13. Coxeter diaries, 1 March 1968.

14. Coxeter–Fuller correspondence, Coxeter Fonds, University of Toronto Archives; Buckminster Fuller Collection, Stanford University Libraries; and Coxeter's copy of Marks's *The Dymaxion World of Buckminster Fuller*.

15. The theorems: "I. Every isometry is a reflection or the product of 2, 3, or 4 reflections. II. Every product of 2 reflections is a translation or rotation. III. Every product of 3 reflections is a glide or a rotary reflection, including a simple reflection as a special case. IV. Every product of 4 reflections is the product of 2 half-turns. V. Every direct isometry is a twist, including a translation or a rotation as a special case. VI. Every similarity is a twist or a glide or a dilative rotation, including a simple dilation as a special case." Coxeter, "Helices and Concho-Spirals," *Proceedings of the 11th Nobel Symposium, Symmetry and Functions of Biological Systems at the Macromolecular Level*.

16. Coxeter–Fuller correspondence, Coxeter Fonds, University of Toronto Archives; and Buckminster Fuller Collection, Stanford University Libraries.

17. Coxeter diaries, 4 September 1968.

18. Thomas, interviews; Schattschneider, interviews.

19. Coxeter diaries, 30 September 1968; "Fuller Sees Housing Service Industry by '72," *Globe and Mail*, October 1, 1968; and "Fuller Advises Students Go Outstairs Not Upstairs," *Telegram*, October 1, 1968.

20. Smith, interviews.

21. Coxeter, interviews.

22. Whiteley, interviews.

23. Coxeter, interviews; Smith, interviews.

24. George Escher, interviews; Schattschneider, interviews; and Schattschneider, interview with George Escher, November 2005.

25. M. C. Escher to Arthur Loeb, 8 June 1964, Escher Archives, Gemeentemuseum, The Hague.

26. Barry Farrell, "The View From the Year 2000," *LIFE*, 70, no. 7 (February 26, 1971), 46–58.

27. Coxeter published his proof, saying: Fuller's theorem ". . . (very nearly) stated: When congruent balls are arranged in cubic close-packing so as to fill a tetrahedron, square pyramid, octahedron, cuboctahedron, truncated tetrahedron or truncated octahedron, with $n+1$ balls along each edge, the total number of peripheral balls is bn^2+2, where $b=2, 3, 4, 10, 14,$ or $30,$ respectively." Coxeter, interviews; Coxeter–Fuller correspondence; Buckminster Fuller Collection, Stanford University Libraries; Coxeter diaries, 29 September 1970; and Coxeter, "Polyhedral Numbers," in *For Dirk Struik, Boston Studies in the Philosophy of Science*, 25–35.

28. Coxeter–Fuller correspondence, Coxeter Fonds, University of Toronto Archives; and Buckminster Fuller Collection, Stanford University Libraries.

29. Coxeter, "Virus Macromolecules and Geodesic Domes," in *A Spectrum of Mathematics*, a collection of essays written as a Festschrift for geometer H. G. Forder, at the University of New Zealand, in Auckland, where Coxeter did a stint as a visiting professor. It was also to become the inspiration for the book *Spherical Models*, by Magnus Wenninger, a Benedictine monk with whom Coxeter corresponded.

30. The common cold virus is comprised of three proteins that swirl together to form an icosahedron. "It self-assembles out of these proteins in your body to your detriment," said Walter Whiteley. According to Whiteley most viruses are icosahedral or dodecahedral (none are tetrahedral, cubic, or octahedral; though some are more tubular than polyhedral in structure). Ibid.; and Whiteley, interviews.

31. Coxeter, in "Virus Macromolecules and Geodesic Domes," compared Fuller's method

in constructing the icosahedral map with the geodesic dome: "When Buckminster Fuller makes a map of the whole earth, he imagines a regular icosahedron inscribed in the geographic globe, transfers the outlines by gnomic (or 'central') projection from the sphere to the surface of the icosahedron, cuts this surface along a suitable set of eleven edges, and unfolds it into a flat 'net.' . . . When Fuller makes a geodesic dome, the process is reversed: he fills a large equilateral triangle with a pattern of small equilateral triangles, repeats this pattern on each face of an icosahedron, and transfers the repeated pattern by gnomic projection to the surface of the circumsphere." Coxeter–Fuller correspondence, Coxeter Fonds, University of Toronto Archives; and Buckminster Fuller Collection, Stanford University Archives.

32. Another word Fuller invented was "vector equilibrium" to supplant "cuboctahedron," a replacement Coxeter took offense to. "[Fuller] thought highly of himself, to the extent of being allowed to invent new words when he felt like it," Coxeter said disapprovingly. As far as the term "tensegrity" was concerned, Coxeter didn't mind it. However, New York artist Kenneth Snelson, a student of Fuller's at the experimental Black Mountain College in 1948, claimed that Fuller invented the term "tensegrity" as a way of appropriating Snelson's discovery of this sticks and string mobile. Snelson calls them either "floating compression structures" or "discontinuous compression, continuous tension structures"—though he mostly refers to them simply as "my structures." Coxeter, interviews; Smith, interviews; Kenneth Snelson, e-mail correspondence "Re: Fuller and Tensegrity," June 2005; Fuller, *Synergetics*.

33. Coxeter, interviews.

34. Smith, interviews.

35. Fuller, *Synergetics*, ix.

36. Whiteley, interviews.

37. Whiteley's first foray into this realm was with an interdisciplinary geometry group. He connected with Janos Baracs, a structural engineer, now retired from the University of Montreal. Baracs has built the largest tensegrity sculpture in the world, in Montreal's St. Hyacinth Park, and over the course of his career he consulted with architects Arthur Erickson and Moshe Safdie on such projects as Expo '67 (of which Safdie was the chief architect). Said Whiteley, "Janos works with people who do unusual designs. He has the capacity to step outside the box and do a direct analysis of these unusual things because he knows a lot of geometry. And why does he know a lot of geometry?" Whiteley asked. "Because he was trained in Hungary . . . Janos is another of these people, like Coxeter, who are keeping with a live geometric tradition, with roots in Europe, bringing along the rest of us from North America." Whiteley, interviews; and Whiteley, "Learning to See like a Mathematician."

38. Whiteley, "Learning to See like a Mathematician."

39. Ibid.; and Howard Wainer, *Visual Revelations*, 58–60.

40. Whiteley; and Edward R. Tufte, *Visual Explanations*, 52.

41. Sir Michael Atiyah articulated the same idea in his 1980s address to the Mathematical Association: "[W]e have to appreciate that mathematics is a human activity and that it reflects the nature of human understanding. Now, the commonest way of indicating that you have understood an explanation is to say 'I see.' This indicates the enormous power of vision in mental processes, the way in which the brain can analyse and sift what the eye sees. Of course, the eye can sometimes deceive and there are optical illusions for the unwary but the ability of the brain to decode two- and three-dimensional patterns is quite remarkable." Whiteley; Atiyah, "What Is Geometry," in *The Changing Shape of Geometry*," 24–29.

42. Whiteley; Stephen M. Kosslyn, *Image and Brain: The Resolution of the Imagery Debate*, 13–14.

43. Whiteley, interviews; Thomas Homer-Dixon, *The Ingenuity Gap: Can We Solve the Problems of the Future?*

44. Whiteley, interviews.

45. Whiteley, interview, Budapest, July 11, 2002; and Henry Petroski, *Design Paradigms: Case Histories of Error and Judgement in Engineering*, 81–179.

46. Whitely, interviews.

47. Petroski, "Past and Future Failures," *American Scientist*, 500–4; and Petroski, *Success through Failure: The Paradox of Design*, 170–1.

48. Coxeter, *Projective Geometry*, 2.

49. Coxeter defined projective geometry mathematically as follows: "The plane geometry of the first six books of Euclid's *Elements* may be described as the geometry of lines and circles: its tools are the straight-edge (or unmarked ruler) and the compasses. A remarkable discovery was made independently by the Danish geometer Georg Mohr (1640–97) and the Italian Lorenzo Mascheroni (1750–1800). They proved that nothing is lost by discarding the straight-edge and using the compasses alone . . ." He went on to say: "It is natural to ask how much remains if we discard the compass instead, and use the straightedge alone . . . Is it possible to develop a geometry having no circles, no distances, no angles, no intermediacy (or 'betweenness'), and no parallelism? Surprisingly, the answer is Yes; what remains is projective geometry: a beautiful and intricate system of propositions, simpler than Euclid's but not too simple to be interesting . . . This geometry of the straight-edge seems at first to have very little connection with the familiar derivation of the name *geometry* as 'earth measurement.' Though it deals with points, lines, and planes, no attempt is ever made to measure the distance between two points or the angle between two lines . . . In the words of D. N. Lehmer: 'As we know nothing experimentally about such things, we are at liberty to make any assumptions we please, so long as they are consistent and serve some useful purpose' " Coxeter, ibid., 1–5.

50. In tracing the genealogy of projective geometry, Coxeter noted that conics were studied by Euclid, Archimedes, and Apollonius, but that the earliest truly projective theorems were discovered by Pappus of Alexandria in AD 3. Kepler realized the important concept of "a point at infinity," as did seventeenth-century French architect Girard Desargues, declaring that "parallel lines have a common end at an infinite distance." Coxeter prefaced his book by quoting historian E. T. Bell. "If Desargues, the daring pioneer of the seventeenth century, could have foreseen what his ingenious method of projection was to lead to, he might well have been astonished. He knew that he had done something good, but he probably had no conception of just how good it was to prove." Coxeter, ibid.

51. Gian-Carlo Rota, "Book Reviews," *Advances in Mathematics*, 263.

52. Whiteley, interviews; Weiss, interviews; Coxeter, "My Graph," *Proceedings of the London Mathematical Society*, 117–36.

53. Coxeter–Hofstadter correspondence, December 1992–July 1993, Coxeter Fonds, University of Toronto Archives; Hofstadter, interviews, August 2001, May and June 2005, e-mail correspondence June 2005–January 2006.

54. Ibid.

55. Ibid.

56. Ibid.

57. Each of a triangle's centers has its own justification, and some are more appealing than others, Hofstadter said. "But in the end all of it coming down to a subjective ranking based purely on esthetics." The Greeks knew about the three centers Hofstadter named, and, he continued: "Then a millennium and a half later came the Fermat point, and then a century or two

later the nine-point center. And then maybe the Gergonne point. They started proliferating in the 19th century—the Nagel point and the Brocard points and there were stunning connections among these points through circles and straight lines they lie on. And then there were conjugacies—transformations that flip points back and forth into each other." Ibid.

58. Hofstadter has since broken his own path, creating two new courses, allowing him to do mathematics and geometry the way he loved. The first, in the early 1990s, was "Circles and Triangles: Diamonds of Geometry"—"I will be giving this course for the first time starting in about a month," he reported to Coxeter, "and am very excited about it. There are two textbooks: your *Geometry Revisited*, naturally, and David Wells's charming new *Penguin Dictionary of Curious and Interesting Geometry*." And in the last few years, he decided to revisit the "thorny" field of group theory that so repelled him at Berkeley, and reclaim it on his own terms. "I came back to group theory and Galois theory some thirty-five years later and I said, 'By God, I'm going to make this loveable. I am going to make this hateful stuff loveable. I'm going to go through it again and find a way of visualizing it. I'm going to chop down those trees, I'm going to clear the prickly jungle, I'm going to make it so that I can see long distances.' And I have done this. I don't mean that I have created a revolution in group theory," he said. "I'm not trying to claim anything grandiose." The course is called "Group Theory and Galois Theory Visualized." And in the process of creating concrete images of groups, one new image presented itself for what's called the "second isomorphism theorem"—and with the image, he said, the truth of the theorem became totally obvious. The image was of two concentric circles sliced radially, and looked like a small pizza sitting inside a larger one, earning the course its nickname—"pizza theory"(reinforced by the fact that Hofstadter would bring along pizzas for the class's evening meetings). The pizza imagery was used in several places, with slight modifications, including in the proof of a rather difficult theorem called Zassenhaus's lemma. However, by far the most important diagrams Hofstadter uses in his course are Cayley diagrams. "Coxeter was very involved with Cayley diagrams," said Hofstadter. "He and Willy Moser used Cayley diagrams in their book *Generators and Relations for Discrete Groups*. It seems to me that the key elements that one uses in visualizing group theory, and thereby Galois theory, are closely connected with the ideas that Coxeter was exploring." Ibid.

59. In his article, Hofstadter pondered the prospects of winning a Fields for his elementary triangular discovery: "Was I truly the first person in decades to find a major new property of the triangle? In an exuberant mood that evening, I described my breakthrough to my wife, and she fantasized that now, some 25 years after having abandoned mathematics, I might be awarded the Fields Medal! We chuckled over this idea, especially the irony that I had quit math in utter despair over its enormous abstraction. Of course, any talk of a Fields Medal was just a silly dream and we knew it, but it did occur to me to wonder just why it was so utterly laughable to think that work in Euclidean geometry, no matter how elegant or new or fundamental, might be considered worthy of a the Fields Medal—or even worthy of serious attention by mathematicians at all." Ibid.

60. Ibid.

61. Ibid.

62. A recent study found that an isolated Amazon group, the Mundurukú in Brazil, intuitively understand spatial geometrical concepts—such as right angles and parallelism—despite the fact that the tribe does not possess words to describe these concepts, and that their intrinsic knowledge is based on no formal education, and certainly no tinkering with a geometry kit. The study indicated that: "The spontaneous understanding of geometrical concepts and maps by this remote human community provides evidence that core geometrical knowledge . . . is a universal constituent of the human mind." Stanislas Dehaene, Véronique Izard, Pierre Pica, Elizabeth Spelke, "Core Knowledge of Geometry in an Amazonian Indigene Group," *Science*, 381–84.

63. Coxeter–Hofstadter correspondence; and Hofstadter, interviews.

64. One of Hofstadter's "most bitter mathematical memories" involves a book with the alluring title *Three Pearls of Number Theory*, by Russian mathematician A. Y. Khinchin. The book was "written for the admirable purpose of helping a wounded soldier-friend of Khinchin's pass several dreary months in a hospital," as Hofstadter describes it, and contained Khinchin's proof of an "absolutely beautiful result in number theory" known as Van der Waerden's theorem. Hofstadter considered this theorem so elegant that he felt he absolutely had to learn and understand it. As a graduate student in physics, when he was "very deeply involved in mathematics on a daily basis," he tried the proof. It was only five pages long and yet after hours of struggle he gave up. This made him think that a better title for the book might be "Three Oysters of Number Theory." He went back at it a year later, "out of shame, or renewed curiosity," he said. "And the second time around for some miraculous reason I was able to plow through it. It took me many hours. And when I finally finished I had converted it into a set of pictures—the canonical pictures, just colored boxes nested inside each other. The pictures were so simple and clear that I was able to explain to non-mathematician friends, people who claimed they didn't like mathematics, not only the theorem but the entire proof in about 15 minutes. All they needed to understand was high school math notions and colored nested boxes." The best part (or the worst, depending on which way you look at it) was that Hofstadter later came upon an article written by Van der Waerden describing the discovery of the original proof. "This article showed the very pictures that I had drawn!" said Hofstadter. "And it showed that Van der Waerden and the other two people who found the proof had done so *only* through pictures. There was not a single formula in the whole article. It confirmed," Hofstadter said, "my intuitive belief that nobody could understand the proof in Khinchin's book without constructing these diagrams, whether on paper or in their mind." Ibid.

65. He named John Conway and William Thurston. Thurston is a "modern" geometer who won the Fields, but also a geometer clearly very fond of the visual; his ideas about visualizing hyperbolic geometry are explained in the much-lauded computer graphics video *Not Knot*. Conway and Thurston, when together at Princeton, taught a "wildly popular" mathematics course for undergraduates, relaying cutting-edge geometric and topological ideas via visual and hands-on exercises more commonly found in grade schools. Ibid.; Conway, interviews.

66. A quick litmus test for the pendulum's recent wobble toward geometry is seen by the proliferation of books on the subject, such as *The Changing Shape of Geometry*. Published in 2003, it contains one of Coxeter's last contributions to the collective geometry oeuvre (his Budapest paper was the last, the conference proceedings published belatedly in 2006). *The Changing Shape of Geometry* culls an archive of articles and essays published over the last century, testimonies to the status of geometry, opening with G. H. Hardy's 1925 essay, "What Is Geometry?" Another entreaty was penned in 1973 by J. V. Armitage, an honorary senior fellow in the department of mathematical sciences at the University of Durham, in northeast England. In pleading geometry's case, Armitage cites a lesson from Euclid, who faced a question from one of his students as to whether there was not an easier, shorter way to study geometry than to learn the salient lessons from *Elements*. "There is no royal road to geometry," Euclid famously replied. Armitage elaborated: "Mathematics is an attractive subject because it is difficult. Obstinacy and the ability to concentrate are a necessary part of the equipment of the mathematician and, in other contexts, they are desirable for life in general . . . Geometry is a good discipline in which to encounter difficulties for the first time, because its problems lend themselves to pictorial exploration and investigation and that combination of intuition and logic which is the essence of mathematics . . . The real answer is Euclid's own; those who expect education to produce a return in kind for financial investment will always seek to weigh and measure its fruits and

ultimately impoverish the human spirit. But if one believes that the intellect and the ability to reason should be developed for their own sakes, then geometry, like the classics, has a value which cannot be measured." Not long ago Armitage reread his article, in between supervising and marking exams for the course "Foundations of Mathematics"—taken mostly by chemistry and natural science students, as well as a few arts majors. After refreshing his memory, Armitage said his argument holds, still relevant today, if not more so. Classical elementary geometry is a timeless archetypal approach to mathematics and the sciences. Armitage, however, isn't as optimistic as Atiyah. He worries that all the pleading has been a losing battle. Coxeter raised the specter of geometry, to be sure, and gave it a grassroots surge, but the subject has not been reinstated in the educational curriculum, at the grade school nor university level, to the extent he believes it should. Increasingly, he finds that students coming to university don't know much geometry at all. "But there are some honourable exceptions," he said. He taught a girl from Manchester not long ago, who had been at an ordinary high school and was keen, keen on classical geometry. Armitage observed that she clearly had a teacher who had just stepped out of a time machine and who, by all indications, had ignored the Euclid-barren syllabus. He encouraged this student to record her favorable experiences with the Royal Society of London, the apogee of all mathematical and scientific wisdom, where a report on the status of geometry education was in the works. J. V. Armitage, "The Place of Geometry in Mathematical Education," in *The Changing Shape of Geometry*, 515–26; Armitage, interview, May 27, 2005; and Atiyah, interviews.

67.   "Teaching and Learning Geometry, 11–19," a report of a Royal Society/Joint Mathematical Council working group, July 2001.

CHAPTER 10—C$_{60}$, IMMUNOGLOBULIN, ZEOLITES, AND COXETER@COXETER. MATH.TORONTO.EDU

1.   Coxeter, "The Problem of Packing a Number of Equal Nonoverlapping Circles on a Sphere," *Transactions of the New York Academy of Science*, 320–31.

2.   Ibid.; Conway and Sloane, *SPLAG*; and Conway, interviews; Sloane, interviews; and George G. Szpiro, *Kepler's Conjecture*.

3.   Ibid.

4.   In 1956, a friend of Coxeter's, Scottish mathematician John Leech, provided a simple proof. Coxeter, *Twelve Geometric Essays*, 179–188.

5.   Ibid., 180–181.

6.   Lang, interviews; O'Connor and Robertson, "Claude Elwood Shannon," http://www-groups.dcs.st-and.ac.uk/~history/Mathematicians/Shannon.html (accessed February 3, 2006); Sloane, interviews.

7.   Ibid.

8.   Lang, interviews.

9.   Sloane, interviews.

10.   Ibid.

11.   Ibid.

12.   Lang, interviews; and Conway, interviews.

13.   In sphere packing, the problem of finding the right answer for all dimensions still is unsolved. In some dimensions, the potential answer to the problem has been narrowed, whittled to within upper and lower bounds. In his paper, Coxeter gave a formula that proposed new bounds. In four dimensions his kissing number bound—the number of billiard balls that fit around a central ball—was between twenty-four and twenty-six (twenty-four had been known as the lower bound for a century or so, but Coxeter produced the new upper bound of twenty-

six). In eight dimensions, Coxeter produced an upper bound of 244, as against the long-known lower bound of 240, from the E8 Lattice. Coxeter's work required difficult calculations, and so this was one of those occasions when he recruited his friend in Scotland, John Leech, whom he called a "computing man," to do the messy computer calculations for him.

In twenty-four dimensions, some progress on this problem was later made by Leech and John Conway—Leech "dangled under a few noses" his structure for a twenty-four-dimensional grid, now known as the Leech lattice—the noses including Conway's and Coxeter's. Leech needed someone to determine the symmetries of his lattice (knowing it would have multiple symmetries, which would form a group), as he didn't have the group theory skills to do the investigation himself. Conway was the first to take the bait. In the book *Kepler's Conjecture* by George G. Szpiro, the author recounts: "Conway told his wife that this was something important and difficult, and that he was going to work on it Wednesdays from six to midnight and Saturdays from noon to midnight. He need not have planned that far ahead. It took him only a single Saturday session to crack the puzzle." What Conway discovered that evening was that the group describing the symmetries of the Leech lattice was none other than one of the sporadic simple groups that had eluded discovery until then." The discovery of the "Conway group" catalyzed the detection of additional sporadic simple groups a breakthrough that catapulted the global classification effort a magnificent leap forward. And Conway, who was then rather down in the dumps about his mathematical productivity, or lack thereof, received a welcome shot to his ego—almost instantaneously he was installed as a Fellow of the Royal Society, and he has never looked back as one of the front-running mathematicians in the world. However, Conway complained recently of suffering from the chronic disease called laziness, playing and puzzling his mathematical days away in the math department common room at Princeton, though occasionally he catches the bug of a heady topic, such as "the free will theorem," which he and Simon Kochen have been developing with a vengeance.

In 2004, sphere-packing's translational case was entirely solved for twenty-four dimensions by Henry Cohn and Abhinav Kumar. In four dimensions, Neil Sloane and Andrew Odlyzko reduced the bound to twenty-five a few years later; and subsequently Ronald Hardin cut it down to twenty-four; Sloane and Odlyzko knew twenty-five was impossible but couldn't prove it. In eight dimensions, Sloane and Odlyzko reduced the bound to 240. Their method was much easier than Coxeter and Leech's crude computing, because they utilized the powers of linear programming. Coxeter, "An Upper Bound for the Number of Equal Nonoverlapping Spheres That Can Touch Another of the Same Size," *Proceedings of the Symposia in Pure Mathematics*, 53–71; Coxeter, interviews; Conway, interviews; Conway and Sloane, *SPLAG*; O'Connor and Robertson, "John Leech," http://www-groups.dcs.st-and.ac.uk/~history/Mathematicians/Leech.html (accessed February 5, 2006); Szpiro, 95; Sloane, interviews; Smith, interviews.

14. Lang submitted his proposal to a precursor of the International Telecommunications Union—a United Nations agency fostering cooperative standards for telecommunications equipment and systems, located in Geneva, Switzerland. Lang, interviews.

15. Lang showed Coxeter his proposition on paper, since it wasn't until 1986 that a practical E8 Lattice Modem was produced. And with technology's ever-forward march, the E8 was soon superseded. "That modem was short-lived as there was a great push on in the industry for further increased modem performance," said Lang. Motorola produced the Leech Lattice Modem in twenty-four dimensions. It became particularly popular in Sweden, put to use by large companies with lots of offices that needed to communicate computer-to-computer. The Leech modem was soon superseded by Motorola's Trellis Modem, and so on. Lang, interviews.

16. Robert Tennent (professor in semantics and design of programming languages at the

School of Computing, Queen's University, Kingston, Ontario), interview, e-mail correspondence "Re Coxeter anecdote," January 2004–June 2005.

17. Conway has worked on the more theoretical side of sphere packing. "I was almost seduced into that against my will," Conway said. "One of the things that made my mathematical name, in fact THE thing that made my mathematical name was studying the symmetry of a certain packing of spheres in 24 dimensions," he said, referring to his collaboration with John Leech (see endnote 13). And a friend of Conway's, Neil Sloane, dragooned him into writing a book, the standard book on the subject—*Sphere Packing, Lattices, and Groups*, also known as *SPLAG*—a book that now generates a nice nest egg of a royalty check annually. "So I'm personally grateful in a way. But I wasn't really interested in the sphere-packing problem myself, as an [applied] problem. I was interested in the symmetry and the beauty of this particular arrangement, which is what Coxeter was interested in." But, as Conway said, almost ruefully, any good piece of mathematics has applications. Conway, interviews.

18. Indeed, Coxeter didn't hold any grudges against Lang and his work, and they kept in touch. Years later, Lang invited Coxeter for a visit at ESE, the Canadian subsidiary of Motorola, where Lang was then vice president of research. Another visitor that day was G. David Forney Jr., now adjunct professor in the department of electrical engineering and computer science at the Massachusetts Institute of Technology, and formerly a Codex Motorola scientist. Forney is regarded as the founder of the modem industry; in 1970 he invented the first reliable high-speed modem, later adopted as an international standard from which modem technology exploded. Forney, whose current interests include coding and decoding for Euclidean-space channels, recalled meeting Coxeter that day. "I made a special trip to Toronto at the invitation of Gord Lang, to be present at Coxeter's visit to ESE Ltd, our sister company," said Forney, who was then group vice president in the Motorola Information Systems Group, located in Mansfield, Massachusetts. The visit entailed a long social lunch, a brief tour of ESE, with a Leech lattice modem to look at. Forney found Coxeter in dandy form: "a courteous, distinguished-looking professorial gentleman with an amiable air and his wits very much about him," Forney says. "He was almost ninety then and he was fascinating." David Forney, e-mail correspondence "Re: Gord Lang, E8, and Coxeter," May 31, 2005; and Robert Gallager, "A Conversation with G. David Forney, Jr.," *IEEE Information Theory Society Newsletter*, http://www.itsoc.org/publications/nltr/97_jun/jdvcon.html (accessed February 4, 2006).

19. Coxeter, "University of Toronto Oral History Project."

20. Coxeter, interviews; Thomas, interviews.

21. C. Davis, interviews.

22. Pat Kerr (Engineering Directorate, NASA Langley Research Center), e-mail correspondence, January 31, 2006; and Kurt Severance, Paul Brewster, Barry Lazos, and Kaniel Keefe, "Wind Tunnel Data Fusion and Immersive Visualization: A Case Study," NASA Langley Research Center.

23. Whiteley, interviews; and Weeks, interviews.

24. Studying physics and then computer graphics at Berkeley in the 1960s—at a time when geometry was "non-constructive, not tangible, abstract"—DeRose learned all the geometry he could, but on his own, browsing through the library stacks for works published in the 1800s, by Hamilton and Caley, and by the anachronistic Coxeter. He also plumbed the works of Austrian and German practitioners. "There is a nice tradition of geometry there," he said. The dearth of geometry in the classrooms, he feels, is a crisis, and one nearer and dearer to his heart as his children get older. "The people teaching mathematics and geometry at the elementary and secondary levels don't understand it. They don't understand the ways in which it can be made relevant," he said. "This turns kids off mathematics in general and geometry in particular, as be-

ing irrelevant and not very interesting. I've got an eleven-year-old and a seven-year-old. We spend lots of time wandering around the world seeing geometry everywhere we look: 'Oh, look, the shape of the outline of that headlight on the ground is a parabola!' " Tony DeRose, Pixar senior scientist, interview, November 22, 2005.

25. DeRose demonstrated by walking through the creation of Bob Parr: "Imagine you've got, say Bob Parr from *The Incredibles*, standing with his arms out. He's a three-dimensional character, and the computer understands him as a three-dimensional hunk of geometry," said DeRose. "Now place a virtual camera in front of him, ten feet away. We're going to model that camera as a point, directed toward Bob. And then we're going to erect in front of the camera a square that is perpendicular to the direction of the camera. That's where the image is formed, the two-dimensional image that people see in the theatres. So what we'll do is conceptually the following thing: if we want to figure out where the tip of his nose goes in the image, we're going to draw a line from that camera center point, out to his nose, and at some point that line is going to cross the image plane—that's where the image of his nose is going to appear. And we do that for all the other points on him and that way we form a two-dimensional projection of Bob on the picture plane. In geometric terms, what we are doing is we are taking a three-dimensional set of geometry and constructing from it a two-dimensional set of geometry on the image plane, and that's a projective transformation." DeRose, interview.

26. Ibid.

27. Ibid.

28. Ibid.; and "The Pixar Process," http://www.pixar.com/howwedoit/index.html (accessed February 3, 2006).

29. Whiteley, interviews.

30. In 1996, ten mathematicians and computer scientists, led by Egon Schulte, at Northwestern University, submitted a proposal to the NSF-funded Geometry Center, in Minneapolis, for a 3-D program analogous to the 2-D Geometer's Sketchpad—allowing play with virtual tiles and blocks. The suggested names for the program were the Coxeter Project, or Project Coxeter. In e-mail correspondence among the group, University of Alberta's Robert Moody pointed out: "There is a possible acronym from his initials: Hands-on Synthetic Manipulation (of 3d graphics). I am pretty sure [Coxeter] would not approve of that!" To which Marjorie Senechal replied: "Coxeter might not approve of his initials being interpreted as 'Hands-on Synthetic Manipulation' but I'll bet his wife would! I've heard her tell him several times that he should get with it and learn to use computers. She could talk him into giving it his blessing even if the HSM part is a joke." Unfortunately, the proposal was not approved, as the Geometry Center's funding was cut later that year and the center shut its doors. Senechal, "Donald and the Golden Rhombohedra," *The Coxeter Legacy*, 159–77.

31. The principal investigator was Eugene Klotz, and the program designer was Nicholas Jackiw.

32. Schattschneider, interview, November 2005.

33. When the program became a success, IBM wanted it as well, but it had been written for Mac. The cost of translating the program to a new platform initially seemed like a deterrent for Key Curriculum, the publisher of Geometer's Sketchpad, until IBM gladly agreed to front the money. In the United States, Sketchpad is used in 60 percent of courses. Elsewhere, its reach can be based only on licensing and distribution agreements. There are national licenses for the software in Malaysia and Thailand; there are strong international cultures of Sketchpad use in South Korea, Singapore, Costa Rica, Taiwan, and a growing interest in China; and there is "substantial penetration and broad familiarity" in Canada (including an Ontario provincial license for all schools and home users, the goal being one free license per student), the United Kingdom, and

Australia. Two similar programs are Cabri and Cinderella. Ibid; and Steven Rasmussen (president, Key Curriculum Press), e-mail to Siobhan Roberts "Re: Sketchpad Stats," June 7, 2005.

34.   When Hofstadter had his reunion with geometry in the early 1990s, it occurred to him that he could probably write a computer program that would enable him to construct interrelated circles and triangles in such a way that he could drag points or lines around on the screen with the geometric consistencies of their constructions adapting in real time. When he mentioned this hope to a visually minded friend, he was surprised to learn the program already existed. Hofstadter tried out Geometer's Sketchpad and immediately was hooked. In one of his letters to Coxeter, Hofstadter asked if he had seen this modern tool, describing how this mind-blowing new way of doing geometry assisted in one of his discoveries: "I ran to my computer and fired up the astonishing program Geometer's Sketchpad, with which you may be familiar. If not, I hope you have a chance to see it. It is absolutely revolutionary. You make any geometric construction on the screen, no matter how complex, and then you can dynamically move points about on the screen and all the constraints will be obeyed. You had praise in your book for the movie *Simson Line* by T. J. Fletcher—well, all I can say is, if you liked that movie, you would go wild about Geometer's Sketchpad. Every single construction you make in GS is as good as the *Simson Line* movie! With GS, you see straight to the heart of geometry . . . I truly wish some of the old-time geometers—Poncelet, Steiner, Coolidge, Brocard—could have seen it. They would have gone crazy." Coxeter–Hofstadter correspondence, Coxeter Fonds, University of Toronto Archives; and Schattschneider, interview.

35.   Whiteley, interviews.

36.   Ibid.

37.   The Math Lab holds several large circular worktables, Mac computers in each corner, and cupboards full of "toys" for geometric modeling. Whiteley makes some of his own models by cutting high-grade plastic strips, with bolt-and-screw joints forming the vertices of shapes (a homemade model made ages ago by Coxeter is still lying around; hand-colored cardboard strips held together with rivets). The geometrical gadget of late is ZomeTool, a kit of struts and nodes that assemble into any of Coxeter's polytopes (Coxeter's signature on the box endorses the product, with his statement: "Zome System considerably simplifies the procedure of construction and unifies the study of space frame structures into one coherent system, of great educational value in teaching of solid geometry, science, art, engineering and architecture"). Other popular modeling systems include UniStrut, an old system produced for engineers; Polydron, with hollow and solid regular polygons in red, blue, green, and yellow, which click together to form hingelike joints and are marketed toward school kids; Molecular Visions, specifically intended for the assembly of the Bucky Ball C_{60}; Tensegritoys; Molymod made for organic chemistry; and "the latest rage," as Whiteley described it, Deluxe Magnetic Kit, with colored plastic struts connected to magnetic balls at the vertices (manufactured by Lee Valley Tools, specialists in woodworking and gardening tools). Whiteley, interviews.

38.   Whiteley, interviews.

39.   Ibid.

40.   Ibid.

41.   Ibid.

42.   Ibid.

43.   Ibid.; and John Andros (professor of chemistry, York University, Ontario), CHEM 3070 Handout #6, "Where Chemistry Meets Society: Current Examples."

44.   Whiteley, interviews.

45.   Material scientists are trying to classify all the possible patterns and give some order to their search. Broadly speaking, this is "reticular science" (a reticulum being a netlike structure).

It is a serendipitous science self-criticized for its "shake and bake," "mix and wait," and "heat and beat" methods. The classification system inherent in Coxeter groups holds potential to provide some much-needed order.

46. Whiteley, interviews.

47. Ibid.

48. Ibid.

49. Kroto, interviews; Moody, interviews.

50. Kroto described it as a "plethora of exotic molecules in a wide range of physico-chemical environments." And these exotic molecules were in addition to the "well-known species such as ammonia, water, and ethanol—enough for 10^{28} bottles of schnapps in Orion alone." Kroto, "C_{60}: Buckminsterfullerene, The Celestial Sphere That Fell to Earth," *Angewandte Chemie*, 111–29.

51. One of the first compounds subjected to a microwave study—trying to find the compound's frequency in outer space using a piece of equipment called a microwave spectroscope—was the extragalactic molecule HC_3N. After this was successfully detected, Kroto decided he wanted to trawl outer space for a slightly larger compound, HC_5N, since he and a group of scientists had determined its frequency. He wrote to his colleague Takeshi Oka, at National Research Council (NRC) laboratory in Ottawa—"the Mecca for spectroscopists." Oka responded that he was "very, very, very, very, very much interested." In 1975, Kroto and a cluster of Canadian astronomers were successful in discovering HC_5N. Then Kroto upped the ante. Why not see if HC_7N was present as well? Since his team's allotted time with the radio telescope was coming to an end, they had to race to find the frequency of the molecule. By the time Kroto and his crew set up the radio telescope apparatus in the dark woods of Algonquin Park, Ontario, they were still awaiting word on the frequency from the colleague calculating it in England. The frequency arrived (via telephone and then radio transmission to the Algonquin hinterland) with not a moment to spare, just as Taurus climbed the horizon of the night sky. Kroto recounted the emotionally charged finale to the experiment in the magazine *Angewandte Chemie*: "The next few hours were high drama. We dashed out to the telescope and tuned the receiver to the predicted frequency range . . . We tracked the extremely weak signals from the cold dark cloud throughout the evening. The computer drove the telescope and stored incoming data, but to our frustration we could not process the data on-line while the telescope was running. The system did, however, display individual ten-minute integrations, and as the run progressed we watched the oscilloscope for the slightest trace of the predicted signal in the receiver's central channel . . . Desperate for even the faintest scent of success, we carried out a simple statistical analysis in order to determine whether the signal level of the channel was greater than the noise. As the night wore on, we became more and more excited, convinced that the signal was significantly more often high than low; we could hardly wait for Taurus to set. By 1:00 a.m. we were too excited and impatient to wait any longer, and shortly before the cloud vanished completely, Avery stopped the run and processed the data. The moment when the trace . . . appeared on the oscilloscope was one of those that scientists dream about and which, at a stroke, compensate for all the hard work and the disappointments which are endemic in life." Ibid.

52. Ibid.

53. Kroto, interviews; and Kroto, "The Celestial Sphere That Fell to Earth."

54. Smalley had never come across the British term "wadge" for a cluster and liked it so much that he started to refer to C_{60} as the "Mother Wadge." The omnipresence of the wadge led Sir Harry to call it the "Godwadge." Ibid.

55. Ibid.

56. Coxeter wasn't thrilled with the name Buckminsterfullerene. "He argued that the

structure already had a beautiful name—the truncated icosahedron," recalled Glenn Smith, who suggested there were any number of names that would have been more appropriate. The truncated icosahedron had been discovered by Archimedes—how about the Archimedene? Or, named after Leonardo da Vinci for his beautiful renderings? Or—although Coxeter would never suggest such a thing—how about the Coxeterene, acknowledging the twentieth century's curator of such shapes? Coxeter, interviews; Smith, interviews.

57.   This work was done together with his colleague Ken McKay in the UK. Kroto, interviews; and Kroto, "The Celestial Sphere That Fell to Earth."

58.   Hansard, *House of Lords*, December 2, 1991–January 9, 1992, 590.

59.   "C Sixty Inc.," http://www.csixty.com/ (accessed February 4, 2006).

60.   Kroto, interviews.

61.   Whiteley, interviews.

CHAPTER 11—"COXETERING" WITH M. C. ESCHER (AND PRAISING OTHER ARTISTS)

1.   Coxeter, "Aspects of Symmetry," Banff, Alberta, August 27, 2001.

2.   One public lecture in Coxeter's repertoire was "The Mathematics of Leonardo da Vinci," in which he stated: "Leonardo became interested in regular polygons and their symmetry groups through his desire to design buildings with symmetrical plans, such as the hexagonal Capella Emiliana in Venice and the octagonal San Maria degli Angeli in Florence . . . Surely Leonardo would have agreed with Professor J. L. Coolidge of Harward, who said (in 1929): 'It is my personal credo that Geometry is a branch of Art.'" Coxeter, "The Mathematics of Leonardo da Vinci," Coxeter Lectures File, Coxeter Fonds, University of Toronto Archives.

3.   Ed Barbeau (emeritus professor of mathematics, University of Toronto), interviews, Toronto, February 26, 2003, February 24, 2005.

4.   Coxeter Diaries, 3 September 1954; Coxeter, interviews.

5.   J. L. Locher et al., *M. C. Escher, His Life and Complete Graphic Work*, 73.

6.   Coxeter diaries, 6 September 1954; Coxeter, interviews.

7.   Locher, 73.

8.   Coxeter to M. C. Escher, 24 October 1954, Coxeter Fonds, University of Toronto Archives.

9.   Locher, 70–71.

10.   J. L. Locher, *The World of M. C. Escher*, 48.

11.   Locher et al., *M. C. Escher: His Life and Complete Graphic Work*, 71–73.

12.   Ibid.

13.   Snow was a Cambridge man of Coxeter's vintage and, like Escher, based his ideas on a meeting of the two cultures in his own experience. As Snow described, he was by training a scientist and by vocation a writer (Escher's father and brothers were scientists, whereas he studied at the School for Architecture and Decorative Arts in Haarlem, and there found a mentor in teacher Samuel Jessurun de Mesquita). In his lecture, Snow recounted an anecdote of Cambridge lore demonstrating the tepid interaction between artistic and scientific minds at the Trinity high table. An artistically inclined visitor tried to engage his neighbor to his left in conversation and "got a grunt" in response. "He then tried the man on his own right hand and got another grunt." The president of Trinity, sitting at the table, explained: "Oh, those are the mathematicians! We never talk to them!" Snow, however, was worried about, as he said, "something more serious. I believe intellectual life of the whole of Western society is increasingly being split into two polar groups . . . I remember G. H. Hardy once remarking to me in mild puzzlement, some time in the 1930s: 'Have you notice how the word "intellectual" is used

nowadays? There seems to be a new definition which certainly doesn't include Rutherford or Eddington or Dirac or Adrian or me. It does seem rather odd.'" Snow agreed. "[I]t is bizarre how very little of twentieth-century science has been assimilated into twentieth-century art." The divide didn't hold with Escher and Coxeter. C. P. Snow, *The Two Cultures*.

14. Coxeter diaries, 14 September 1954.

15. Coxeter focused the opening section of his paper, "The Mathematical Implications of Escher's Prints," on Escher's depiction of the Platonic solids: "Like Leonardo da Vinci and Albrecht Dürer, Escher has a strong appreciation of the five Platonic solids . . . 'They symbolize,' he says, 'man's longing for harmony and order, but at the same time their perfection awes us with a sense of our own helplessness. Regular polyhedra are not inventions of the human mind, for they existed long before mankind appeared on the scene.'" Coxeter, "The Mathematical Implications of Escher's Prints," 49, in *The World of M. C. Escher* (citing Escher, in *The Graphic Work of M. C. Escher*, 9, Ballentine Books, New York, 1967).

16. Doris Schattschneider, *M. C. Escher: Visions of Symmetry*, 2.

17. Coxeter, "Coloured Symmetry," in Coxeter et al. *M. C. Escher: Art and Science*, 16.

18. Schattschneider, interviews, Toronto, May 16, 2004, Banff, August 3, 2005 and June 15, 2005; e-mail correspondence May 2004–January 2006; Schattschneider, "Coxeter and the Artists: Two-Way Inspiration," at the Coxeter Legacy Conference, May 15, 2004, and in *The Coxeter Legacy*, 255–80; and "Coxeter and the Artists: Two-Way Inspriation, Part 2," at the Renaissance Banff Bridges Conference, August 3, 2005, and in *Renaissance Banff Bridges Proceedings 2005*, 473–80.

19. Schattschneider, *M. C. Escher: Visions of Symmetry*, 2 (as cited from Escher's book *Regular Division of the Plane*, 1958 – Stichting de Roos, Utrecht).

20. Ibid., 9–10, 17–18; and Locher, *M. C. Escher: His Life and Complete Graphic Work*, 23–24, 50.

21. Coxeter wrote a paper, "Escher's Fondness for Animals," wherein he quoted Escher's practical explanation for choosing the animals he did: "My experience has taught me that the silhouettes of birds and fish are the most gratifying shapes of all for use in the game of dividing the plane. The silhouette of a flying bird has just the necessary angularity, while the bulges and indentations in the outline are neither too pronounced nor too subtle." Coxeter, "Escher's Fondness for Animals," in *M. C. Escher's Legacy: A Centennial Celebration*, 1–4; and Coxeter, *Introduction to Geometry*, 58.

22. "Mathematics is like a pig—you never throw anything out," said Michele Emmer, a mathematician-filmmaker at the University of Rome, with whom Coxeter made a film *The Fantastic World of M. C. Escher*. Michele Emmer, interview, Toronto, May 16, 2004; information about the film, *The Fantastic World of M. C. Escher*, and another film in which Coxeter participated, *Platonic Solids*, can be found at http://www.mat.uniroma1.it/people/emmer/.

23. Coxeter, *Introduction to Geometry*, 50–66.

24. Schattschneider, interviews.

25. Lightman, *The Discoveries*; and Kroto, interviews.

26. It was not lost on Coxeter the coincidence that there are seventeen 2-D crystallographic groups and that he had seventeen Ph.D. students. Schattschneider, interviews. Thomas Banchoff (geometer, Brown University), interview, March 3, 2006.

27. Schattschneider, interviews; Coxeter, *Introduction to Geometry*, 57, 59; Gardner; and M. C. Escher, correspondence to George Escher, 29 January and 30 July 1961, National Gallery of Canada Library.

28. Escher held his first American exhibit at the Whyte Gallery, in Washington, in October 1954, the month after he and Coxeter had intersected in Amsterdam. One-third of his prints

were snapped up before the show opened, and more than one hundred prints sold in all. Locher et al., *M. C. Escher: His Life and Complete Graphic Work*, 73–74.

29. Schattschneider, interviews.

30. O'Connor and Robertson, "Maurits Cornelius Escher," http://www-groups.dcs.st-and.ac.uk/~history/Mathematicians/Escher.html (accessed February 1, 2006); for a slightly different translation, see Locher et al., 168.

31. Schattschneider, interviews.

32. Ibid; and Emmer, interview.

33. George Escher, "M. C. Escher at Work," in Coxeter et al. *M. C. Escher: Art and Science*, 1–2.

34. M. C. Escher, "The Regular Division of the Plane," in *Escher on Escher: Exploring the Infinite*, 42.

35. M. C. Escher to Coxeter, 5 July 1958, M. C. Escher Archives, National Gallery of Art, Washington, D.C.

36. Breaking the news of his discovery, Escher explained to Coxeter that most of his letter had been too learned and technical "for a simple, self-made plane pattern-man like me," but the mathematical illustrations were just what he had been searching for: "Since a long time I am interested in patterns with 'motives' getting smaller and smaller till they reach the limit of infinite smallness. The question is relatively simple if the limit is a point in the centre of a pattern. Also a line-limit is not new to me, but I was never able to make a pattern in which each 'blot' is getting smaller gradually from a centre towards the outside circle-limit, as shows your Figure 7." Ibid.

37. M. C. Escher to George Escher, 9 November 1958, National Gallery of Canada Library.

38. Schattschneider, interviews; "Coxeter and the Artists: Two-Way Inspiration"; and "Coxeter and the Artists: Two-Way Inspiration, Part 2."

39. M. C. Escher to George Escher, 9 November 1958. Copyright © 2024 The M. C. Escher Company-The Netherlands. All rights reserved. www.mcescher.com.

40. M. C. Escher to George Escher, 7 December 1958: Conway, interviews, op. cit.; Weeks, interviews.

41. M. C. Escher to George Escher, ibid.

42. Coxeter to M. C. Escher, 29 December 1958, M. C. Escher Archives, National Gallery of Art, Washington, D.C.

43. Locher et al., *M. C. Escher: His Life and Complete Grahic Work*, p. 91.

44. M. C. Escher to George Escher, 4 January 1959, National Gallery of Canada Library.

45. M. C. Escher to George Escher, 15 February 1959, National Gallery of Canada Library.

46. Schattschneider, interviews.

47. John Conway is a bit like Coxeter in this way, though their personalities are extreme opposites. "Conway will button-hole somebody on the street and try to explain knot theory to them," said Schattschneider. Indeed, meandering through the Princeton campus, Conway occasionally gives a tour narrated by a paper he wrote called "How to Stare at a Brick Wall"—bricks being another practical manifestation of tiling the plane. Conway can provide proper names for all the brickwork on campus, discussing the period between the "headers" and the "stretchers" that vary to form the patterns of the different brick "bonds" (there are twenty-six different bonds on the Princeton campus: Flemish bond, English bond, Dutch bond, Narrow American bond, Running bond, and so on). Ibid., and Conway, interview, June 22, 2004.

48. Schattschneider has conducted an extensive study, a two-part paper, titled "Coxeter and the Artists: Two-Way Inspiration." She examined Coxeter's interactions with the model-makers Alicia Boole Stott, Michael Longuet-Higgins, Paul Donchian, Father Magnus Wen-

ninger (see endnote 53), and Marc Pelletier (a sculptor, and the inventor, with Paul Hildebrandt, of the polyhedral modeling kit called Zometool—"zome" being a hybrid of "zone" and "dome"). And Schattschneider observed Coxeter's productive rapport with the artists George Odom, Peter McMullen (a mathematician at the University College, London, who as an undergrad drew projections of the four-dimensional polytopes that greatly impressed Coxeter), Tony Bomford (an Australian surveyor, adventurer, and maker of geometrically inspired woolen hooked rugs), English sculptor John Robinson, George Hart, and the Netherlands' Rinus Roelofs (an applied mathematician cum sculptor who often produces digital [virtual] sculpture, and rapid prototyping sculpture, in which a "3-d printer" produces a wax, metal, nylon or plastic model directly from his computer data file). Scattschneider, interviews.

49. Schattschneider, interviews; and Senechal, interviews.

50. Coxeter, interviews; Coxeter, "Math and Aftermath," CBC *Ideas*.

51. George Odom (artist and geometer), interview, 20 May 2005.

52. Upon hearing of Coxeter's death, Odom wrote to Magnus Wenninger: "I don't know what I would have done without you, Soccarides, and Coxeter. The three of you have been my only contact with the human race." Odom–Wenninger correspondence, personal files; Odom, interview.

53. Wenninger, a master model-maker, got to know Coxeter in the late 1960s on the international mathematics conference circuit, where Wenninger was an eye-catching mainstay. He would set up table in a prominent hallway and make polyhedral models, like a balloon twister at a child's birthday party. People flocked to his table, Coxeter being one of them—watching a monk, more a magician really, cutting and gluing a rainbow array of colored paper into intricate polyhedra, was naturally quite entrancing. Wenninger amassed a considerable congregation of polyhedra beside him over the course of a conference (giving the occasional model away). The time spent on each model ranged from three to ten hours, and he has made two models a day, roughly five hundred a year, for the last forty years. Father Magnus first started making polyhedra using the instructions in Coxeter's *Mathematical Recreations and Essays*, chap. 5. His success at modeling gave him the "courage" to write to Coxeter, who encouraged him to publish his own instruction book. Coxeter provided an introduction, praising his new friend's "infectiously enthusiastic style." His first book, *Polyhedron Models*, became an authoritative reference and spawned two sequels, *Spherical Models* and *Dual Models*. The latter book, Father Magnus prefaced by saying, in words that could easily be Coxeter's own, "Only when you handle a model yourself will you see the wonders that lie hidden in this world of geometrical beauty and symmetry." Father Magnus Wenninger interview, Budapest, August 2003.

54. Odom, interview.

55. Discussing Odom's suicide attempt some years later, and Odom's last wish to simply "be a memory in the mind of God," Coxeter wrote in a letter: "Somehow this reminds me of a passage in Lewis Carroll's *Through the Looking Glass*, about the Red King (just after the The Walrus and The Carpenter): 'He's dreaming now,' said Tweedledee: 'and what do you think he's dreaming about?' Alice said: 'Nobody can guess that.' 'Why, about you!' Tweedledee exclaimed, clapping his hands triumphantly. 'And if he left off dreaming about you, where do you suppose you'd be?' 'Where I am now, of course,' said Alice. 'Not you!' Tweedledee retorted contemptuously. 'You'd be nowhere. Why, you're only a sort of thing in his dream!' 'If that there King was to wake,' added Tweedledeedum, 'you'd go out—bang!—just like a candle!'" Coxeter continued to point out the relevance: "As Martin Gardner remarks, in his [*Annotated Alice*] (p. 238) this idea was first posed by Socrates (in the same work by Plato that introduces the five Platonic solids): 'A question which I think that you must often have heard persons ask: how can you determine whether, at this moment, we are sleeping and all our thoughts are a dream, or whether

we are awake and talking to one another in the waking state?' Is your life more enjoyable these days? Best wishes, Donald Coxeter." Coxeter–Odom correspondence, 1974–2000, Coxeter Fonds, University of Toronto Archives, and personal files of Magnus Wenninger (courtesty of George Odom).

56. Coxeter–Odom correspondence.

57. Coxeter–Odom correspondence; and Odom, interview.

58. Initially Coxeter told Odom his construction was incorrect. "I owe you a profound apology for my careless remark about Drawing No. 2 ('. . . not even a close approximation'!) As Father Magnus kindly pointed out, I was mistaken: This new construction for the golden section is quite correct. I must have been tired from overwork that day. Very sorry." In 1983, problem E 3007 in the *American Mathematical Monthly*, proposed by George Odom, stated: "Let *A* and *B* be the midpoints of the sides *EF* and *ED* of an equilateral triangle *DEF*. Extend *AB* to meet the circumcircle (of *DEF*) at *C*. Show that *B* divides *AC* according to the golden section." Odom, interview; Coxeter, interviews; Schattschneider, interviews, "Coxeter and the Artists: Two-Way Inspiration"; and George Odom, Problem E3007, *American Mathematical Monthly*, 482.

59. Coxeter may have treated Odom as an equal and given him due credit for his discoveries, but Coxeter also made sure Odom did not get too big for his britches: "Dear George, Our phone conversation made me suspect that you are concerned as to whether you have made any real contribution to mathematics. I said Yes, but neither your contributions nor mine are as important as Pythagoras. You illustrated the octahedral group by your 4 hollow triangles, you gave a new construction for the golden section, and you discovered a significant compound of ten cubes. These are important contributions, for which you have been recognized . . . I hope you have been able to reduce your smoking and coffee drinking: nicotine and caffeine are not good for one's health. Exercise is important, so I hope you have opportunities for walking. Yours, HSMC." Coxeter–Odom correspondence.

60. Odom, interview; Coxeter–Odom correspondence; Schattschneider, interviews, and "Coxeter and the Artists: Two-Way Inspiration."

61. Robinson connected with Coxeter upon the recommendation of Ronald Brown, professor of mathematics and director of the Centre for the Popularisation of Mathematics at the University of Wales, Bangor. John Robinson (sculptor, honourary fellow, University of Bangor, Wales), interview, Galhampton, September 10–11, 2003; Coxeter–Robinson correspondence, courtesy of Robinson, and Coxeter Fonds, University of Toronto Archives; and Robinson, *Symbolic Sculpture*.

62. Ibid.

63. Ibid.

64. Robinson also cited the sculptor Aristide Maillol: "My point of departure is always Geometric figure—square, lozenge, triangle—for those are the shapes that stand up best in Space." And: "I consider that I make statues as an apple tree gives apples. It is in me and I give what I have." Robinson, *Symbolic Sculpture*, 103–104.

65. Ibid., 104.

66. Robinson, *Symbolic Sculpture*, 66, 67; Fields Institute, Toronto; Isaac Newton Institute for Mathematical Sciences, Cambridge University.

67. Coxeter to John Robinson, 11 November 1992, Coxeter Fonds, University of Toronto Archives.

68. Robinson, "Symbolic Sculptures by John Robinson—Intuition," http://www.popmath .org.uk/sculpture/pages/2intuiti.html (accessed February 2006), Centre for the Popularisation of Mathematics, University of Wales, Bangor.

69. From a photo of Robinson's *Intuition*, Coxeter assumed the outer edges of the hollow

triangles were twice the length of the inner edges, and that an inner vertex of any triangle co-incided with the midpoint of an outer edge of another. According to Schattschneider, "He also noted Odom's and Robinson's sculptures had an essential difference: while every two triangles in Odom's sculpture were interlocked, Robinson's three triangles were interlocked in the man-ner of Borromean rings." Coxeter relayed all this in a letter to Robinson, closing with a re-search request: "P.S. One important question: Is your structure inherently rigid, or would it collapse if placed on a slippery floor?" Robinson replied: " 'Intuition' is extremely rigid . . . It would make a self-supporting roof for a building." Robinson enclosed an architect's drawing of the proposed structure, which he still hopes will be built. Coxeter continued his research on *Intuition* by bringing a model of Robinson's sculpture to a workshop at the Regional Geome-try Institute, at Smith College, in the summer of 1993. He showed workshop participants how it could indeed collapse. He then asked if anyone could determine the point at which it became rigid, as Robinson insisted it did (participants were surprised that Mr. Geometry didn't know the answer). The answer was that the ratio of outer to inner edge was 1.9663265: 1, not 2:1 as Coxeter previously assumed (though Coxeter still held that the sculpture would collapse of its own weight; and Robinson kept to his belief in its inherent rigidity). Robinson interview; Cox-eter–Robinson correspondence, Coxeter Fonds, University of Toronto Archives; and Schattschneider, "Coxeter and the Artists: Two-Way Inspiration."

70. Coxeter, "Symmetrical Combinations of 3 or 4 Hollow Triangles," *The Mathematical Intelligencer*, vol. 16, no. 3, 25–30.

71. Robinson–Coxeter correspondence.

72. M. C. Escher to George Escher, 17 May 1960, National Gallery of Canada Library.

73. Ibid., 28 May 1960.

74. Ibid., 15 February 1959.

75. M. C. Escher, "Approaches to Infinity," in *Escher on Escher: Exploring the Infinite*, 123–127.

76. Coxeter diaries, 23 November 1968.

77. Coxeter diaries, 18 November 1974.

78. Coxeter, interviews.

79. Escher, "Approaches to Infinity."

80. Schattschneider, interviews; and Coxeter–Escher correspondence, National Gallery of Canada Library and National Gallery of Art, Washington, D.C.

81. M. C. Escher, *Circle Limit III*, currently with Susan Thomas.

82. M. C. Escher to Coxeter, 1 May 1960, National Gallery of Art, Washington, D.C.

83. Coxeter to M. C. Escher, 16 May 1960.

84. Ibid.

85. M. C. Escher to Coxeter, 28 May 1960.

86. Coxeter diaries, 9 and 10 November 1960.

87. Coxeter diaries, 3 July 1962, 4 October 1965, 30 June 1966.

88. Schattschneider, interviews; and Schattschneider, "Coxeter and the Artists: Two-Way Inspiration."

89. Coxeter, "The Mathematical Implications of Escher's Prints," Escher retrospective catalog (1968), 87–89; Schattschneider, interviews; and Schattschneider, "Coxeter and the Artists: Two-Way Inspiration."

90. Coxeter, "The Mathematical Implications of Escher's Prints," in *The World of M. C. Escher* (1991), 49–52.

91. Coxeter, "The Non-Euclidean Symmetry of Escher's Picture *Circle Limit III*," in *Leonardo*, 19–25.

92. Coxeter, interviews; Schattschneider, interviews; and Schattschneider, "Coxeter and the Artists: Two-Way Inspiration."

93. Ibid.

94. Ibid.

95. M. C. Escher to George Escher, 30 July 1961, National Gallery of Canada Library.

96. Coxeter also kept in his file the letters to the editor the following day, condemning this critique as "lower scribbling." "Writing of the Escher exhibit, [Mays] says, 'It's enough to make a grown-up art critic weep.' How would he know?" and another offering: "it reminds me of something that was said of a hugely successful Broadway show many years ago: 'Everyone hated it except the public.' Miscellaneous Escher File, Coxeter Fonds, University of Toronto Archives.

97. G. W. Locher, "Structural Sensation," in *The World of M. C. Escher*, 47–48; and Emmer, interview.

98. Schattschneider, "Escher's Metaphors," *Scientific American*, 48–53; and Emmer, Ibid.

99. M. C. Escher to George Escher, 15 February 1959, National Gallery of Canada; and J. L. Locher et al., *M. C. Escher, His Life and Complete Graphic Work*, 92–93.

100. The smallest of the spheres is steel, with four wooden balls turned on a lathe, and the largest is only a hemisphere. The top sphere opens along its circumference, with Coxeter's letter and Soddy's equation stashed in a compartment inside. When Robinson was in Toronto planning the installation of *Intuition*, Coxeter suggested he might like to try a sculpture of five mutually tangent spheres with radii governed by Soddy's geometric theorem ("The square of the sum of all five bends is thrice the sum of their squares"). Robinson was intrigued and couldn't get the idea out of his head, ultimately making it for Coxeter as a more personal ninetieth-birthday present, which he could keep at home. Coxeter, interviews; Robinson, interviews; Coxeter–Robinson correspondence, Coxeter Fonds, University of Toronto Archives.

101. Robinson, "Symbolic Sculptures by John Robinson—Firmament," http://www.popmath .org.uk/sculpture/pages/5firm.html (accessed February 2, 2006).

CHAPTER 12—THE COXETERIAN SHAPES OF THE COSMOS

1. There is a minor planet named after Coxeter—Coxeter asteroid No. 18560—confirmed by the Royal Astronomical Society of Canada. It was discovered by the Italian Paul G. Comba, observed on "1997 03 07" photographically from his home on Galaxy Lane, in Prescott, Arizona. A former mathematics professor at the University of Hawaii and later a software developer at IBM, Comba retired to become an amateur astronomer who, to date, has discovered 1,145 asteroids (some using the old method of using mirrors for detection). He began naming his discoveries after his hit parade composers, his wife, and kids. When he realized he needed a hefty store of names, he began going through the alphabet of famous mathematicians, which soon brought him to Coxeter. Paul G. Comba, interview, June 4, 2005; Peter Jedicke (president, Royal Astronomical Society of Canada), e-mail correspondence, "Re: Coxeter Asteroid," June 3, 2005; for profile information, http://www.rasc.ca/faq/asteroids/home.htm; and to track the Coxeter Asteroid, http://neo.jpl.nasa.gov/cgi-bin/db?name=18560 (accessed January 31, 2006).

2. Coxeter diaries, 16 December 1933.

3. Coxeter, interviews; Thomas, interviews; Weiss, interviews.

4. Thomas, interviews.

5. Conway, interviews; Smith, interviews; Marc Pelletier (geometric sculptor), interview, Toronto, May 13 and 14, 2004, June 2 and October 18, 2005.

6. Coxeter, interview, February 2002.

7. Conway, interviews.

8. Bezdek, interview.

9. Coxeter, interview, Budapest, July 9, 2002.

10. János Bolyai Conference on Hyperbolic Geometry, Programme.

11. Jeff Weeks (freelance geometer, Canton, New York), interview, Budapest, July 10, 2002, October, November 2003, and May 2005.

12. Weeks also echoed other Coxeter fans in raving about the visual element of Coxeter's work: "You see it and it's a joy to behold, to just appreciate these images," Weeks said. "Coxeter gets away from the overly abstract axiomatic approach to mathematics and shows you something that has real substance. You can see it. In that sense, Coxeter is an influence on my own work, and the work of people doing similar things—promoting the concrete and visual approach to things. I'm tempted to say that the fact Coxeter worked with fundamental things in this way is likely what's responsible for his ideas having found various applications, and in many ways unanticipated applications. I think that's the sign of good mathematics—it's something really basic and then shows up in lots of places along the way." Ibid.

13. It used to be that once a decade or so, scientists asked, "What is the shape of the universe?" A hypothesis would arise—for example, that the universe was flat and infinite—followed by a spurt of research, and then our collective cosmic curiosity was sated with enough information to last us a while on the space-time odometer. Since the early 1990s, however, cosmology is where much of the exciting science has been happening. "We've been looking for the shape of the universe like Columbus did the shape of the Earth," said Glenn Starkman, an astrophysicist at Case Western Reserve University, in Cleveland. The "crisis" in cosmology these days, according to Starkman, speaking only somewhat with tongue in cheek, is that time is running out. "If we don't figure out the shape of the universe soon," he said, "the universe will hide this secret from us forever." This is because the research depends on data salvaged from the microwave background, the echoes of the big bang that created the universe in the first place. And as Starkman explained, "Today, the place from which the echoes come to us is moving away from us faster than the speed of light, which means we can't receive light from that place any more—we can no longer see or learn about that place, never mind any farther away. We have enough data now to be able to determine the shape of the universe if the shortest distance around the universe is less than the distance across the microwave sphere of the big bang. But we do not have enough data if the universe is any bigger." Max Tegmark, an astrophysicist and professor at Princeton's Institute for Advanced Study, likened it to trying to figure out the shape of the Earth if you're not able to see beyond the walls of your bedroom. "Nature has a censorship where we can only see so far," Tegmark said. "We can't see anything from farther than 14 billion light-years. This limits us in what we can see and what data we can gather." The only way there will be enough data, according to Starkman, is if the universe ceases with its expansion. "It might stop its accelerated expansion, but probably not for many billion years, which doesn't help us much." Starkman, interview, Toronto, March 19, 2004 and April 22, 2004; Tegmark, interview, April 2004.

14. Weeks's collaborators are Jean-Pierre Luminet at the Paris Observatory, Alain Riazuelo and Roland Lehoucq at Service d'Astrophysique du Commissariat à l'Energie Atomique (CEA) de Saclay, and Jean-Philippe Uzan at Institut d'Astrophysique de Paris. Weeks, interviews.

15. Ibid.

16. Weeks provided a good description of the process by which he applies pure geometry to cosmology. He encounters a question—such as "What is the shape of the universe?"—and then embarks on understanding how things work, "by looking and exploring, either in your mind, on a sheet of paper, or with computer simulation. You're exploring a situation in the same way that you'd explore a forest," he says. "You go in and start kicking around, you determine what the patterns are and what's going on. You get a better feel for what you're looking at, trying to find the theorem that summarizes it all. And working with the visual approach, a theorem

is not just a statement on a piece of paper with a bunch of lines and proofs supporting it. Ideally, the theorem is also a visual element that gives you an insight into how to look at something. You struggle, you're exploring around these woods, and eventually you see the pattern—in much in the same way that when exploring in an orchard you find the viewpoint where the trees all line up. You have that AHA! moment. If you're looking at the orchard from the wrong angle, the trees are just all over the place. When you look in the right direction they line up and you understand the pattern and what makes it work, and that's often the case with geometrical mathematics. You look at a lot of examples, you're confused, but eventually you find some patterns and you get to that moment where you say AHA! Ok, this is all so simple. Why didn't I see this two months ago?!" Ibid.

17. Ibid.

18. J.-P. Luminet, J. Weeks, A. Riazuelo, R. Lehoucq, and J.-P. Uzan, "Dodecahedral Space Topology as an Explanation for Weak Wide-Angle Temperature Correlations in the Cosmic Microwave Background," *Nature*, 593–595.

19. Weeks, interviews.

20. The circle searching is being conducted by another international medley of cosmologists: Canadian Glenn Starkman, based at Case Western in Cleveland; Neil Cornish, an Australian at Montana State; David Spergel, at Princeton; and Japan's Eiichiro Komatsu, at the University of Texas. Starkman et al. have not yet found any circles. They have looked where the circles should be with no luck. In the search, they see there is something odd going on—the fluctuations on the microwave sky are aligning themselves in strange ways; they seem to be aligning themselves with the plane of the solar system. "This just shouldn't be," Weeks said. "What goes on in deep space and the distant past should not be affected by the path the planets follow around the sun." One possibility for this odd behavior is a miscalculation, which could also explain the failed circle-searching. A German team, meanwhile, is suggesting that outer-spatial noise, galactic contamination, or a cloud of dark matter could obscure results and make finding circles impossible. Tegmark is currently calculating just how much data distortion could result from these galactic pollutants." The most amazing thing of all," he said, "is that we humans can address these questions in a scientific way; that these philosophical questions—like is space infinite?—have become scientific questions." Though, the answer to these questions is still philosophical—the "So what?" factor. The answers mainly just serve to satisfy the age-old innate human curiosity, our egocentric pondering about our local place in the universal scheme of existence. There is always a chance that the scientific answers will lead to more questions and then more answers again, but these subsequent Q&As are in areas of science that are currently hidden and unfathomable, until we find the initial answers. Weeks, interviews; and Starkman, interviews.

21. Weeks, interviews.

22. Ibid.

23. Coxeter and Weeks, Budapest conference, July 9, 2002. Delivering a talk to the Royal Astronomical Society Club in 1972, Coxeter had borrowed an observation from fellow geometer László Fejes-Tóth, who claimed that the geometer is "able to create an infinite set of new universes, the laws of which are within our reach, though we can never set foot in them." Coxeter, "Royal Astronomical Society Club," March 10, 1972, Lectures File, Coxeter Fonds, University of Toronto Archives; László Fejes-Tóth, *Regular Figures*, 125 – Macmillan, New York, 1964.

24. Coxeter, interview, January 2003.

25. J. Richard Gott (professor of astrophysics, Princeton) correspondence to Siobhan Roberts, May 1, 2005, and e-mail correspondence May–June 2005.

26. "I had adopted a less stringent criterion of regularity," said Gott, "namely that these are networks of congruent regular polygons such that each vertex is surrounded by the same number and arrangement of polygons, the sum of the face angles at a vertex is greater than 360 degrees, two polygons meet at each edge, and polygons meet at only one edge. Coxeter had in addition the requirement that all dihedral angles between faces should be the same. Therefore he found only three examples. Because I allowed different dihedral angles, I found four additional polyhedra (all of which I called pseudopolyhedra) beyond those found by Petrie and Coxeter." Gott found pentagons-five-around-a-point, squares-five-around-a-point, triangles-eight-around-a-point, and triangles-ten-around-a-point. Ibid.

27. Chemist A. F. Wells later found additional regular skew polyhedra, using regularity criteria identical to Gott's (Wells was unaware of Gott's paper but knew of Coxeter's). He pinpointed squares-five-around-a-point, triangles-seven-around-a-point, triangles-eight-around-a-point, triangles-nine-around-a-point, triangles-ten-around-a-point, and triangles-twelve-around-a-point, which he discussed in his book *Three-Dimensional Nets and Polyhedra*. Ibid.

28. Ibid.

29. Gott has since continued to be intrigued by geometrical topics. He found an exact solution to Einstein's field equations for the geometry around one cosmic string, and for two moving cosmic strings. "This solution is of particular interest," he said, "because it allows time travel to the past . . . My solution thus joins a list of time travel solutions in general relativity including a rotating universe solution found by K. Gödel in 1949, and wormhole solutions found by Kip Thorne, Mike Morris, and Ulvi Yurtsever in 1988." Gott has written a popular account of his work in *Time Travel in Einstein's Universe*, published in 2001. Ibid.

30. Leonard Mlodinow, *Euclid's Window* (Greene's dust jacket blurb).

31. Moffat, interview.

32. Conway, interviews; Greene, *The Elegant Universe*.

33. Witten.

34. Coxeter, interview, March 14, 2003.

35. The first strings revolution occurred in 1984, inflating the field from a few pioneers to hundreds of zealots. The next revolution, the duality revolution (relating different strands of string theory through "duality transformations"), occurred in 1995. The feeling is that it is about time another revolution bolstered the string theorists' spirits; the arrival of this third revolution has been rather more agonizing than many had hoped. This think-tank session, however, raised only a sinkhole of questions. Raphael Bousso, a professor of particle physics at the University of California–Berkeley, put it slightly more hopefully with his "wish list of questions to be answered some time soon." His and others' list of questions included: "Is quantum mechanics more fundamental than general relativity?" "Is there a wavelength function of the universe? How do we pick it?" The doozy: "Will the classical notion of time survive?" And a question one would not expect string theorists to be asking, "What is string theory?" The highlights of a rat-a-tat spiel by Eva Silverstein, also from Stanford, were uplifting as found poetry, printed with colored markers on her overhead transparencies—"no man's land"—"big crunch"—"potentially interesting/crazy estimate"—"light degrees of freedom"—and, alas, "no smoking-gun signature." In closing her presentation, Silverstein mock-lamented: "My phenomenological friends make fun of these sessions where we sit around and talk about our feelings." Strings05, Toronto, July 11–16, 2005.

36. According to another audience member, hobby-string theorist Carl Feinberg, from New York, Witten is "the brightest light in the firmament. He's incorporeal. It's like this disembodied intelligence. The lucidity of his thought and expression is so unique—it's pure thought."

Feinberg wasn't quite so impressed with how the panel session proceeded. He found it at once disappointing and scintillating. "I was struck by the way the panel was all sitting like this"—arms crossed and clenched (Feinberg sat the same way for much of the evening), angst-ridden either at the lack of revolutionaries or at the lack of productivity during the discussion. Both were apropos to Witten's point. "In 1990, if we had had too much of a preconception, we would never have . . ." he said, trailing off. "If we'd had a panel, opinions would have been all over the map." He clarified later by saying, "If you have too many preconceptions, about what problems you want to solve, you often don't get there. If you decide you want to solve problem X, you might not see problems Y and Z, and problems Y and Z might be easier and might be necessary as examples before you solve problem X." In other words, all this stewing will only tangle a string theorist's creativity into knots. Witten, interview, Toronto, July 2005.

37. Witten, e-mail correspondence.

38. Greene and Plesser's mirror manifolds grew from their contemplation of the Dixon-Lerche-Vafa-Warner conjecture. And simultaneously, Greene learned, complementary discoveries exposing the mirror symmetry of string theory were made by Philip Candelas, Monika Lynker, and Rolf Schimmrigk, at the University of Texas. Briane Greene, *The Elegant Universe*, 255–62.

39. Greene, 260.

40. Witten, op. cit.

41. Lederman, interview; Greene, *The Elegant Universe*.

42. Marc Henneaux, "Platonic Solids and Einstein Theory of Gravity: Unexpected Connections," general seminar February 26, 2003, available at http://tena4.vub.ac.be/cgi-bin/seminars.pl?datafile=seminars20022003.data&which=all; Henneaux and his collaborators have written more scholarly papers on the subject, namely "Cosmological Billiards," *Classical and Quantum Gravity*, "Einstein Billiards and Overextensions of Finite Dimensional Simple Lie Algebras," *Journal of High Energy Physics*, and "E(10) and BE(10) and Arithmetical Chaos in Superstring Cosmology," *Physical Review Letters*; and the more recent seminar, "From Platonic Solids to Infinite Coxeter Groups and Lorentzian Kac-Moody Algebras: The Road to Uncover the Symmetries of Gravity?" at the Theoretical Physics Colloquium, May 11, 2005.

43. Ibid.

44. Henneaux's collaborators are Thibault Damour (IHES, France), Bernard Julia (ENS, Paris), and Hermann Nicolai (MPI, Germany). Henneaux is also director of the International Solvay Institutes, and associated with Centro de Estudios Cientificos, in Valdivia, Chile.

45. Marc Henneaux (physicist, Free University, Brussels), e-mail correspondence "Re: Platonic Solids and Einstein Theory of Gravity: Unexpected Connections," November 8, 2004; and interviews, November 16, 2004 and September 28, 2005.

46. Ibid.

47. Ibid.

48. Witten.

49. Atiyah, interviews.

50. Ibid.

51. Ibid.

52. Ibid..

53. Coxeter, Budapest, July 9–13, 2002. Coxeter often quotes this passage from Chesterton, such as in his book *Regular Complex Polytopes*, on p. 21.

54. Gyorgy Darvas (director, Symmetrion, and CEO International Symmetry Association), interview, Budapest, July 7, 2002.

55.   Coxeter, with Darvas, Budapest, July 7, 2002.

56.   Coxeter, Budapest, July 13, 2002.

CHAPTER 13—FULL CIRCLE SYMMETRY

1.   Coxeter, *Regular Polytopes*, 289.

2.   John Derbyshire, *Prime Obsession: Bernhard Riemann and the Greatest Unsolved Problem in Mathematics*, 378.

3.   The Dutch Consulate in Canada, in particular, kept close tabs on Coxeter. Each year he had to submit what he called his "still alive form," since in 1938 he had made one of his best financial decisions in declining a 100-pound payment for a lecture he had given in Utrecht, opting instead for a lifetime 2.5 percent annually on that fee. Coxeter diaries, 10 May 1938; and Coxeter, interviews.

4.   Coxeter–Pólya correspondence, Coxeter Fonds, University of Toronto Archives.

5.   Coxeter–Syne correspondence, Coxeter Fonds, University of Toronto Archives; and Cathleen Synge Morawetz, family papers.

6.   Clifford Zvengrowski to Coxeter, 12 September 1998, Coxeter Fonds, University of Toronto Archives.

7.   Coxeter, interviews.

8.   C. Davis, interviews.

9.   Another disappointment was the void left upon his retirement. The greatest living classical geometer, who had put the University of Toronto on the international mathematical map, was not replaced; there is now no classical geometer, not even a classically inclined geometer, on staff. Mathematical traditions usually do not reside in isolated individuals. Certainly, one of Coxeter's legacies is the community of his students who still come together in celebration of their work, and share his classical spirit with up and coming generations. And, quite conspicuously, Coxeter's students are the sort of mathematicians who take more pleasure in their work than most; they live and breathe and truly love the art. But for many, the decision taken by the University of Toronto, and other institutions internationally, not to invest in the future of classical geometry is alarming, if not tragic. For Walter Whiteley, this is an expression of the geometry gap. Coxeter, interviews; Whiteley, interviews; Thomas, interviews.

10.   Michael Longuet-Higgins, interviews; Coxeter–Longuet-Higgins correspondence, Coxeter Fonds, University of Toronto Archives, and Longuet-Higgins, personal papers; and Schattschneider, "Coxeter and the Artists: Two-Way Inspiration, Part 2."

11.   Coxeter, interviews; Thomas, interviews.

12.   Ibid.

13.   John Conway, Fields Institute, February 15, 2002; Conway, interviews.

14.   This is a simple statement of Coxeter's "Murder Weapon": If "$A^p = B^q = C^r = ABC = 1$" defines a finite group, then "$A^p = B^q = C^r = ABC = Z$" implies $Z^2 = 1$. Conway; and Conway, interview, November 26, 2005; Conway with Coxeter and G. C. Shepherd, "The Centre of a Finitely Generated Group," in *Tensor*, 1972 (contains the proof for "The Murder Weapon").

15.   Conway, Fields Institute.

16.   John Galsworthy, *The Forsyte Saga*, 318–19. Coxeter also ritualistically dipped into the Bible. When Rien died, he stopped attending church entirely (he went just to please her), but continued to appreciate biblical truths as he did mathematical ones. He referred to his Bible as he did his *Encyclopedia of Mathematics*. During afternoon tea with his first PhD student Willy Moser, their last visit, Coxeter began reading from Ecclesiastes 11:12—"Remember now thy Creator in the days of thy youth, while the evil days come not, nor the years draw nigh, when

thou shalt say, I have no pleasure in them." He parsed it more for the vocabulary than anything else. "What do you think is meant by 'grinders'?" he queried, continuing, "In the day when the keepers of the house shall tremble, and the strong men shall bow themselves, and the grinders cease because they are few, and those that look out of the windows be darkened." He guessed grinders might be teeth, seeing as even his dentures no longer fit properly. ". . . when they shall be afraid of that which is high, and fears shall be in the way, and the almond tree shall flourish, and the grasshopper shall be a burden, and desire shall fail: because man goeth to his long home, and the mourners go about the streets . . ."

17. Thomas and Coxeter, Budapest, July 13, 2002.
18. Thomas, interviews.
19. Ibid.
20. Coxeter, interviews, Toronto, January–March 2003.
21. Coxeter, interview, Toronto, March 29, 2003.
22. Coxeter, interviews; Thomas, interviews.
23. Abraham.
24. Conway, interviews.
25. Sandra Witelson (neuroscientist, McMaster University), interview, Hamilton, March 2003.
26. Ibid.
27. Coxeter, interview, Hamilton, March 2003.

APPENDIX 1—FIBONACCI AND PHYLLOTAXIS

1. Coxeter, *Introduction to Geometry*, 160–72; and Conway, interviews.
2. Coxeter and Ball, *Mathematical Recreations and Essays*, 57; and Weisstein, "Phyllotaxis," http://mathworld.wolfram.com/Phyllotaxis.html (accessed February 5, 2006).
3. Coxeter, "Chirality and Phyllotaxis," Coxeter Lectures File, Coxeter Fonds, University of Toronto Archives.
4. Coxeter, *Introduction to Geometry*, 172; and Conway, interviews. A more detailed explanation of the phyllotaxis mechanism appears in *The Book of Numbers* by John Conway and Richard Guy. And Coxeter's elder friend and colleague, Sir D'Arcy Thompson, with whom he corresponded, provided a timeless account of why living things and physical phenomena take the shape they do in his book, *On Growth and Form*. See also Coxeter's papers, "The Role of Intermediate Convergents in Tait's Explanation for Phyllotaxis," *J. Algebra* 167–75, and "The Golden Section, Phyllotaxis, and Wythoff's Game," *Scripta Mathematica*, 135–43.

APPENDIX 2—SCHLÄFLI SYMBOLS OF THE 3-D AND 4-D REGULAR POLYTOPES

1. Schläfli symbols charts, courtesy of Doris Schattschneider.

APPENDIX 3—COXETER DIAGRAMS

1. Conway, interviews; Coxeter diagrams, courtesy of Doris Schattschneider and John Conway.

APPENDIX 4—COXETER GROUPS

1. Monson, interviews; and Conway, interviews.
2. Ibid.
3. Coxeter, *Regular Polytopes*, 290–305; chart of multiple reflected images and diagram, courtesy of Doris Schattschneider.

APPENDIX 5—MORLEY'S MIRACLE

1. Coxeter, *Introduction to Geometry*, 23–25.
2. Ibid., 23.
3. Conway, interviews. See "Morley's Theorem" at the Mathworld site for a historical reference of constructions and proofs of Morley's theorem, http://mathworld.wolfram.com/MorleysTheorem.html.
4. O'Connor and Robertson, "Frank Morley," http://www-groups.dcs.st-and.ac.uk/~history/Mathematicians/Morley.html (accessed February 5, 2006).
5. Coxeter, *Introduction to Geometry*, 24; and Coxeter and Greitzer, *Geometry Revisited*, 47.
6. Conway, interviews.
7. All three of Morley's sons, Christopher, Frank V., and Felix, were Rhodes scholars. Christopher became a novelist, and his books include *Thunder on the Left*, *The Trojan Horse*, *Kitty Foyle*, and *The Old Mandarin*; Felix became editor of the *Washington Post*; and Frank became a director of the publishing firm Faber and Faber, and collaborated with his father on mathematics. O'Connor and Robertson.
8. O'Connor and Robertson.
9. Coxeter, *Introduction to Geometry*; Coxeter and Greitzer, *Geometry Revisited*; and Conway, interviews.
10. "John Conway's Morley Proof," http://paideiaschool.org/TeacherPages/Steve_Sigur/resources/TB%20pictures/morley-dissected.html (accessed March 14, 2006). Conway's proof and diagrams, courtesy of John Conway and Steve Sigur.
11. Conway, interviews.

APPENDIX 6—FREEMAN DYSON ON "UNFASHIONABLE PURSUITS"

1. Freeman Dyson (polymath, emeritus professor of physics, Institute for Advance Study), interview, October 26, 2004; and Coxeter–Dyson correspondence, Coxeter Fonds, University of Toronto Archives; and Coxeter–Dyson correspondence, courtesy of Freeman Dyson.
2. Freeman Dyson, "Unfashionable Pursuits," Coxeter–Dyson correspondence, Coxeter Fonds, University of Toronto Archives; reprinted here with the permission of Freeman Dyson (originally published in *The Mathematical Intelligencer*, 47–54, and Dyson, *From Eros to Gaia*).
3. In 1992, Dyson's prophecy came to pass when the Monstrous Moonshine conjecture was proved by Richard Borcherds (University of California–Berkeley), revealing a deep connection between elliptic curves, the Monster group, and string theory—a certain strain of string theory has the Monster group symmetries. The Monster group, so named by Conway because it is the most humongous, mysterious, and fascinating of the sporadic groups, had been predicted to exist in 1973 by Robert Griess (University of Michigan) and Bernd Fischer (University of Hamburg). Fischer had visited Cambridge looking for help in defining the group, and Conway took up the challenge. In 1978, an observation by John McKay (Concordia University) brought to light possible connections between the Monster group and the elliptic modular function, which Conway and Simon Norton (University of Cambridge) developed into a proposition called the Monstrous Moonshine conjecture in 1979. Borcherds worked on the proof of the Moonshine conjecture for eight years. He later said, "I was over the moon when I proved the moonshine conjecture. If I get a good result I spend several days feeling really happy about it. I sometimes wonder if this is the feeling you get when you take certain drugs. I don't actually know, as I have not tested this theory of mine." He won a Fields Medal for his work, which, compared to actually proving the conjecture, left him nonplussed. Conway, interviews; John McKay, interview,

May 2004; Conway and Norton, "Monstrous Moonshine," Bulletin of the London Mathematical Society, 308–39; "Monstrous Moonshine Conjecture," http://www.daviddarling.info/encyclopedia/M/Monstrous_Moonshine_conjecture.html (accessed February 5, 2006); and Weisstein, "Monstrous Moonshine," http://mathworld.wolfram.com/MonstrousMoonshine.html (accessed February 5, 2006).

4. Coxeter to Freeman Dyson, 4 October 1981, Coxeter–Dyson correspondence, Coxeter Fonds, University of Toronto Archives.

5. Dyson, "Unfashionable Pursuits."

APPENDIX 7—CRYSTALLOGRAPHY AND PENROSE TOILET PAPER

1. Schattschneider, interviews.

2. Coxeter, Introduction to Geometry, 50.

3. Weisstein, "Tiling," http://mathworld.wolfram.com/Tiling.html (accessed February 5, 2006); for a definitive reference on tiling, see Tilings and Patterns, by Branko Grünbaum and G. C. Shephard.

4. Penrose, interviews.

5. Ibid.

6. Ibid.

7. Ibid.

8. Matthew Rose, "Mathematician Sues Kimberly-Clark Unit over Its Toilet Paper—At Heart of the Messy Issue Is Tissue's Quilted Design; Is It the 'Penrose Pattern'?" The Wall Street Journal, April 14, 1997.

9. Penrose, interviews.

10. Marion Walter (emeritus professor of mathematics education, University of Oregon), interview, August 2, 2005.

APPENDIX 8—THE MATHEMATICAL PUBLICATIONS OF H. S. M. COXETER

1. Reprinted (with updates) from Kaleidoscopes (eds. Sherk, McMullen, Thompson, Ivić Weiss), with permission.

BIBLIOGRAPHY

For a full list of Publications by H. S. M. Coxeter, see appendix 8.

Abbott, Edwin A. *Flatland: A Romance of Many Dimensions*. New York: Dover Publications, 1952.

Abbott, Edwin A. With notes by Ian Stewart. *The Annotated Flatland*. Cambridge, MA: Perseus Publishing, 2002.

Abercrombie, W.E. "Geometry—An Interlude or an Essential," *Ontario Mathematics Gazette* 5, no. 3 (1967): 17–24.

Abraham, Carolyn. *Possessing Genius*. Toronto: Penguin Canada, 2001.

Albers, Donald, ed. "Is Geometry Dead?" *The Two-Year College Mathematics Journal* 11, no. 1 (January 1980).

Albers, Donald J., and G. L. Alexanderson, eds. *Mathematical People: Profiles and Interviews*. Boston: Birkhäuser, 1985.

Aldrich, John. "The Mathematics PhD in the United Kingdom." http://www.economics.soton.ac.uk/staff/aldrich/PhD.htm

Anderson, Marlow, Victor Katz, and Robin Wilson, eds. *Sherlock Holmes in Babylon and Other Tales of Mathematical History*. Washington, D.C.: Mathematical Association of America, 2004.

Armitage, J. V. "The Place of Geometry in a Mathematics Education," *The Changing Shape of Geometry*. New York: Cambridge University Press, 2003, 515–26.

Aspray, William. "The Emergence of Princeton as a World Center for Mathematical Research, 1896–1939," in *A Century of Mathematics in America*, vol 3. Edited by Peter Duren. Providence: American Mathematical Society, 1989, 195–215.

Baake, M., and R. V. Moody, eds. *Directions in Mathematical Quasicrystals*. Providence: American Mathematical Society, 2000.

Baake, M., Uwe Grimm, and Robert V. Moody. "What Is Aperiodic Order?" *Spektrum der Wissenschaften*, February 2002, 64–74.

Ball, W. W. Rouse. *A Short Account of the History of Mathematics*. London: Macmillan, 1912.
———. *Cambridge Papers*. London: Macmillan, 1918.
———. *Mathematical Recreations and Essays,* Tenth Edition. London: Macmillan, 1931.

Ball, W. W. Rouse. Revised by H. S. M. Coxeter. *Mathematical Recreations and Essays,* 11th ed. New York: Macmillan, 1939.

Ball, W. W. Rouse, and H. S. M. Coxeter. *Mathematical Recreations and Essays*, 13th ed. New York: Dover Publications, 1987.

Banchoff, Thomas F. *Beyond the Third Dimension: Geometry, Computer Graphics, and Higher Dimensions*. New York: Scientific American Library, 1990.

Banchoff, Thomas F. "Interview with Fr. Magnus J. Wenninger," *Symmetry: Culture and Science* 13, nos. 1–2 (2002): 63–70.

Barrow-Green, June, and Jeremy J. Gray. "Geometry at Cambridge, 1863–1940," *Historia Mathematica*, June 2006, 42.

Barwell, Noel. *Cambridge.* London: Blackie & Son Limited, n.d.

Batterson, Steven L. *Stephen Smale: The Mathematician Who Broke the Dimension Barrier.* Providence: American Mathematical Society, 2000.

Beaulieu, Liliane. "A Parisian Café and Ten Proto-Bourbaki Meetings (1934–1935)," *The Mathematical Intelligencer* 15, no. 1 (1993): 27–35.

———. "Bourbaki's Art of Memory," *Osiris* 2, second series, vol. 14, *Commemorative Practices in Science: Historical Perspectives on the Politics of Collective Memory*, 1999, 219–51.

———. "Dispelling a Myth: Questions and Answers about Bourbaki's Early Work, 1934–1944," *The Intersection of History and Mathematics.* Basel: Birhäuser, 1994, 241–52.

Beckwith, Philip. "Paul S. Donchian Opens Door to Fairyland of Pure Science: His Wire and Cardboard Models Explain Highest Mathematics," *The Hartford Daily Courant*, January 20, 1935, E3.

Bell, E. T. *The Development of Mathematics.* New York: McGraw-Hill, 1945.

———. *Men of Mathematics.* London: Victor Gollancz, 1937.

Bell, Jordan. "The Geometrical Foundations of Coxeter Groups." Honours Project, Carleton University, School of Mathematics and Statistics, January 14, 2005.

Benn, John. "Universities and Public Life: A Contrast between England and America in Which Room for Improvement Is Noted," *Princeton Alumni Weekly*, March 17, 1933.

Berberian, S. K. "Bourbaki, the Omnivorous Hedgehog: A Historical Note?" *The Mathematical Intelligencer* 2 (1980): 104–5.

Blackburn, David, and Jonathan Rosenhead. "Cambridge Mathematics," *The Eagle*, EL 8, 46–49.

Blay, Michel. *Reasoning with the Infinite.* Chicago: University of Chicago Press, 1998.

Boas, R. P. "Bourbaki and Me," *The Mathematical Intelligencer*, no. 4 (1986): 84.

Bodanis, David. $E = mc^2$. Toronto: Random House, 2001.

Bogomolny, Alexander. http://www.cut-the-knot.org.

Bonpunt, Louis. "The Emergence of Symmetry Concepts by the Way of the Study of Crystals (1600–1900)," *Symmetry: Culture and Science*, 10, no. 1–2 (1999), 127–41.

Borel, Armand. "Twenty-Five Years with Nicolas Bourbaki, 1949–1973," *Notices of the AMS* 45, no. 3 (March 1998): 373–80.

———. "The School of Mathematics at the Institute for Advance Study," *A Century of Mathematics in America*, vol 3. Edited by Peter Duren. Providence: American Mathematical Society, 1989, 119–47.

Borwein, Jonathan M., and David H. Bailey. *Mathematics by Experiment: Plausible Reasoning in the 21st Century.* Natick: A. K. Peters, 2004.

Bourbaki, Nicolas. "The Architecture of Mathematics," *American Mathematical Monthly* 57 (April 1950): 221–32.

———. *Groupes et Algèbres de Lie, 4–6.* Paris: Hermann, 1968 (reprinted New York: Masson, 1981).

———. *Elements of the History of Mathematics.* New York: Springer, 1999.

Brannan, David A., Matthew F. Esplen, and Jeremy J. Gray. *Geometry.* New York: Cambridge University Press, 1999.

Brewster, Sir David. *The Kaleidoscope: Its History, Theory, and Construction.* Holyoke: Van Cort Publications, 1987.

Brown, Ronald. "Sculptures by John Robinson at the University of Wales, Bangor," *The Mathematical Intelligencer* 16, no. 3 (1994): 62–64.

Bruhn, Jörn. "Mathematics Education and Comparative Studies: Two Examples," Reflections on Educational Achievement, Papers in Honour of T. Neville Postlethwaite. New York: Waxmann-Verlag, 1995, 69–74.

Brunés, Tons. *The Secrets of Ancient Geometry, and Its Use*. Copenhagen: Rhodos, 1967.

Bryden, John. *Best-Kept Secret: Canadian Secret Intelligence in the Second World War*. Toronto: Lester Publishing, 1993.

Bursill-Hall, Piers. "Why Do We Study Geometry? Answers Through the Ages," Lecture delivered at the opening festivities of the Faulkes Institute for Geometry, University of Cambridge, May 2002.

Calaprice, Alice. *The New Quotable Einstein*. Princeton: Princeton University Press, 2005.

Cambridge University. *Cambridge University Reporter*. Cambridge: Cambridge University Press, October 8, 1926.

Canadian Press. "Drew Demands Gov't Investigate U of T Professor," *The Daily Press*, March 17, 1950.

Carroll, Lewis. *Euclid and His Modern Rivals*. New York: Dover Publications, 1973.

Carroll, Lewis. With introduction and notes by Martin Gardner. *The Annotated Alice*. New York: Clarkson N. Potter, 1960.

Cartan, Henri. "Nicolas Bourbaki and Contemporary Mathematics," *The Mathematical Intelligencer* 2 (1980): 175–80.

Chaplin, Virginia. "Princeton Mathematics: A Notable Record," *Princeton Alumni Weekly*, May 1958, 9, 6–15.

Church, A. H. *On the Relation of Phyllotaxis to Mechanical Laws*. London: Williams & Norgate, 1904.

Coleman, A. J. "Algebraic Methods," *Encyclopedia of Applied Physics*, vol. 1. VCH Publishers, 1991, 515–37.

Conway, John Horton, and Neil J. A. Sloane. *Sphere Packings, Lattices, and Groups*. New York: Springer, 1999.

Conway, John H., and Richard K. Guy. *The Book of Numbers*. New York: Copernicus, 1996.

Coxeter Fonds, B2004–0024, University of Toronto Archives.

Coxeter, Harold Scott Macdonald. "Some Contributions to the Study of Regular Polytopes," PhD. diss., December 18, 1931, Cambridge University Library.

———. University of Toronto Library Oral History Project. B1986–0088, University of Toronto Archives.

———. Rockefeller Fellowship Record Card, Rockefeller Archive Center, Sleepy Hollow, New York.

———. Student File, Alumni Records, Trinity College, Cambridge.

———. "Chirality and Phyllotaxis," Coxeter Lectures File, Coxeter Fonds, University of Toronto Archives.

———. "The Mathematics of Leonardo da Vinci," Coxeter Lecture Files, Coxeter Fonds, University of Toronto Archives.

———. "Royal Astronomical Society Club," March 10, 1972, Coxeter Lectures File, Coxeter Fonds, University of Toronto Archives.

Dakers, Caroline. *The Holland Park Circle: Artists and Victorian Society*. New Haven: Yale University Press, 1999.

Davenport, Guy. "The Kenner Era," *National Review*, December 31, 1985.

Davies, John D. "The Curious History of Physics at Princeton," *Princeton Alumni Weekly*, October 1973, 2, 8–11.

Davis, Chandler. "The Purge," in *A Century of Mathematics in America*, vol. 1. Edited by Peter Duren. Providence: American Mathematical Society, 1988, 413–28.

———. "Where Did Twentieth-Century Mathematics Go Wrong?" *The Intersection of History and Mathematics*. Edited by Sasaki Chikara et al., Boston: Birkhäuser, 1994.

Davis, Chandler, and Erich W. Ellers, eds. *The Coxeter Legacy: Reflections and Projections.* Toronto/Providence: American Mathematical Society/Fields Institute, 2006.

Davis, Chandler, Branko Grünbaum, F. A. Sherk, eds. *The Geometric Vein, The Coxeter Festschrift.* New York: Springer-Verlag, 1981.

Davis, Chandler, and Marjorie Senechal, eds. "The World of Coxeter," *The Mathematical Intelligencer* 26, no. 3 (Summer 2004).

Davis, Philip J. "The Rise, Fall, and Possible Transfiguration of Triangle Geometry: A Mini History," *American Mathematical Monthly* 102 (1995): 204–14.

———. *Thomas Gray, Philosopher Cat.* Boston: Harcourt, Brace, Jovanovich, 1988.

Dehaene, Stanislas, et al. "Core Knowledge of Geometry in an Amazonian Indigene Group," *Science* 20 (January 2006): 381–84.

Derbyshire, John. *Prime Obsession: Bernhard Riemann and the Greatest Unsolved Problem in Mathematics.* Washington, D.C.: Joseph Henry Press, 2003.

Devlin, Keith. *The Language of Mathematics: Making the Invisible Visible.* New York: W. H. Freeman, 2000.

Devlin, Keith. *Mathematics: The Science of Patterns.* New York: Scientific American Library, 1997.

Dickson, Paul. *Sputnik: The Launch of the Space Race.* New York: Walker & Company, 2001.

Dieudonné, Jean. "New Thinking in School Mathematics," in *New Thinking in School Mathematics.* Organisation for European Economic Co-operation, 1961, 31–45.

———. "The Work of Bourbaki during the Last Thirty Years," *Notices of the AMS* 29 (1982): 618–23.

———. "The Work of Nicholas Bourbaki," *American Mathematical Monthly* 77 (February 1970): 134–45.

Dunne, J. W. *An Experiment with Time.* Charlottesville: Hampton Roads Publishing Company Inc., 2001.

Duren, Peter, et al., eds. *A Century of Mathematics in America*, vols. 1–3. Providence: American Mathematical Society, 1988–89.

Dyson, Freeman. *From Eros to Gaia.* New York: Pantheon Books, 1992.

Einstein, Albert. *Sidelights on Relativity.* New York: Dover Publications, 1983.

Ellers, Erich W., Branko Grünbaum, Peter McMullen, and Asia Ivić Weiss. "H. S. M. Coxeter (1907–2003)," *Notices of the AMS* 50, no. 10 (November 2003): 1234–40.

Emmer, Michele, ed. *Mathematics and Culture*, vol 1. New York: Springer, 2000.

———. *Mathland: From Flatland to Hypersurfaces.* Basel: Birkhäuser, 2004.

———, ed. *The Visual Mind: Art and Mathematics.* Cambridge, MA: The MIT Press, 1995.

Eriksson, Tommy et al. "Enantiomers of Thalidomide: Blood Distribution and the Influence of Serum Albumin on Chiral Inversion and Hydrolysis," *Chirality* 10 (1998): 223–28.

Ernst, Bruno. *The Magic Mirror of M. C. Escher.* New York: Random House, 1976.

Escher, M. C. Correspondence, Library and Archives, National Gallery of Canada, Ottawa, Ontario.

———. Correspondence, Haags Gemeentemuseum, The Hague, the Netherlands.

———. Archives, National Gallery of Art, Washington, D.C.

———. *Escher on Escher: Exploring the Infinite.* New York: Harry N. Abrams, 1989.

Farrell, Barry. "The View from the Year 2000," *LIFE*, February 26, 1971, 46–58.

Ferguson, Kitty. *Tycho & Kepler.* New York: Walker Books, 2002.

"The Fields Medal," http://www.fields.utoronto.ca/aboutus/jcfields/fields_medal.html.

Fillmore, Peter, ed. *Canadian Mathematical Society, 1945–1995.* Ottawa: Canadian Mathematical Society, 1995.

Fitzgerald, Penelope. *The Gate of Angels*. New York: Houghton Mifflin Company, 1998.

Fosdick, Harry Emerson. *On Being a Real Person*. London: Student Christian Movement Press, 1954.

Fosdick, Raymond B. *The Story of the Rockefeller Foundation*. New York: Harper & Brothers, 1952.

Friedman, Martin L. *The University of Toronto: A History*. Toronto: University of Toronto Press, 2002.

Freudenthal, Hans. "Geometry Between the Devil and the Deep Sea," *Educational Studies in Mathematics* (1971): 413–35.

Fuller, R. Buckminster, Papers, M1090, Department of Special Collections and University Archives, Stanford University Libraries.

———. *Critical Path*. New York: St. Martin's Press, 1981.

———. *Synergetics: Explorations in the Geometry of Thinking*. New York: Macmillan, 1975.

———. *Synergetics 2: Further Explorations in the Geometry of Thinking*. New York: Macmillan, 1979.

Galison, Peter. "Images Scatter into Data, Data Gather into Images," in *Iconoclash*. Edited by Bruno Latour and Peter Weibel. Cambridge, MA: MIT Press, 2002, 300–23.

Gallager, Peter. "A Conversation with G. David Forney, Jr.," *IEEE Information Theory Society Newsletter*, February 2006.

Galsworthy, John. *The Forsyte Saga*. Middlesex: Penguin Books Canada, 1986.

Gardner, Howard. *Intelligence Reframed*. New York: Basic Books, 1999.

Gardner, Martin. *Martin Gardner's New Mathematical Diversions from* Scientific American. New York: Simon & Schuster, 1966.

———. *The New Ambidextrous Universe: Symmetry and Asymmetry from Mirror Reflections to Superstrings*. New York: W. H. Freeman Company, 1990.

Ghyka, Matila. *The Geometry of Art and Life*. New York: Dover Publications, 1977.

Gleick, James. *Isaac Newton*. Toronto: Random House Canada, 2003.

———. *Chaos: Making a New Science*. New York: Penguin Books, 1988.

———. "Rethinking Clumps and Voids in the Universe," *New York Times*, November 9, 1986, A1.

Goggin, P. M. and J. N. Gordon. "Thalidomide and Its Derivatives: Emerging from the Wilderness," *Postgraduate Medical Journal*, 79, no. 929 (March 2003), 127–32.

Gombrich, E. H. et al. *Art, Perception, and Reality*. Baltimore: The Johns Hopkins University Press, 1972.

Gombrich, E. H. *Art and Illusion: A Study in the Psychology of Pictorial Representation*. Princeton: Princeton University Press, 1969.

Goodman, Roe. "Alice Through Looking Glass after Looking Glass: The Mathematics of Mirrors and Kaleidoscopes," *Mathematical Association of America Monthly* 111 (April 2004): 281–98.

Gosset, Thorold. "The Hexlet," *Nature* 139 (January 1937): 62.

Gott, J. R. III. "Pseudopolyhedrons," *American Mathematical Monthly* 74, no. 5 (May 1967): 497–504.

Gott, J. R. III, and A. Melott. "The Spongelike Topology of Large Scale Structure in the Universe," *Astrophysical Journal*, 1986.

Gott, J. R. III, A. Melott, and M. Dickinson. "The Spongelike Topology of Large Scale Structure in the Universe," *Astrophysical Journal* 306 (1986): 341–57.

Gray, George W. *Education on an International Scale*. New York: Harcourt, Brace and Company, 1941.

Gray, Jeremy J. *The Hilbert Challenge*. New York: Oxford University Press, 2000.

———. *János Bolyai, Non-Euclidean Geometry, and the Nature of Space*. Cambridge, MA: Burndy Library Publications, 2004.

———, ed. *The Symbolic Universe: Geometry and Physics 1890–1930*. Toronto: Oxford University Press, 1999.

Green, Christopher. "Classics in the History of Psychology, Timaeus." http://psychclassics .yorku.ca/Plato/Timaeus/timaeus2.htm, York University.

Greene, Brian. *The Elegant Universe: Superstrings, Hidden Dimensions, and the Quest for the Ultimate Theory*. New York: Vintage Books, 1999.

———. *The Fabric of the Cosmos: Space, Time and the Texture of Reality*. New York: Alfred A. Knopf, 2004.

Guedj, Denis. "Nicholas Bourbaki, Collective Mathematician: An Interview with Claude Chevalley," *The Mathematical Intelligencer* 7, no. 2 (1985): 18–22.

Hadamard, Jacques. *The Psychology of Invention in the Mathematical Field*. New York: Dover Publications, 1954.

Halmos, Paul R. *I Want to Be a Mathematician: An Automathography*. New York: Springer-Verlag, 1985.

———. " 'Nicolas Bourbaki,' " *Scientific American* 196, no. 5 (May 1957): 88–99.

Hancock, Geoff. "The Many Sides of Donald Coxeter," *Graduate* 7, no. 1 (September/October 1979): 10–12.

Hansard, Parliamentary Debates, House of Lords, Official Records. London: HMSO, December 10, 1991, 590.

Hardy, G. H. "What Is Geometry?" *The Changing Shape of Geometry*. New York: Cambridge University Press, 2003, 13–23.

———. *A Mathematician's Apology*. Cambridge: Cambridge University Press, 1940.

Hardy, G. H. With a foreword by C. P. Snow. *A Mathematician's Apology*. Cambridge: Cambridge University Press, 2004.

Hargittai, István, "Lifelong Symmetry: A Conversation with H. S. M. Coxeter," *The Mathematical Intelligencer* 18, no. 4 (1996): 35–41.

———. "John Conway—Mathematician of Symmetry and Everything Else," *The Mathematical Intelligencer* 23, no. 2 (November 2001): 6–14.

Hargittai, István, and Magdolna Hargittai. *Symmetry, A Unifying Concept*. Bolinas: Shelter Publications, 1994.

Hargittai, István, and T. C. Laurent, eds. *Symmetry 2000*, parts 1 and 2. Wenner-Gren International Series, vol. 80. London: Portland Press, 2002.

Hart, George. *Virtual Polyhedra: An Encyclopedia of Polyhedra*, http://www.georgehart.com/virtual-polyhedra/vp.html

Heath, Sir Thomas L., trans. *Euclid: The Thirteen Books of the Elements*, vols. 1–3. New York: Dover Publications, 1956.

Heilbron, J. L. *Geometry Civilized: History, Culture, and Technique*. Oxford: Clarendon Press, 2000.

Henderson, David W., and Daina Taimina. *Experiencing Geometry: Euclidean and Non-Euclidean with History*. Upper Saddle River: Pearson Prentice Hall, 2005.

Henderson, Linda Dalrymple. *The Fourth Dimension and Non-Euclidean Geometry in Modern Art*. Princeton: Princeton University Press, 1983.

Henneaux, Marc. "Platonic Solids and Einstein Theory of Gravity: Unexpected Connections," Francqui Chair Seminar, February 26, 2003.

Henneaux, Marc, et al. "Cosmological Billiards," *Classical and Quantum Gravity* 20 (2003).

———. "E(10) and BE(10) and Arithmetical Chaos in Superstring Cosmology," *Physical Review Letters* 86 (2001): 4,749–52.

———. "Einstein Billiards and Overextensions of Finite Dimensional Simple Lie Algebras," *Journal of High Energy Physics*, June 2002.

Herbst, Patricio G. "Establishing a Custom of Proving in American School Geometry: Evolution of the Two-Column Proof in the Early Twentieth Century," *Educational Studies in Mathematics* 49 (2002): 283–312.

Hermann, Robert. "Mathematics and Bourbaki," *The Mathematical Intelligencer* 8, no. 1, 1986): 32–33.

Hilbert, David and S. Cohn-Vossen. *Geometry and the Imagination.* New York: Chelsea Publishing Company, 1952.

Hoffman, Donald D. *Visual Intelligence: How We Create What We See.* New York: W. W. Norton, 1998.

Hoffman, Paul. *The Man Who Loved Only Numbers.* New York: Hyperion, 1998.

Hofstadter, Douglas R. "From Euler to Ulam: Discovery and Dissection of a Geometric Gem," Center for Research on Concepts and Cognition, 1992 (also published in a shorter version: "Discovery and Dissection of a Geometric Gem," *Geometry Turned On!* Edited by J. King and D. Schattschneider. Mathematical Association of America, 1997, 3–14).

———. *Gödel, Escher, Bach: An Eternal Golden Braid.* New York: Basic Books, 1999.

Holden, Alan. *Shapes, Space, and Symmetry.* New York: Columbia University Press, 1971.

Homer-Dixon, Thomas. *The Ingenuity Gap: Can We Solve the Problems of the Future?* Toronto: Vintage, 2001.

Horne, R. W., "The Structure of Viruses," *Scientific American* 208, no. 1 (January 1963): 48–56.

Howson, Geoffrey. "Geometry: 1950–1970," in *One Hundred Years of L'Enseignement Mathématique*, edited by Daniel Coray et al. Geveva: L'Enseignement Mathématique, 2003, 115–31.

Howson, Geoffrey. "Milestone or Millstone?" in *The Changing Shape of Geometry.* New York: Cambridge University Press, 2003, 505–14.

Infeld, Leopold. "He Renounced Canada for Poland," *The Globe and Mail*, February 26, 1951.

———. *Quest: The Evolution of a Scientist.* New York: Doubleday, Doran & Co., Inc., 1941.

Infeld, Leopold. *Whom the Gods Love: The Story of Evariste Galois.* New York: McGraw-Hill Book Company, Inc., 1948.

Ivins, William M. Jr. *Art & Geometry: A Study in Space Intuitions.* New York: Dover Publications, 1964.

Jackson, Allyn. "Interview with Henri Cartan," *Notices of the AMS* 46, no. 7 (August 1999): 782–88.

James, Ioan. *Remarkable Mathematicians: From Euler to von Neumann.* Cambridge University Press/Mathematical Association of America, 2002.

Jay, Martin. *Downcast Eyes: The Denigration of Vision in Twentieth-Century French Thought.* Berkeley: University of California Press, 1994.

Jones, C. Sheridan. *London in War-Time.* London: Grafton & Co., 1917.

Jowett, Benjamin. "Selections from Plato's Timaeus," in *The Collected Works of Plato.* Princeton: Princeton University Press, 1980.

(K-13) Geometry Committee. *Geometry, Kindergarten to Grade Thirteen,* Toronto: Ontario Institute for Studies in Education, 1967.

Kanigel, Robert. *The Man Who Knew Infinity: A Life of the Genius Ramanujan.* Toronto: Washington Square Press, 1992.

Kaplan, Robert, and Ellen Kaplan. *The Art of the Infinite*. New York: Oxford University Press, 2003.

Kaplansky, Irving. *Linear Algebra and Geometry*. New York: Chelsea Publishing Company, 1974.

Kaku, Michio. *Parallel Worlds: A Journey Through Creation, Higher Dimensions, and the Future of the Cosmos*. Toronto: Doubleday, 2005.

Kepler, Johannes. *The Six-Cornered Snowflake*. London: Oxford University Press, 1966.

Kline, Morris. *Mathematics for the Nonmathematician*. New York: Dover Publications, 1967.

Kosslyn, Stephen M. *Image and Brain: The Resolution of the Imagery Debate*. Cambridge, MA: MIT Press, 1994.

Kostant, Bertram. "The Graph of the Truncated Icosahedron and the Last Letter of Galois," *Notes of the AMS* 42, no. 9 (September 1995): 959–68.

Kroto, Harold, J. R. Heath, S. C. O'Brien, R. F. Curl, and R. E. Smalley. "C_{60}: Buckminsterfullerene," *Nature* 318 (November 1985, 14): 162–63.

Kroto, Harold W. "C_{60}: Buckminsterfullerene, The Celestial Sphere that Fell to Earth," *Angewandte Chemie* 31, no. 2 (February 1992): 111–29.

———. "Space, Stars, C_{60}, and Soot," *Science* 242 (November 1988, 25): 1139–45.

Lai, Jonathan. "Finite Coxeter Groups." PhD thesis, Carleton University, Department of Mathematics and Statistics, January 16, 1984.

Laurence, William I. "The Week in Science: Measuring the Earth's Age . . . Projecting Hyper-Cubes," *The New York Times*, July 1935 21, 6 (Science).

Lederman, Leon, and Christopher Hill. *Symmetry and the Beautiful Universe*. Amherst: Prometheus Books, 2004.

Lehrer, Tom. "New Math," *The Year That Was*, http://www.lyricsfreak.com/t/tom-lehrer/.

Leitch, Alexander. *A Princeton Companion*. Princeton: Princeton University Press, 1978.

Lightman, Alan. *The Discoveries: Great Breakthroughs in Twentieth-Century Science*. Toronto: Alfred A. Knopf Canada, 2005.

Littlewood, J. E. *Littlewood's Miscellany*. New York: Cambridge University Press, 1986.

Livio, Mario. *The Equation That Couldn't Be Solved: How Mathematical Genius Discovered the Language of Symmetry*. Toronto: Simon & Schuster, 2005.

Locher, J. L. *The World of M. C. Escher*. New York: Harry N. Abrams, Inc., 1971.

———, ed. *M. C. Escher: His Life and Complete Graphic Work*. New York: Harry N. Abrams, Inc., 2000.

Logothetti, Dave. "An Interview with H. S. M. Coxeter, the King of Geometry," *The Two-Year College Mathematics Journal* 11, no. 1 (January 1980): 2–18.

Longuet-Higgins, Michael S. "Encounters with Polytopes," *Symmetry: Culture and Science* 13, nos. 1–2 (2002): 17–31.

Loomis, Elisha S. *The Pythagorean Proposition*. Berea: Ohio Mohler Print Co., 1927.

Lord, E. A., and S. Ranganathan. "Sphere Packing, Helices and the Polytope {3,3,5}," *European Physical Journal* 15 (2001): 335–43.

Luminet, Jean-Pierre, Jeffrey R. Weeks, Alain Riazuelo, Roland Lehoucq, and Jean-Philippe Uzan. "Dodecahedral Space Topology as an Explanation for Weak Wide-angle Temperature Correlations in the Cosmic Microwave Background," *Nature* 425 (October 9, 2003): 593–95.

Lundy, Miranda. *Sacred Geometry*. New York: Walker & Company, 2001.

MacHale, Desmond. *George Boole: His Life and Work*. Dublin: Boole Press, 1985.

MacGillavry, Caroline. *Fantasy and Symmetry—The Periodic Drawings of M. C. Escher*. New York: Harry N. Abrams, 1976.

MacLane, Saunders. "Topology and Logic at Princeton," in *A Century of Mathematics in America*, vol 2. Edited by Peter Duren. Providence: American Mathematical Society, 1989, 217–21.

"Magpie & Stump," *The Trinity Magazine*, December 1928.

Malatyh, George. "Geometry for All and for Elite," Tenth International Congress on Mathematical Education, Copenhagen, July 2004.

Malkevitch, Joseph, ed. *Geometry's Future*. Arlington: COMAP, 1991.

Mammana, Carmelo, and Vinicio Villani. *Perspectives on the Teaching of Geometry for the 21st Century, An ICMI Study*. Boston: Kluwer Academic Publishers, 1998.

Mandelbrot, Benoit. "Chaos, Bourbaki, and Poincaré," *The Mathematical Intelligencer* 11, no. 3 (1989): 10–12.

Mandelbrot, Benoit. *Fractals, Form, Chance, and Dimension*. San Francisco: W. H. Freeman and Company, 1977.

———. *The Fractal Geometry of Nature*. New York: W. H. Freeman and Company, 1983.

Maor, Eli. *To Infinity and Beyond*. Boston: Birkhäuser, 1986.

Marks, Robert W. *The Dymaxion World of Buckminster Fuller*. Carbondale: Southern Illinois University Press, 1960.

Mathias, A. R. D. "The Ignorance of Bourbaki," *The Mathematical Intelligencer* 14, no. 3 (1992): 4–13.

McMullen, Peter, and Egon Schulte. *Abstract Regular Polytopes*. New York: Cambridge University Press, 2002.

Merzback, Uta C. "The Study of the History of Mathematics in America: A Centennial Sketch," in *A Century of Mathematics in America*, vol 3. Edited by Peter Duren. Providence: American Mathematical Society, 1989, 639–66.

Miyazaki, Koji. *An Adventure in Multidimensional Space: The Art and Geometry of Polygons, Polyhedra, and Polytopes*. Toronto: John Wiley & Sons, 1986.

———. *Science of Higher-Dimensional Shapes and Symmetry*. Kyoto: Kyoto University Press, 2004.

Mlodinow, Leonard. *Euclid's Window: The Story of Geometry from Parallel Lines to Hyperspace*. Toronto: Simon & Schuster, 2002.

Monk, Ray. *Ludwig Wittgenstein: The Duty of Genius*. London: Vintage Books, 1991.

Monson, Barry. *Geometry in a Nutshell, Notes for Math 3063*. Fredericton: Department of Mathematics & Statistics, 2000.

Moon, Bob. *The New Maths Curriculum Controversy, An International Story*. London: The Falmer Press, 1986.

Muir, Jane. *Of Men and Numbers*. New York: Dover Publications, 1996.

Nasar, Sylvia. *A Beautiful Mind*. Toronto: Simon & Schuster, 1998.

Ne'eman, Yuval. "Symmetry as the Leitmotif at the Fundamental Level in Twentieth Century Physics," *Symmetry: Culture and Science*, 10, no. 1–2 (1999), 143–62.

Nebeker, Frederik. *The Princeton Mathematics Community in the 1930s: An Oral History Project*. Princeton, NJ: The Trustees of Princeton University, 1985.

Neiger, Brad L. "The Re-emergence of Thalidomide: Results of a Scientific Conference," *Teratology*, 62, no. 6 (2000), 432–35.

O'Connor, John J., and Edmund F. Robertson. *The MacTutor History of Mathematics Archive*. http://www-history.mcs.st-andrews.ac.uk/history/.

Odom, George. "Problem E 3007," *American Mathematical Monthly*, Vol. 90, 1983, 482.

Okumura, Hiroshi. "Geometries in the East and the West in the 19th Century," *Symmetry: Culture and Science*, 10, no. 1–2, 189–97.

Osserman, Robert. "The Geometry Renaissance in America: 1938–1988," in *A Century of Mathematics in America*, vol. 2. Providence: American Mathematical Society, 1989, 513–26.

Pantalony, David. "H. S. M. Coxeter's Unusual Collection of Geometric Models," *Rittenhouse, Journal of the American Scientific Enterprise* 15, no. 1: 11–20.

Pauling, Linus. *No More War!* New York: Dodd, Mead & Company, 1958.

Pedersen, Jean J. "Geometry Is Alive and Well: The Coxeter Symposium in Toronto," *The Two-Year College Mathematics Journal* 11, no. 1 (January 1980): 19–24.

Pedoe, Daniel. "In Love with Geometry," *The College Mathematics Journal* 29, no. 3 (May 1998): 170–88.

Pendergrast, Mark. *Mirror Mirror: A History of the Human Love Affair with Reflection*. New York: Basic Books, 2004.

Penrose, Roger. *The Road to Reality: A Complete Guide to the Laws of the Universe*. London: Jonathan Cape, 2004.

Perkins, Sarah B., and P. J. Rowley. "Bad Upward Elements in Infinite Coxeter Groups," *Advances in Geometry*, 4 (2004), 497–511.

———. "Evil Elements in Coxeter Groups," *Journal of Algebra*, 251 (2002), 538–59.

Petroski, Henry. *Design Paradigms: Case Histories of Error and Judgment in Engineering*. New York: Cambridge University Press, 2000.

———. "Past and Future Failures," *American Scientist* 92 (November–December 2004): 500–4.

———. "Predicting Disaster," *American Scientist* 81, March–April 2003): 110–13.

———. *Success through Failure: The Paradox of Design*. Princeton: Princeton University Press, 2006.

"Poll on Campus," *Princeton Alumni Weekly*, November 4, 1932.

Polo, Irene. *Alicia Boole Stott, a Geometer in High Dimension*. Preprint.

Pritchard, Chris, ed. *The Changing Shape of Geometry*. New York: Cambridge University Press, 2003.

Punke, H. H. "The Family and Juvenile Delinquency," *Peabody Journal of Education*, 1955, 98.

R.E. "Obituaries: Henry Frederick Baker," *The Eagle*, EL 7: 80–83.

Read, A. H. *A Signpost to Mathematics*. London: C. A. Watts & Co., 1951.

Reist, Marianne, et al. "Chiral Inversion and Hydrolysis of Thalidomide: Mechanisms and Catalysis by Bases and Serum Albumin, and Chiral Stability of Teratogenic Metabolites," *Chemical Research in Toxicology*, 11 (1998), 1521–28.

Rider, Robin E. "Alarm and Opportunity: Emigration of Mathematicians and Physicists to Britain and the United States, 1933–1945," *Historical Studies in the Physical Sciences* 15 (1984): 107–76.

Roberts, Siobhan. "Figure Head," *Toronto Life*, January 2003, 82–88.

———. "I've Been Coxetering Today," *National Post*, September 2001, 1, B3.

Roberts, Siobhan, and Asia Ivić Weiss. "Donald in Wonderland: The Many-Faceted Life of H. S. M. Coxeter," *The Mathematical Intelligencer* 26, no. 3 (2004): 17–25.

Robinson, Floyd G. "Book Review, *Geometry: Kindergarten to Grade Thirteen*," *Ontario Journal of Educational Research* 10, no. 1 (Autumn 1967): 67–70.

Robinson, Gilbert de Beauregard. *The Mathematics Department in the University of Toronto, 1827–1978*. Toronto: University of Toronto Press, 1979.

———. *Recollections: 1906–1987*. Toronto: University of Toronto Press, 1987.

Robinson, John. *Symbolic Sculpture*. Carouge-Geneva: Edition Limitée, 1992.

Rose, Matthew. "Mathematician Sues Kimberly-Clark Unit over Its Toilet Paper—At Heart of the Messy Issue Is Tissue's Quilted Design; Is It the 'Penrose Pattern'?" *The Wall Street Journal*, April 14, 1997.

Rosen, Joe. *Symmetry Discovered: Concepts and Applications in Nature and Science*. New York: Cambridge University Press, 1975.

Rota, Gian-Carlo. "Book Reviews," *Advances in Mathematics* 77 (1989), 263.

———. "Fine Hall in Its Golden Age: Remembrances of Princeton in the Early Fifties," in *A Century of Mathematics in America*, vol 1. Edited by Peter Duren. Providence: American Mathematical Society, 1989, 223–26.

———. *Indiscrete Thoughts*. Boston: Birkhäuser, 1997.

Rowe, David. "Coxeter on People and Polytopes," *The Mathematical Intelligencer* 26, no. 3 (2004): 26–30.

Ruiz, Ángel. "A Bridge Across the Americas: The History of the Inter-American Committee on Mathematics Education," *The Mathematical Educator* 9, no. 2 (Spring 1999): 50–53.

Ruiz, Ángel, and Hugo Barrantes. *The History of the Inter-American Committee on Mathematics Education*. Bogotá: Colombian Academy of Exact, Physical, and Natural Sciences, 1998.

Sadoc, J. F., and N. Rivier. "Boerdijk-Coxeter Helix and Biological Helices," *European Physical Journal*, Vol. B12, November, 1999, pp. 309–18.

Sawyer, W.W. *Mathematician's Delight*. Suffolk: Penguin Books, 1944.

———. *Prelude to Mathematics*. Middlesex: Penguin Books, 1955.

Schattschneider, Doris. "Coxeter and the Artists: Two-Way Inspiration," in *The Coxeter Legacy: Reflections and Projections*. Providence/Toronto: American Mathematical Society/Fields Institute, 2006, 255–80.

———. "Coxeter and the Artists: Two-Way Inspiration, Part II," in *Renaissance Banff Bridges: Mathematical Connections in Art, Music, and Science, Conference Proceedings*. Edited by Reza Sarhangi. Kansas: Winfield, 2005, 473–80.

———. "Escher: A Mathematician in Spite of Himself," *Structural Topology*, No. 15, 9–42.

———. "Escher's Metaphors," *Scientific American*, November 1994, 48–53.

———. "In Praise of Amateurs," in *The Mathematical Gardner*. Boston: Prindle, Weber & Schmidt, 1981.

———. *M. C. Escher, Visions of Symmetry*. New York: W. H. Freeman Company, 1999.

Schattschneider, Doris, and Michele Emmer, eds. *M. C. Escher's Legacy*. New York: Springer, 2003.

Schechter, Bruce. *My Brain Is Open*. New York: Touchstone, 2000.

Schmidt, Lawrence E. "Simone Weil's Characterization of Algebra as a 'Monster of Contemporary Civilization'—A Paper Presented to the Annual Colloquy of the American Weil Society," May 2, 2003.

Scholz, Erhard. "Hermann Weyl's Contribution to Geometry, 1917–1923," in *The Intersection of History and Mathematics*. Basel: Birkhäuser, 1994, 203–29.

Schrag, Lex. "Mathematician Hard to Figure, Prof. Infeld Stays in Poland but Doubt He Took A-Secrets," *The Globe and Mail*, September 22, 1950.

Sen, Amartya. *Trinity College: Cambridge, Annual Record 2001*. Cambridge: Cambridge University Press, 2001.

Senechal, Marjorie. "The Continuing Silence of Bourbaki—An Interview with Pierre Cartier, June 18, 1997." *The Mathematical Intelligencer* 20, no. 1, 1998: 22–28.

———. "Coxeter and Friends," *The Mathematical Intelligencer* 26, no. 3 (Summer 2004): 16.

———, ed. *The Cultures of Science*. Commack: Nova Science Publishers, 1994.

———. "Donald and the Golden Rhombohedra," in *The Coxeter Legacy: Reflections and Projections*. Providence/Toronto: MAA/FI, 2006, 159–77.

———. *Patterns of Symmetry*. Amherst: University of Massachusetts Press, 1977.

———. *Quasicrystals and Geometry*. New York: Cambridge University Press, 1995.

Senechal, Marjorie, and George Fleck, eds. *Shaping Space: A Polyhedral Approach*. Boston: Birkhäuser, 1988.

Senechal, Marjorie, and George Fleck. *A Workbook of Common Geometry*. Northhampton: Clark Science Center, Smith College, 1988.

Severance, Kurt, Paul Brewster, Barry Lazos, and Kaniel Keefe. "Wind Tunnel Data Fusion and Immersive Visualization: A Case Study," NASA Langley Research Center, 2001.

Shakespeare, William. *Complete Works of William Shakespeare*. New York: Avenel Books, 1975.

Shaplen, Robert. *Toward the Well-Being of Mankind: Fifty Years of the Rockefeller Foundation*. New York: Doubleday & Company, 1964.

Shell, Barry. "Donald Coxeter, Mathematician and Geometer," http://www.science.ca/scientists/scientistprofile.php?pID=5.

Shenitzer, Abe, and John Stillwell, eds. *Mathematical Evolutions*. Washington, D.C.: Mathematical Association of America, 2002.

Sherk, Arthur. "Remembering Donald Coxeter," *CMS Notes*, September 2003, 4–6.

Sherk, F. Arthur, Peter McMullen, Anthony C. Thompson, and Asia Ivić Weiss. *Kaleidoscopes: Selected Writings of H. S. M. Coxeter*. Toronto: John Wiley & Sons, 1995.

Shubnikov, A. V., and V. A. Koptsik. *Symmetry in Science and Art*. New York: Plenum Press, 1974.

Sieden, Lloyd Steven. *Buckminster Fuller's Universe: His Life and Work*. Cambridge MA: Perseus Publishing, 2000.

"A Simple Story," *The Trinity Magazine*, December 1926, 18.

Sinclair, Lister. "Explorations: Space, Time, and the Atom Bomb," CBC Television, February 26, 1957.

———. "Magic Number," CBC *Ideas*, March 12, 1996.

———. "Math and Aftermath," CBC *Ideas*, May 13, 14, 1997.

Sinclair, Nathalie. *History of the Geometry Curriculum of the United States* (Center for the Study of Mathematics Curriculum Monograph Series, volume 2). Lansing, MI: Information Age Publishing Company.

Singh, Simon. *Fermat's Enigma*. Toronto: Penguin Books, 1998.

Sion, Maurice, ed. *A Pictorial Record of the International Congress of Mathematicians, Vancouver, Canada, August 21–29, 1974*. Vancouver: Canadian Mathematical Congress, 1977.

Smith, Kenneth B. "Fuller Sees Housing Service Industry by '72," *Globe and Mail*, October 1, 1968.

Smith, Michael. "Geometric Progression," *University of Toronto Magazine* 24, no. 3 (Spring 1997): 12–15.

Snow, C. P. *The Two Cultures*. New York: Cambridge University Press, 2000.

Soddy, Frederick. "The Kiss Precise," *Nature* 137 (June 1936) 1021.

Stekel, W. *Disguises of Love: Psycho-Analytical Sketches*. London: Kegan Pau, Trench, Trubner & Co., 1922.

Stekel, W. *Sexual Aberrations: Disorders of the Instincts and the Emotions*. New York: Liveright Publishers, 1940.

Stern, Beatrice M. *A History of the Institute for Advanced Study, 1930–1950*. Princeton, May 1964.

Strauss, Stephen. "Art Is Math Is Art for Professor Coxeter," *The Globe and Mail*, May 9, 1996.

Sutton, Daud. *Platonic and Archimedean Solids*. New York: Walker & Company, 2002.

Swift, Jonathan. *A Modest Proposal and Other Satires*. New York: Prometheus Books, 1995.

Szpiro, George G. *Kepler's Conjecture*. Hoboken: John Wiley & Sons, 2003.

Taylor, Kevin. *Central Cambridge: A Guide to the University and Colleges*. Cambridge: Cambridge University Press, 1994.

Taylor, A. E. *Plato: The Man and His Work*. London: Methuen & Co., 1929.

Thom, René. "'Modern' Mathematics: An Educational and Philosophic Error? A Distinguished French Mathematician Takes Issue with Recent Curricular Innovations in His Field," *American Scientist* 59 (November–December 1971): 695–99.

Thompson, D'Arcy. *On Growth and Form*. New York: Cambridge University Press, 2000.

Thurling, Peter. "Fuller Advises Students Go Outstairs Not Upstairs," *Telegram*, October 1, 1968.

Tierney, John. "Paul Erdös Is in Town. His Brain Is Open," *Science*, October 1984, 40–47.

Tits, Jacques. "Groupes et Géométries de Coxeter," *Wolf Prize in Mathematics* 2 (2001): 740–54.

Toole, John Kennedy. *A Confederacy of Dunces*. New York: Grove Press, 1980.

Tóth, Imre. *Palimpseste: Propos avant un triangle*. Paris: Presses Universitaires de France, 2000.

Trevelyan, G. M. *Trinity College: An Historical Sketch*. Cambridge: Master and Fellows of Trinity College, Cambridge, 1990.

Trinity College Council. *Trinity College, Cambridge, Ordinances*. Cambridge: Cambridge University Press, 1931.

Trudeau, Richard. J. Introduction by H. S. M. Coxeter. *The Non-Euclidean Revolution*. Boston: Birkhäuser, 1987.

Tufte, Edward R. *Envisioning Information: Narratives of Space and Time*. Connecticut: Graphics Press, 2003.

———. *Visual Explanations: Images and Quantities, Evidence and Narrative*. Connecticut: Graphics Press, 2003.

Veblen, Thorstein. *The Higher Learning in America*. http://socserv2.socsci.mcmaster.ca/~eeon/ugcm/3113/veblen/higher.

Wainer, Howard. *Visual Revelations: Graphical Tales of Fate and Deception from Napoleon Bonaparte to Ross Perot*. New York: Copernicus, 1997.

Washburn, Dorothy K., and Donald W. Crowe. *Symmetries of Culture: Theory and Practice of Plane Pattern Analysis*. Seattle: University of Washington Press, 1988.

Weeks, Jeffrey R. *The Shape of Space: How to Visualize Surfaces and Three-Dimensional Manifolds*. New York: Marcel Dekker, Inc., 1985.

Weil, André. *The Apprenticeship of a Mathematician*. Boston: Birkhäuser, 1992.

———. "History of Mathematics: Why and How," in *Proceedings of the International Congress of Mathematicians, 1978*. Helsinki: Academia Scientiarum Fennica, 1980, 229–36.

Weil, Simone. *Gravity and Grace*. London: Routledge & Kegan Paul, 1952.

———. *Notebooks I*. London: Routledge & Kegan Paul, 1956.

Weiss, Asia Ivić. "Dedication of Sculpture in Honour of H. S. M. Coxeter's 95th Birthday," *The Fields Institute Newsletter*, June 2002.

Weisstein, Eric. W. *MathWorld—A Wolfram Web Resource*. http://mathworld.wolfram.com.

Wells, David. *The Penguin Dictionary of Curious and Interesting Geometry*. Toronto: Penguin Books Canada, 1991.

Wells, H. G. *The Time Machine*. New York: Tor Books, 1986.

Wenninger, Magnus J. *Polyhedron Models*. Cambridge: Cambridge University Press, 1996.

——. *Spherical Models*. New York: Cambridge University Press, 1979.

Wenninger, Magnus J. *Dual Models*. New York: Cambridge University Press, 1983.

Weschler, Lawrence. *A Wanderer in the Perfect City*. St. Paul: Hungry Mind Press, 1998.

Weyl, Hermann. *Symmetry*. New Jersey: Princeton University Press, 1952.

——. "Discrete Groups Generated by Reflections," by H. S. M. Coxeter in *The Structure and Representation of Continuous Groups*. Princeton: Institute for Advanced Study, 1935.

Whitehead, A. N. "Geometry," *The Encyclopaedia Britannica*, 11th ed., 1910.

Whiteley, Walter. "The Decline and Rise of Geometry in 20th Century North America," *Proceedings of the Canadian Mathematics Education Study Group, June 1999*. CMESG, 1999.

——. "Learning to See like a Mathematician," *Visual Representation and Interpretation*. Elsevier, 2004.

——. "Working with Visuals in the Mathematics Classroom: Why It Is Needed, Hard and Possible," 2005 OAME Conference Paper.

Wilde, Oscar. *The Works of Oscar Wilde*. London: Collins, n. d.

Worth, Ruth. "U. of T. Heads Seeking End to Nuclear Testing," *The Globe and Mail*, November 23, 1959, A1.

Yaglom, I. M. "Elementary Geometry, Then and Now," in *The Geometric Vein: The Coxeter Festschrift*. New York: Springer, 1981, 253–69.

——. *Geometric Transformations*, vols. 1–3. Toronto: Random House, 1962, 1968, 1973.

Young, Grace Chisholm, and W. H. Young. *Beginner's Book of Geometry*. New York: Chelsea Publishing Company, 1970.

Zwicky, Jan. *Wisdom and Metaphor*. Kentville: Gaspereau Press, 2003.

EPIGRAPH AND ART CREDITS

Chapter 1   Plato, *The Republic*

Chapter 2   J. L. Synge, *Kandelman's Krim*

Chapter 4   Hermann Weyl, *Symmetry*

Chapter 5   Frederick Soddy, "The Kiss Precise" (*Nature* 139, 62). © 1936 Nature Publishing Group. Reproduced with permission.

Chapter 6   Henri Poincaré, *Oeuvres*

Chapter 7   Thorold Gosset, addendum to "The Kiss Precise"

Chapter 9   Henri Poincaré, *La Science et l'Hypothèse*

Chapter 11 Lewis Carroll, *Through the Looking-Glass*

Chapter 12 J. L. Synge, *Kandelman's Krim*

Chapter 13 Shakespeare, "Sonnet 18"

Page 5 Created by by David Logothetti. Courtesy of Faith Logothetti.

Page 8 Nathaniel Kahn, from the film, *My Architect* © Louis Kahn Project, Inc. All rights reserved. Used by permission.

Page 10 Courtesy of Doris Schattschneider

Page 11 Courtesy of the University of Toronto Archives

Page 13 Courtesy of Susan Thomas

Page 14 Created by Jerome Snyder as seen in "'Nicolas Bourbaki,'" by Paul R. Halmos (*American Magazine*, Vol. 196 No. 5).

Page 15 Photograph by Siobhan Roberts

Page 18 Courtesy of Asia Ivic´ Weiss

Page 24 Image file courtesy of Thomas Fisher Rare Books Library, University of Toronto

Page 25 Courtesy of Susan Thomas

Page 31 Courtesy of Canadian Broadcasting Corporation, "Explorations Space, Time and the Atom Bomb," with Lister Sinclair, February 26, 1957

Page 32 From documentary footage by Siobhan Roberts

Page 34 Courtesy of the University of Toronto Archives

Page 39 Courtesy of the University of Toronto Archives

Page 42 Courtesy of the University of Toronto Archives

Page 43 Courtesy of the University of Toronto Archives

Page 48 NASA's Imagine the Universe

Page 53 Courtesy of Dover Publications

Page 57 Courtesy of J. R. Gott III

Page 61 Image file courtesy of Trinity College Library, Cambridge

Page 63 Image file courtesy of Thomas Fisher Rare Books Library, University of Toronto

Page 65 Image file courtesy of Thomas Fisher Rare Books Library, University of Toronto

Page 66 Courtesy of the University of Toronto Archives

Page 72 Courtesy of the University of Toronto Archives

Page 73 From *George Boole His Life and Work*, by permission of the author, Desmond MacHale

Page 76 Courtesy of the University of Toronto Archives

Page 78 Photograph by Seymour Schuster. Courtesy of Mathematical Association of America. © Mathematical Association of America, 2024. All rights reserved.

Page 91 Courtesy of Doris Schattschneider

Page 92 (all) Courtesy of Doris Schattschneider

Page 93 (top) Courtesy of Doris Schattschneider

Page 93 (bottom) Courtesy of Amina Buhler

Page 94 (top) Courtesy of the University of Toronto Archives

Page 94 (bottom) Courtesy of Doris Schattschneider

Page 107 Courtesy of Wittgenstein Archive Cambridge. Copyright © Wittgenstein Archive Cambridge. Reproduced with permission.

Page 109 Courtesy of the University of Toronto Archives

Page 111 Courtesy of the University of Toronto Archives

Page 114 Courtesy of the University of Toronto Archives

Page 121 Created by Jerome Snyder as seen in "'Nicolas Bourbaki,'" by Paul R. Halmos (*American Magazine*, Vol. 196 No. 5).

Page 129 Created by David Logothetti, courtesy of Faith Logothetti

Page 132 Courtesy of Doris Schattschneider

Page 135 Courtesy of the Tee and Charles Addams Foundation "Barber Shop Creature," by Charles Addams, from *The New Yorker*, February 23, 1957, Copyright © Tee and Charles Addams Foundation

Page 144 Courtesy of the University of Toronto Archives

Page 151 Courtesy of the University of Toronto Archives

Page 155 Unknown

Page 162 Courtesy of the University of Toronto Archives

Page 164 Photograph by Seymour Schuster. Courtesy of Mathematical Association of America. © Mathematical Association of America, 2024. All rights reserved.

Page 165 Photograph by Seymour Schuster. Courtesy of Mathematical Association of America. © Mathematical Association of America, 2024. All rights reserved.

Page 166 Photograph by Seymour Schuster. Courtesy of Mathematical Association of America. © Mathematical Association of America, 2024. All rights reserved.

Page 169 Courtesy of Mathematical Association of America. © Mathematical Association of America, 2024. All rights reserved.

Page 170 Created by David Logothetti, courtesy of Faith Logothetti

Page 174 Photograph copyright © Michel Proulx

Page 175 Courtesy of the University of Toronto Archives

Page 177 Bettmann / Contributor, January 01, 1967. Reproduced with permission by Getty.

Page 185 Courtesy of Istvan Orosz

Page 186 (all) Courtesy of Doris Schattschneider

Page 188 Courtesy of Doris Schattschneider

Page 191 Courtesy of the University of Toronto Archives

Page 197 By permission of George Szpiro, copyright © Italy Almog

Page 201 Courtesy Pixar, copyright © 1997 Pixar

Page 205 Courtesy of Michael Treacy and the IZA Database of Zeolite Structures

ACKNOWLEDGMENTS

Whenever I dared mention I was writing a book about the man who saved classical geometry from near extinction—when the topic came up at a dinner party, for instance—the conversation usually skipped a beat or ten. Expressions around the room took on various shades of stupefaction as people entertained flashbacks of math class anxieties, fumbling with compasses and protractors and memorizing the Pythagorean theorem. The conversation resumed only when one brave individual broke the silence with some variation on this rejoinder: "This man saved geometry?!? *Why, on Earth, did he do that?* He would have saved us all a lot of misery if he had let it die!"

My formative experience with geometry in grade six didn't make me phobic, nor a Coxeterian prodigy. But I was very fond of my geometry kit, and as a result I did well in math class that year. So well that I skipped grade seven math and went on to take all the math courses I could in high school. That said, I think Donald Coxeter would agree that I owe my first debt of thanks to my many math teachers along the way who made me love the challenge of math, and my two math-whiz cohorts at high school, Rob Ord and Jenny Michol, who is missed.

King of Infinite Space first began to take shape with John Geiger at the *National Post* (and Ken Whyte who installed me there in the first place). It evolved and transformed early on with Gary Ross and Sarah Fulford at *Toronto Life* magazine, and at The Banff Centre for the Arts with Moira Farr, Alberto Ruy Sánchez, and Ian Pearson, who gamely applied his keen editorial eye as a reader at a later stage as well.

Many thanks to Jackie Joiner, who spotted the book in the Coxeter story in the first place, and Denise Bukowski, who found the book the best homes. Sarah MacLachlan and Lynn Henry at House of Anansi Press are fabulous collaborators and supporters. And the indefatigable George Gibson at Walker & Company astounds with his endless energy, his impeccable insight, and his raging passion for books, which inspires me daily.

The generous outpouring of anecdotes and expertise from mathematicians and scientists throughout the world was astonishing—a testament to their abiding love and admiration for Donald Coxeter.

I am grateful to all of Coxeter's students for their vivid memories and expositions, particularly William Moser (and Beryl Moser), Seymour Schuster, Asia Ivić Weiss, John Coleman, Barry Monson, Chris Fisher, Ed Barbeau, Arthur Sherk, Cyril Garner, Donald Crowe, Norman Johnson, Joseph Sunday, and Robert Moody.

Chandler Davis made himself available for numerous interviews, fielded a regular peppering of questions, and was an incisive reader of the book's first draft. Walter Whiteley and Marjorie Senechal also shared layer upon layer of understanding, and were valued readers.

Capturing Coxeter's spirit, and piecing together historical details of his life as a mathematician, was only possible through the perspectives of Branko Grünbaum, Sir Michael Atiyah, Freeman Dyson, Benoit Mandelbrot, Sir Roger Penrose, Irving Kaplansky, Israel Halperin, Lee Lorch, Emma Castelnuovo, Ron Graham, Simon Kochen, Douglas Hofstadter, Michael Longuet-Higgins (and thank-you to Joan Longuet-Higgins for her hospitality), Eric Infeld, and Cathleen Synge Morawetz.

On the multidimensional subject of geometry, its history, many genres, and Coxeter's contributions, there is a long list of people whose knowledge enriched my perspective: Jeremy Gray, June Barrow-Green, Ravi Vakil, David Mumford, Peter Galison, Martin Jay, Glenn Smith, George Hart, Father Magnus Wenninger, George Odom, David Henderson, Daina Taimina, John Ratcliffe, Roe Goodman, Philip Davis, Martin Gardner, Robert Gunning, Thomas Banchoft, Koji Miyasaki, Jörg Wills, J. Vernon Armitage, Michele Emmer, Gyorgy Darvas, András Prékopa, Karoly Bezdek, Irene Polo, Angel Ruiz Zúñiga, David Pantalony, Alexander Bogomolny, Bill Richardson, Don Albers, Gerald Alexanderson, Brenda Fine, Geoffrey Hinton, Egon Schulte, Douglas Dunham, John McKay, Peter McMullen, Barry Shell, Ken Davidson, Larry Schmidt, Peter Taylor, Bob Erdahl, and William Higginson.

Jean-Pierre Serre, Pierre Cartier, Michel Broué, Jacques Tits, and Liliane Beaulieu, helped in my attempt to get to the bottom of the Bourbaki myth.

On the intersection between Coxeter's geometry and scientific applications, I am grateful to Jeff Weeks, Neil Sloane, Sir Harry Kroto, Marc Henneaux, J. R. Gott III, Leon Lederman, Gordon Lang, Bob Tennent, László Lovász, John Moffat, Edward Witten, Tony DeRose, Laurence Johnson, and on a quirky neurological tangent, Sandra Witelson. The expertise of Robert Craig and Charles Small informed my consideration of Coxeter's musical compositions.

Coxeter's story could not have been told without pictures. Thank you to Faith Logothetti for allowing the reproduction of David Logothetti's cartoon Coxeters; Marc Pelletier for his 120-cell sculpture (and Amina Allen for her photographs) and for the Donchian images; Stan Sherer and Marion Walter for photographs of Coxeter at work and play; Michel Proulx for the Fuller geodesic dome; Eden Robbins for the cover photo; and Nigel Dickson, who gallantly traded a bottle of Bordeaux for my jacket photo (and who took the last portrait of Coxeter, which appears in chapter 13).

Marnee Gamble and Garron Wells at the University of Toronto Archives were very patient with my process; Ken Rose at the Rockefeller Archive Center obliged beyond the ordinary requests; copies of the Fuller-Coxeter correspondence were sent from the Department of Special Collections at the Stanford University Libraries; and Escher's letters to his family from the National Gallery of Canada, as well as the Coxeter-Escher exchange housed at the National Gallery of Art in Washington, D.C. I was lucky as well to find George Escher and get his memories of his father's friendship with Coxeter.

During research in the UK, Trinity College, Cambridge, kindly let me stay in residence for three weeks during the summer of 2003. Many people at the university aided my archival dig, tunneling seventy-five years into the past: Catherine Boyle, Raymond Lickorish, Samantha Pinner, Jonathan Smith, Jacqueline Cox, Adam Perkins, Malcolm Underwood, and many others.

Nesta Coxeter opened her stash of family archives in London, and Eve Coxeter invited me to stay in Liverpool (and Joan Coxeter reminisced over the phone from California). Jane and Royston Carpenter allowed a complete stranger to sleep in one of Coxeter's childhood homes; Timothy Prus agreed to let me to wander through 34 Holland Park Road. John and Margie Robinson were gracious hosts at Agecroft, Galhampton, and proffered all sorts of sustenance, then and since.

Coxeter's children, Susan Thomas and Edgar Coxeter, eagerly read the manuscript, and offered a steady stream of personal reflections that deepened the portrait of Coxeter, the man.

Infinite thanks to Doris Schattschneider, for her intricate reading of the manuscript, and for providing many of the book's illustrations (and more).

Monster-sized thanks to Princeton's John Horton Conway, who—his daunting erudition notwithstanding—is a joy to know. His contribution was invaluable in every respect, and, in particular, is valued to a magnitude of a very high order for his vetting of the manuscript not once, but thrice (and thank you to Diana and Gareth Conway, who welcomed me in their home during those three extended visits).

Both the book and I were sustained with financial support from the Ontario Arts Council, the Canada Council for the Arts, and the Writers' Trust Woodcock Fund.

Many fellow writers, editors, and colleagues, have inspired and encouraged me along the way: Stuart McLean, David Hayes, Tim Falconer, Virginia

Smart, Sarah Galashan, Maya Gallus, Justine Pimlott, Don Obe, Lynn Cunningham, Lindalee Tracey, David MacFarlane, Ian Brown, Rick Boychuk, Ken Alexander, Paul Wilson, Noah Richler, Jeet Heer, and many others.

Shannon Black is a strong and steady friend with sharp advice.

Douglas Bell is brilliant and funny and, like water, essential in every way.

And I owe it all, really, to H. S. M. Coxeter, who tolerated me chasing him around for two years, asking at once simple and complex questions. He launched me on an all-consuming journey that opened my perspective on the world and took me places I otherwise would never have found—all of which, looking back, was a huge amount of fun.

With special thanks for the new edition to Diana Gillooly, Whitney Rauenhorst, Terri O'Prey, Mark Bellis, and company at Princeton University Press.

INDEX

NOTE: DC refers to Donald Coxeter

Also by
SIOBHAN ROBERTS

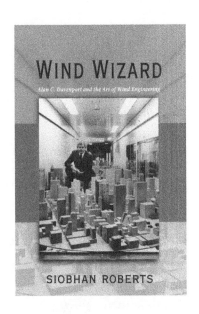
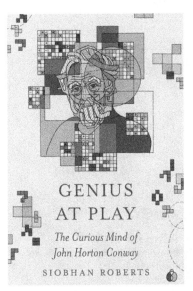

 PRINCETON UNIVERSITY PRESS

Available wherever books are sold.
For more information visit us at www.press.princeton.edu